CHEMICAL SPECTROSCOPY

SOLE DISTRIBUTORS FOR THE UNITED STATES AND CANADA
AMERICAN ELSEVIER PUBLISHING COMPANY, INC.
52 Vanderbilt Avenue, New York 17, N.Y.

CHEMICAL SPECTROSCOPY

by

R. E. DODD

Lecturer in Chemistry,
King's College,
University of Durham,
Newcastle-upon-Tyne (England)

ELSEVIER PUBLISHING COMPANY
AMSTERDAM – NEW YORK
1962

Library of Congress Catalog Card Number 61–13127

With 145 illustrations and 21 tables

PRINTED IN THE NETHERLANDS BY
N.V. DIJKSTRA'S DRUKKERIJ V.H. BOEKDRUKKERIJ GEBR. HOITSEMA, GRONINGEN

PUBLISHER'S PREFACE

Only a few weeks after completing the manuscript of this book the author died suddenly from a heart attack. During the time when the book was being produced the usual co-operation between author and publisher was therefore not possible. Some of his colleagues, who prefer to remain anonymous, very kindly assisted the publisher in that period by taking over the author's task of answering questions about minor details in the text or figures, correcting the proofs, filling in the cross-references and compiling the Index. Not only the publisher, but also the relatives of the author and the future users of the book owe them much gratitude.

We hope that the book will prove to be what the author envisaged when writing it, a guide to students and all those who wish to become conversant with the applications of spectroscopy. May it be a worthy memorial to a man who had great didactic gifts and specialist knowledge of the field treated.

Amsterdam, April 1962 — ELSEVIER PUBLISHING COMPANY

CONTENTS

LIST OF TABLES

INTRODUCTION

Chemical spectroscopy today covers so many techniques that it is unusual for either chemist or spectroscopist to have a working knowledge of all of them. Nevertheless, there is a common basis to the subject which may form the starting point of our discussion. It is simply that *all spectra arise from transitions between energy states.*

It is perhaps worth noting that it took many years to come to this simple conclusion. If spectroscopy is said to date from Sir Isaac Newton's publications in 1672, and spectrometry from Joseph Fraunhofer's measurements between 1814 and 1823, it was not until 1859 that emission spectra were clearly seen to be characteristic of the elements present in the emitter and Kirchoff and Bunsen set about mapping characteristic spectra. In 1885 Balmer announced his discovery of the simple mathematical relationship between the wavelengths of the lines of the hydrogen spectrum: in 1900 Planck explained the intensity distribution of black-body radiation with the revolutionary hypothesis that radiative energy was quantized; and the stage was set for Bohr to demonstrate in 1913 the quantized nature of the potential energy of the hydrogen atom. Although fundamental modifications were yet to be made to the system of mechanics in terms of which the energy states could be satisfactorily described, it was from this point in time that it came to be accepted that atomic and molecular species (and not only the hydrogen atom) can only exist in certain quite characteristic states of energy (although in certain regions these may merge into continuity); that these energy states are accessible to observation by virtue of the equally characteristic differences in energy between them; and that a transition between states leads to emission or absorption of a characteristic quantum of energy which, by virtue of Planck's law, also has characteristic frequency or wavelength.

We are interested in the information of chemical value which can be gleaned from the observation of spectra. We shall therefore state but not prove such formulae as are available from quantum mechanics for the quantitative interpretation of spectra. These formulae are derived — and thus in the mathematical sense proved — from the basic postulates of quantum mechanics: the derivations will be found in standard works to which references are given. In a wider sense, the proof of the formulae and therefore of the validity of the postulates from which they are derived lies in their consistent success in the explanation of the vast accumulation of spectrometric data.

Broadly the subject falls into four parts. In the first chapter some attention is given to the instrumentation of spectroscopy, in sufficient detail to show the sort of results which are obtained; how the essential quantities, frequency and intensity, are measured; and what instrumental limits are placed on such measurements. Secondly, we discuss in Chapters 2 and 3 the application of spectroscopy to the many cases where the chemical species under observation are simple enough for a more-or-less exact description of energy states to be given. The spectra can be interpreted, the relative values of the characteristic states can be assessed and conclusions can be drawn of such direct chemical significance as, for instance, the electronic configurations of atoms and ions, the shapes and dimensions of molecules, or the energies required to dissociate molecules. There are, however, many molecules for which, because of their complexity, a detailed theoretical interpretation is lacking. Here the empirical correlations of spectroscopic features with chemical features are valuable in qualitative analysis and structural diagnosis and we may regard this as the third part of the subject (Chapter 4). Finally there is the question of the intensity of emission or absorption, which depends both upon the inherent probability of the transition between states and upon the number of atoms or molecules in which the transition may occur. There are obvious quantitative applications here leading to chemical information which, in certain circumstances, cannot be got in any other way.

Chapter 1

EXPERIMENTAL METHODS OF SPECTROSCOPY

Spectroscopic measurements are essentially measurements of the intensity and the wavelength of radiative energy. An emitter radiates energy at many different wavelengths according to the number of energy levels between

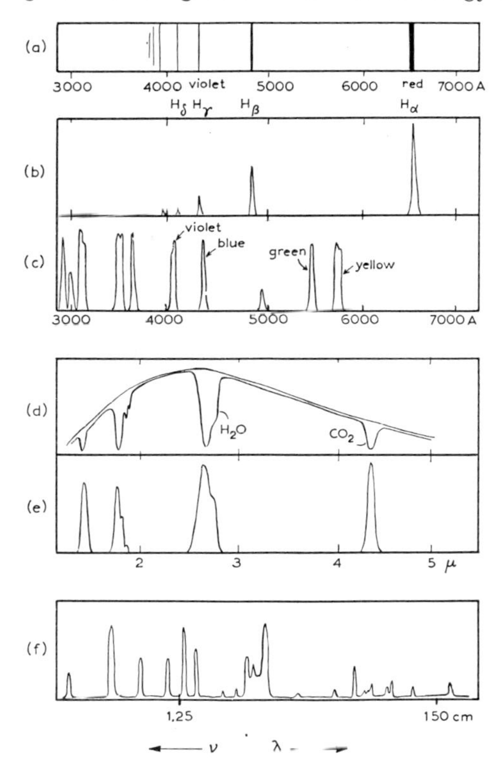

Fig. 1. (a) Photographic record of emission spectrum from a discharge tube containing hydrogen (vibible and ultraviolet). (b) Photometric record of the spectrum in (a). (c) Photoelectric record of mercury discharge emission spectrum (visible and ultraviolet). (d) Absorption spectrum of atmospheric water and carbon dioxide superimposed on continuous background, as observed (infra-red). (e) Absorption spectrum in (d) presented in terms of fraction of energy absorbed. (f) Absorption spectrum of ammonia (microwave).

References p. 64–66

which transitions are possible. A pattern of wavelengths at which energy is radiated is an *emission spectrum* and it will generally also indicate the intensity of the emission at each wavelength. Alternatively radiation varying continuously in wavelength may be passed through the sample of material. It is then found that energy has been absorbed by the material at certain wavelengths. A pattern of these wavelengths, with an indication of the proportion of energy absorbed at each wavelength, is an *absorption spectrum*. Examples of emission and absorption spectra are given in Figure 1.

The purpose of this chapter is to survey the methods by which such spectra are obtained so that the chemist who uses spectroscopy will have some idea of the significance of his measurements and be aware of some of the limitations.

Wavelength and Frequency

The spectroscopist is generally concerned with both *wavelength* and *frequency* as characteristic of the quality of electromagnetic radiation, for it is generally wavelength which he measures and always frequency which is more significant for interpretation of the spectra. In so far as the radiation behaves as a wave phenomenon having periodicity both in space and time, the wavelength λ measures the periodicity in space and the frequency ν the periodicity in time. These are related to the velocity of propagation of the wave by the equation:

$$c = \nu\lambda \tag{1.1}$$

The velocity c in vacuum is constant and does not vary with wavelength:* it is therefore referred to as the *velocity of light* without qualification. It has the value[1] $2.9979304 \pm 0.0000017 \times 10^{10}$ cm sec^{-1} (3.0×10^{10} cm sec^{-1} is of sufficient accuracy in many cases). In transparent media other than vacuum the velocity c' is less than c by a factor which is the inverse of the refractive index, n, of the medium; *i.e.*

$$nc' = c \tag{1.2}$$

c', and therefore n, vary with the wavelength and such variation is referred to as the *dispersion* of the medium.

The frequency of light is unaffected by the medium through which it travels. Since wavelength, frequency and velocity are still connected by the relationship:

* This is axiomatic and consistent with experiment to an accuracy of 1 part in 10^5. It is evidently not possible to verify its constancy to the accuracy with which it has been measured at one wavelength.

$$c' = \nu\lambda' \tag{1.3}$$

the wavelength, λ', in the medium is less than it is in vacuum. In interpretation of the spectra the frequency is more significant than the wavelength but what is measured is usually, in absolute terms, the wavelength in air. (In optical spectroscopy ultimately it is always wavelength which is measured and related to the standard of length, the wavelength of the red cadmium line or, more recently, a red krypton line. In microwave and radiospectroscopy the measurement of frequency is directly related to a frequency standard and through it to the astronomical clock.) It is therefore necessary, when the precision warrants it, to correct the wavelength to the vacuum value before using equation (1.1) to obtain the frequency. Equations (1.1), (1.2) and (1.3) lead to

$$\lambda = n\lambda' = \lambda' + (n - 1)\lambda' \tag{1.4}$$

so that precise determinations of ν from λ' may be made either with equation (1.3) in conjunction with values of the velocity of light in air at various wavelengths[2] (see also Ref. 1, p. 76) or with equation (1.1) after correcting the wavelength to vacuum by addition of the term $(n - 1)\lambda'$. The corrections are small, being of the order of 0.03 %. They are nevertheless significant corrections in measurements which may be made to an accuracy of about 0.001 Å. Some values of $n - 1$ and the appropriate wavelength correction are listed in Table 1: more detail is to be found in published tables[2].

TABLE 1

CORRECTION TO VACUUM OF WAVELENGTH OR WAVENUMBER

λ	$(n - 1)10^6$ at 15°C	$\delta\lambda$	$\delta\omega$ (cm^{-1})
2000 Å	325	0.6512 Å	16.27
3000	290.7	0.8721	9.69
4000	281.7	1.127	7.04
5000	278.1	1.391	5.56
6000	276.3	1.658	4.60
7000	275.3	1.927	3.93
8000	274.7	2.197	3.43
9000	274.2	2.468	3.05
10,000	273.9	2.739	2.74
1 μ	273.9	0.000274 μ	2.74
2	272.3	0.000545	1.36
4	271.0	0.001084	0.68
8	270.7	0.002166	0.34
13	269.8	0.003510	0.21

n = refractive index of air; vacuum $\lambda = \lambda_{air} + \delta\lambda$; vacuum $\omega = \omega_{air} - \delta\omega$

References p. 64–66

It is often the case that a spectrometer has been calibrated (p. 42) against a standard spectrum in which the wavelengths are already corrected to vacuum. In this case wavelengths are obtained by graphical or algebraic interpolations of a semi-empirical kind and will not require any further correction. Conversion to frequency is then directly by equation (1.1).

The range of wavelengths which spectroscopy covers extends from 10^{-8} to 10^{6} cm. Because of the different techniques which are required for detection and measurement this range is divided into regions which are named accordingly, as in Table 2. For convenience different units are employed. These are Ångstroms (Å), millimicrons (mμ) and microns (μ) related thus:

$$\frac{\lambda}{(\text{cm})} = 10^{-8}\frac{\lambda}{(\text{Å})} = 10^{-7}\frac{\lambda}{(\text{m}\mu)} = 10^{-4}\frac{\lambda}{(\mu)} \tag{1.5}$$

The more familiar larger metric units are also used in radiospectroscopy.

Frequency is quoted in $\sec^{-1}$ (cycles per second) or, in the microwave and radiowave regions, in terms of kilocycles per second (kc/s), megacycles per second (Mc/s) and kilomegacycles per second (kMc/s). These are related thus:

$$\frac{\nu}{(\sec^{-1})} = 10^{3}\frac{\nu}{(\text{kc/s})} = 10^{6}\frac{\nu}{(\text{Mc/s})} = 10^{9}\frac{\nu}{(\text{kMc/s})} \tag{1.6}$$

Wave number

A more convenient measure of frequency, used particularly in the infrared region, is the *wave number*. Defined as the inverse of the wavelength in centimetres and designated in various books by the symbols ω, ν or σ, it is the number of waves per centimetre. The unit is 1 cm^{-1}. (The name of H. Kayser is commemorated in the suggested alternative name for this unit but the symbol K for cm^{-1} has not yet come into general use. It has the advantage that subdivision and multiplication are possible, *e.g.* $1\text{K} = 10^{3}\text{mK}$, $10^{3}\text{K} = 1\text{kK}$).

Wave number is related to frequency by the equation

$$\omega = 1/\lambda = \nu/c \tag{1.7}$$

and thus, so long as the wavelength used is corrected to vacuum (when c is constant), ω is directly proportional to ν. Wave number is thus a measure of frequency even though it is incorrect to speak of "a frequency of so many reciprocal centimetres". It is also incorrect to speak of "a band at 3000 wave numbers".

As with wavelength so with wave number, it often happens that a spectrometer is calibrated against a standard spectrum in which wave numbers are given, properly derived from corrected wavelengths. In this case wave numbers obtained by interpolation are correct, in so far as vacuum correction is not required. Some wave-number corrections are listed in Table 1.

The frequencies and wave numbers corresponding to various wavelengths are given in Table 2.

Energy and Frequency

For radiation of frequency ν, wavelength λ, and wave number ω, the transmitted quantum of energy, ε, is:

$$\varepsilon = h\nu = hc/\lambda = hc\omega \tag{1.8}$$

With Planck's constant, $h = 6.6252 \times 10^{-27}$ erg sec, ν in sec^{-1} and λ in cm, ε is calculated in ergs. The value in electron volts is obtained from the following conversion:

$$\frac{\varepsilon}{(\text{eV})} = \frac{6.6252 \times 10^{-27}}{1.601864 \times 10^{-12}} \frac{\nu}{(\text{sec}^{-1})}$$

$$= 4.135038 \times 10^{-15} \frac{\nu}{(\text{sec}^{-1})}$$

$$= 4.135038 \times 2.99793 \times 10^{-5} \frac{\omega}{(\text{cm}^{-1})}$$

$$= 1.239644 \times 10^{-4} \frac{\omega}{(\text{cm}^{-1})} \tag{1.9}$$

Conversely:

$$\frac{\nu}{(\text{sec}^{-1})} = 2.41812 \times 10^{14} \frac{\varepsilon}{(\text{eV})} \tag{1.10}$$

and

$$\frac{\omega}{(\text{cm}^{-1})} = 8065.98 \frac{\varepsilon}{(\text{eV})} \tag{1.11}$$

The energy transmitted by N quanta is called the *gram-molecular quantum*, $Nh\nu$, and this energy is conveniently quoted (for comparison with chemical bond energy terms) in joules or calories per mole

$$\frac{Nh\nu}{(\text{J mole}^{-1})} = 3.990026 \times 10^{-10} \frac{\nu}{(\text{sec}^{-1})}$$

$$= 11.96170 \frac{\omega}{(\text{cm}^{-1})}$$

$$= 1.196170 \times 10^{9} \frac{(\text{Å})}{\lambda} \tag{1.12}$$

$$\frac{Nh\nu}{(\text{cal mole}^{-1})} = 9.5364 \times 10^{-11} \frac{\nu}{(\text{sec}^{-1})}$$

$$= 2.8589 \frac{\omega}{(\text{cm}^{-1})}$$

$$= 2.8589 \times 10^{8} \frac{(\text{Å})}{\lambda} \tag{1.13}$$

Energies per quantum and per gram-molecular quantum are recorded for various wavelengths and frequencies in Table 2.

References p. 64–66

TABLE 2

WAVELENGTH, ENERGY AND FREQUENCY

	λ		ν		ω	ε	$Nh\nu$
	cm	other units	sec^{-1}	other units	cm^{-1}	eV	kcal/mole
X-rays	10^{-8}	1 Å	3×10^{18}		10^{8}	1.2×10^{4}	2.9×10^{5}
Vacuum	10^{-6}	100 Å, 10 mμ	3×10^{16}		10^{6}	124	2.9×10^{3}
ultraviolet	2×10^{-5}	2000 Å, 200 mμ	1.5×10^{15}		50,000	6,2	143
Ultraviolet	4×10^{-5}	4000 Å, 400 mμ	7.5×10^{14}		25,000	3.1	72
Violet		4200 Å	7.1×10^{14}		23,800	3.0	68
Blue		4900 Å	6.1×10^{14}		20,400	2.5	58.3
Green		5300 Å	5.7×10^{14}		18,900	2.3	54
Yellow		5900 Å	5.1×10^{14}		17,000	2.1	49
Orange		6500 Å	4.6×10^{14}		15,400	1.9	44
Red		7500 Å	4.0×10^{14}		13,300	1.6	38
	10^{-4}	10000 Å, 1 μ	3×10^{14}		10,000	1.2	29
Infra-red	10^{-3}	10 μ	3×10^{13}		1,000	0.12	2.9
	10^{-2}	100 μ	3×10^{12}		100	0.012	0.29
	10^{-1}	1 mm	3×10^{11},	300 kMc	10	1.2×10^{-3}	2.9×10^{-2}
Microwave	1	10 mm	3×10^{10},	30 kMc	1	1.2×10^{-4}	2.9×10^{-3}
	10	100 mm	3×10^{9},	3000 Mc	0.1	1.2×10^{-5}	2.9×10^{-4}
	10^{2}	1 m	3×10^{8},	300 Mc	10^{-2}	1.2×10^{-6}	2.9×10^{-5}
Radiowave	10^{4}	100 m	3×10^{6},	3 Mc	10^{-4}	1.2×10^{-8}	2.9×10^{-7}
	10^{6}	10 km	3×10^{4},	30 kc	10^{-6}	1.2×10^{-10}	2.9×10^{-9}

General Experimental Arrangement

The basic requirements for all spectroscopic measurements are a source, a dispersion element and a detector. The source may be one whose emission is to be measured, or it may be one chosen to emit all wavelengths (a *continuum*) within a certain range so that absorption by material in the light path may be observed. In either case the radiation has to be sorted into its constituent parts at different wavelengths (dispersion) and means have then to be found of detecting radiant energy at those wavelengths and of measuring its intensity. These and other essentials are described in the following sections but first we may take note of some of the ways in which the components are arranged.

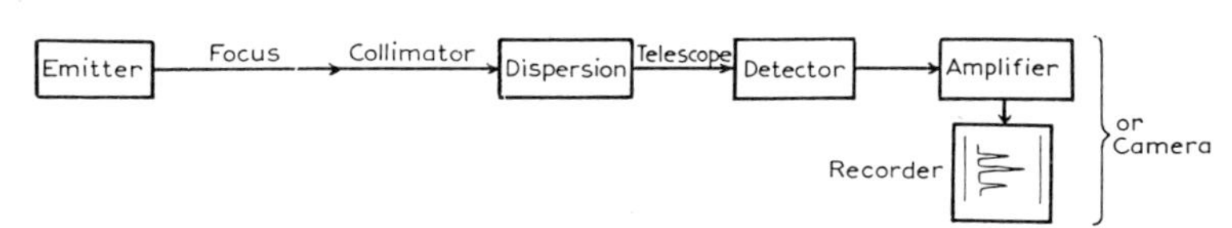

Fig. 2. Arrangement for observation of emission spectra.

Fig. 2 shows the simplest arrangement which can be employed for emission spectra such as Figs 1(a)—(c). The emission from, say, a gaseous discharge is focussed onto the entrance slit of the dispersion unit (the *spectroscope* or *monochromator*) within which the light is normally collimated on to the dispersion element and is then focussed on to the detector. The dispersion element may be controlled manually, or it may be moved automatically in a controllable manner, to scan the wavelength range: and the appropriate response comes at the wavelengths at which emission occurs. For absorption spectra it is only necessary to substitute a source which emits all wavelengths over a useful interval of wavelength and to focus the radiation on to the entrance slit of the monochromator through a convenient thickness of an absorbing sample — Fig. 3. The recorded spectrum is then, as in Fig. 1(d), a graph showing the energy reaching the detector at each wavelength. To determine the absorption the wavelength interval is covered again without the absorbing sample, care being taken to ensure that the source and detector have remained stable over the two scans.

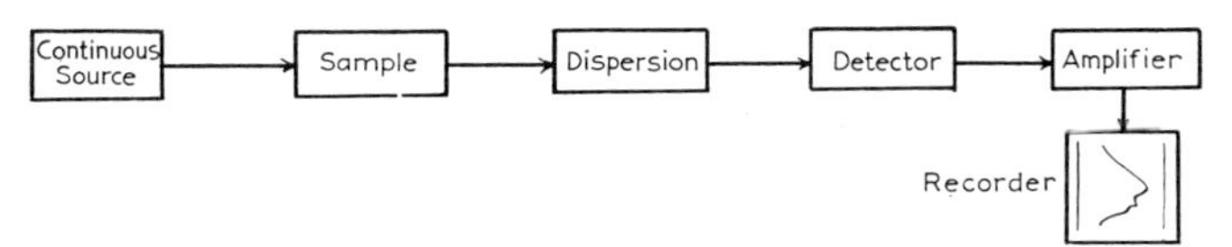

Fig. 3. Arrangement for observation of absorption spectra — single beam.

An absorption spectrum showing just the fraction of energy absorbed, as in Fig. 1(e), can be plotted from Fig. 1(d) by making a point-to-point comparison of the absorption and background curves. Alternatively the spectrometer can be arranged — Fig. 4 — to make this comparison at every

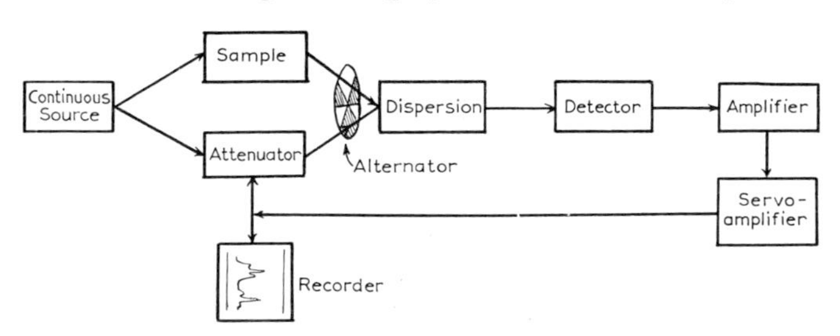

Fig. 4. Arrangement for observation of absorption spectra — double beam comparison.

wavelength by dividing the radiation from the source into two beams. One of these passes through the sample, the other through a variable and calibrated attenuator: an oscillating or rotating mirror presents the two beams alternately to the monochromator and detector. As the monochromator scans the wavelength range the detector-and-amplifier system notes and discriminates between the two signals. Any difference between them, as will

References p. 64–66

occur when there is absorption by the sample, is used by the servo-amplifier to actuate a motor to drive the attenuator into the reference beam. The energy of the comparison beam is thus reduced until there is no difference between the two beams. The servo-mechanism is arranged to move in such a way as always to reduce the difference between the two beams and thus to maintain a balance. The absorption by the sample is given by the position of the attenuator at each wavelength: a mechanical or electrical connection between the attenuator and a recorder pen enables a curve such as Fig. 1(e) to be drawn on paper which is moved forward as the wavelength is changed. This double-beam facility has been generally available in commercial infra-red spectrometers since about 1948 but has, until recently, been less widely available in spectrophotometers covering the visible and ultraviolet regions.

Many instruments are in use which are, in principle, similar to the double-beam spectrometer. Those absorptiometers which are operated manually (and are therefore cheaper) are generally provided with a sliding carriage which enables the sample to be moved in and out of the beam. At a chosen wavelength the energy transmitted by the sample is noted. The sample is then removed (or, if it is a solution, replaced by a cell containing the pure solvent) and an attenuator is adjusted to reduce the energy to the same value. The attenuator is usually a diaphragm calibrated directly in terms of percentage absorption or of optical density (see p. 47): in this case the eye and the hand holding the pencil make the connection between the attenuator and the record. To record a spectrum the wavelength setting is changed by hand and the measurement is repeated as often as is necessary. This can be a tedious business but it is the method by which most absorption spectra were obtained until comparatively recently.

For Raman and fluorescence spectroscopy the arrangement is very similar to that used for emission spectra but with modification at the source. These are discussed separately on pp. 60 and 63.

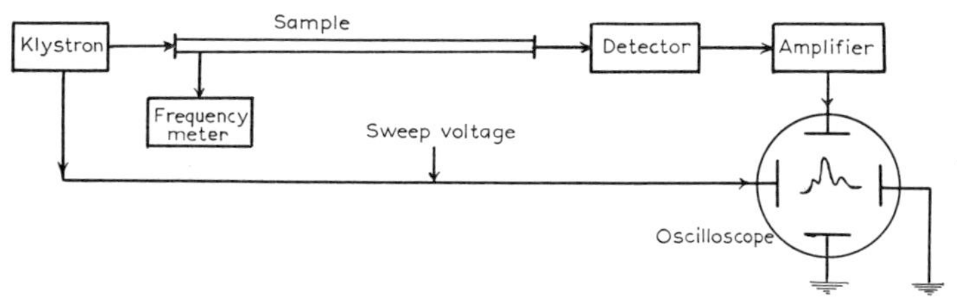

Fig. 5. Arrangement for observation of microwave absorption.

In microwave spectroscopy (Fig. 5) the experimental arrangement is simplified to the extent that the source is also the dispersion element. That is because the source is monochromatic but, within limits, its frequency can be varied. It is therefore only necessary to vary the source frequency by changing the applied voltage to find at which frequencies the sample absorbs energy. A balance between power input and power received at the detector is upset by absorption by the sample: the out-of-balance is a measure of the absorption and can be presented as such by the recorder. Generally the source frequency is swept through a small range by a relatively slowly changing voltage (say at 50 c/s) which is also applied to the X-plates of a cathode-ray oscilloscope. The amplifier output goes to the Y-plates and there is drawn on the CRO screen a graph of absorption against frequency.

Electron spin resonance (p. 103) involves transitions between energy levels whose separation is caused by and depends upon a magnetic field. With the magnetic fields available in the laboratory, around 10,000 gauss, the transitions between levels fall in the microwave region and hence the technique is similar to microwave absorption. However, the dependence of the energy level separation on the magnetic field enables the absorption to be more conveniently observed by holding the source frequency constant and varying the magnetic field by means of subsidiary sweep coils, Fig. 6.

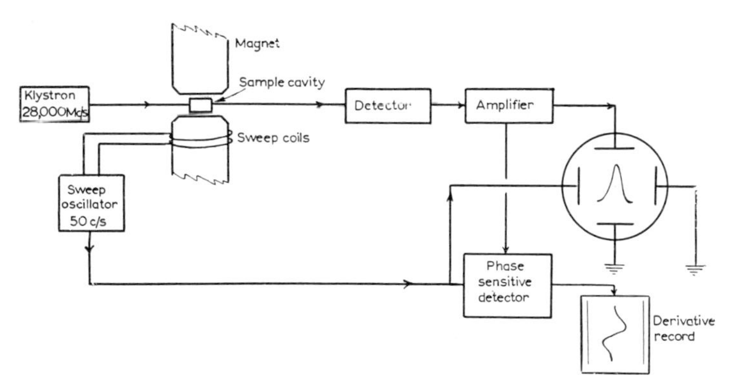

Fig. 6. Arrangement for observation of electron spin resonance absorption (microwave region).

Absorption then occurs at that value of the magnetic field which separates the levels by an amount corresponding to the fixed oscillator frequency. The sweep voltage controlling the variation in the field is now applied to the X-plates of the oscilloscope and a graph is drawn of absorption against magnetic field. This is entirely equivalent to a graph against frequency since frequency and field are directly proportional. For electron spin resonance,

$$1 \text{ gauss} = 2.80 \text{ Mc/s} \tag{1.14}$$

so that in a 10,000 gauss field absorption occurs at approximately 28,000 Mc/s.

References p. 64–66

Nuclear magnetic resonance involves similar principles but a different range of frequencies. Energy-level separations arising from the interaction of laboratory magnetic fields and nuclear magnets (p. 104) are smaller than those for electrons: corresponding frequencies lie in the radiofrequency (RF) range, say 2—60 Mc/s. in a 10,000 gauss field. The source is a RF signal generator and the sample is contained in a small tube within an induction coil whose axis is perpendicular to the magnetic field, Fig. 7. Variation of

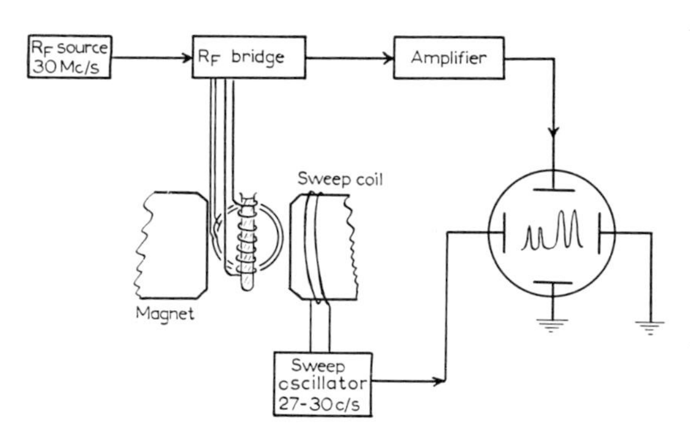

Fig. 7. Arrangement for observation of nuclear magnetic resonance absorption.

the magnetic field changes the energy level separation and at one value of the field strength the separation corresponds to the RF signal. Energy is absorbed by the sample from the RF field set up by the coil. The magnetic susceptibility of the coil is altered in consequence, the inductance of the coil changes, and the RF bridge of which the coil is a part is thrown out of balance. This allows a signal to get through to the amplifier and thence to the Y-plates of the oscilloscope. The magnetic sweep voltage is again applied to the X-plates and thus absorption is plotted against magnetic field. Again this is equivalent to plotting against frequency but the proportionality factor is different for each nucleus (see Table 8, p. 106).

Sources

Emission spectra are, for the most part, concerned with transitions from upper electronic levels in atoms and in simple molecular species. Sources vary in the means of excitation of the upper levels. Temperatures reached in a flame are sufficient to excite upper electronic levels in neutral atoms (that is, without ionization) and in molecules. Flame spectra are therefore either line spectra of neutral atoms or bands of very closely spaced lines arising from molecules. The diatomic or polyatomic species responsible for emission bands may be the molecule as introduced into the flame but frequently it is a more stable fragment. For instance, a common component of the spectrum of hydrocarbon flames is a band due to the "spectroscopic molecule", C_2: there are many other diatomic species which are only observed spectroscopically under conditions necessary to excite emission.

The technique of production of atomic flame spectra (of metallic elements) is to inject a fine spray of a solution of a reasonably volatile salt of the metal into a hot flame such as the oxy-acetylene flame. Chlorides are generally preferred. Where the emission is relatively simple and perhaps contains only one strong line in the visible region the result is a colour characteristic enough for visual identification of the element. This is the case for the alkali metals and alkaline earths. With careful control of the flame and of the admission of the sample it is possible to use this source for quantitative estimation of a number of elements: this is *flame photometry*.

More vigorous excitation comes from electric discharges of various forms. A *discharge* can be excited in a gas or vapour at low pressure with little contribution from the material of the electrode. At higher voltages and closer approach of the electrodes an *arc* may be formed in which electrode material is volatilized and excited to emission. The rapid passage of a high current for a short time across a small electrode gap at a very high voltage is a *spark*. The violence of excitation, increasing in the order in which the methods are mentioned, leads to increasing ionization of the emitters so that different systems of spectral lines appear. In classification these are distinguished by Roman numerals after the symbol of the element: thus Zn I denotes the spectrum arising from neutral zinc, Zn II from Zn^+, Zn III from Zn^{2+}, etc. The notation is easy to describe: sorting out the systems on a photographic plate is another matter. The highest system to be identified is Sn XXIV.

Discharge tubes

The gaseous-discharge tube is one in which the current is carried both by electrons and by positive ions which arise from electron bombardment of the gas at low pressure (a few millimetres). If, as in many tubes, the positive column in the discharge is being observed — sometimes end-on down the length of the tube for greater intensity — the emission is predominantly atomic. Not only can these tubes be used for permanent gases (hydrogen, helium, oxygen, nitrogen, neon, argon, etc.) but also for some of the more volatile metals. If the metal is sufficiently volatile the heat of the discharge produces metal vapour which takes over the major portion of the current from the gas in which the discharge is started. Sodium, cadmium and mercury are examples of elements whose spectra can be excited in this way. Sodium

References p. 64–66

street lamps glow red on first lighting up: the discharge is initially carried by neon and gradually the red neon emission gives way to the characteristic sodium yellow as the heat of the discharge volatilizes metallic sodium.

In the hollow-cathode tube, Fig. 8, it is the negative glow of the gaseous discharge which is under observation. Here the excitation is almost completely by electron bombardment and the same means is used to volatilize

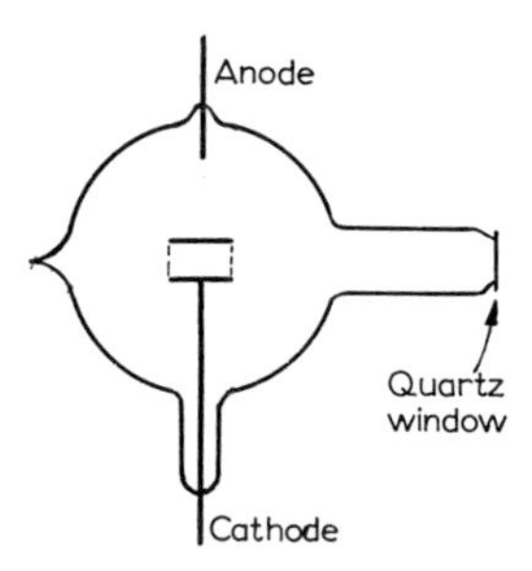

Fig. 8. Hollow cathode discharge lamp.

the metal lining of the hollow tube which constitutes the cathode. The emission, almost entirely from the hollow cathode, contains lines from the ionized species (spark lines) in addition to those from neutral atoms (arc lines): this depends upon the carrier gas, argon (ionization potential, 11.5 V) favours arc lines, helium (19.7 V) favours spark lines. As a source of atomic lines the hollow-cathode discharge tube containing a particular element is ideal for the observation of absorption by atoms of the same element, and this has obvious analytical application. It is for this purpose (atomic-absorption spectroscopy) that hollow-cathode lamps have been much developed recently: they are available for elements Cu, Fe, Mn, Ni, Cr, Co, Al/Mg, Au, Ag, Rh, Pd, Pt, Cu/Zn, Ca, Sr, Ta, Ti, W, Mo, Pb.

Discharge tubes containing mercury are available in a wide variety of shapes, sizes and running conditions. Distinction between arcs and discharge tubes is vague and the term *mercury vapour arc* is frequently used. Operated at low pressures with water-cooled pools of mercury as electrodes, the discharge produces a spectrum of sharp lines rich in the green, blue and ultraviolet regions. Such an arc, wound in the form of a spiral up to a metre in length and capable of carrying direct currents of 20—30 A across a voltage drop of about 100 V, is the so-called Toronto arc used for excitation of Raman spectra[3–5]. For that purpose (p. 60) the two requirements (which are not easily fulfilled simultaneously) are high intensity and sharp lines. A low pressure of mercury may also be combined with a small pressure of inert gas and the electrodes are of tungsten coated with oxide. Some such lamps can be operated with alternating current and, because of the low pressures used, the emitted lines are sharp enough for Raman spectroscopy. For very high intensities of ultraviolet light high-pressure mercury arcs are used but the emission is of greatly broadened lines with considerable self-absorption and some continuous background. Hence, although they are of considerable use for photochemistry, high-pressure arcs have limited spectroscopic value.

Another means of excitation which has possible advantages for Raman sources is that which places a tube close to a high-power very high frequency radio emitter or in a resonant cavity[6]. The tube is without electrodes and can therefore be charged with the element whose spectrum is required, sealed off and kept without attachment to a cumbrous vacuum system. Mercury, helium, and cadmium are examples of elements which behave well in electrodeless discharges.

Arcs and sparks

Open arcs and sparks are perhaps the most widely used sources for elementary analysis by emission spectrography. Metals for analysis can, of course, be employed as electrodes themselves. Carbon, copper and aluminium are frequently used as electrodes for detection of other elements: a solution containing the unknown element can be evaporated on to these electrodes or a powdered sample may be packed into holes drilled into the carbon electrodes. Intense local heating at the electrode serves to vapourize the material. In general the arc spectrum, except near the electrodes, is of un-ionized atoms and the spark spectrum contains lines from ionized species.

Incandescent sources

Absorption spectroscopy requires a continuous variation of frequency over the appropriate frequency range. This may be achieved either by use of a source which emits a continuum in the required range or of one which is monochromatic and continuously variable. Except in the microwave and radio-frequency regions the only available general sources are those emitting a continuum. Principal among these are the sources which are essentially incandescent solids. The emission from a hot body has a distribution of energy against wavelength which is plotted in Fig. 9. The shape of the curves, first determined by O. Lummer and E. Pringsheim in 1899, was, of course, the major piece of experimental evidence upon which Planck's quantum theory was founded in 1900. The essential features to note are that the total energy emitted (W, the area under the curve) increases with temperature and that the wavelength of the maximum λ_m, moves to shorter wavelength

References p. 64–66

at high temperature. These characteristics are expressed in the formulae:

$$\frac{W}{(\text{watts cm}^{-2})} = 5.74 \times 10^{-12} \left(\frac{T}{°\text{K}}\right)^4 \qquad (1.15)$$

and

$$\frac{\lambda_m}{(\text{Å})} \cdot \frac{T}{(°\text{K})} = 2.88 \times 10^7 \qquad (1.16)$$

The first is the Stefan-Boltzmann law and the second Wien's law. These relationships, and the curves of Fig. 9, are strictly for "black-body" radiation. Practical forms of incandescent sources are not ideal in this respect and show various irregularities but they conform reasonably well.

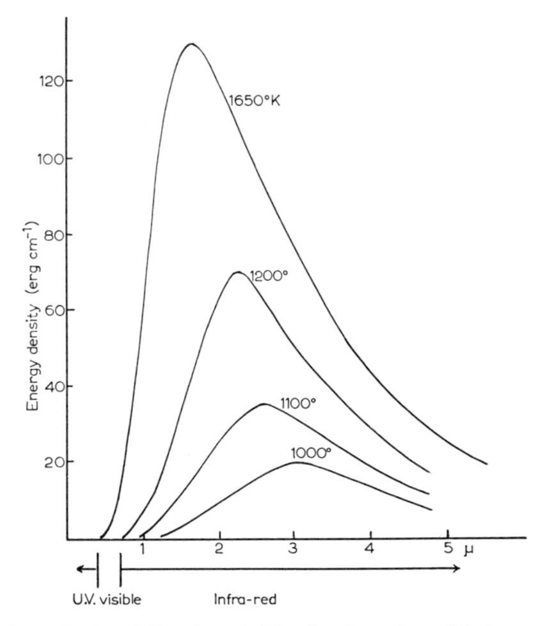

Fig. 9. Black-body radiation. Wavelength distribution of equilibrium energy density at 1000, 1100, 1200 and 1650°K.

For use in the infra-red region, mainly beyond 2μ, the hot body takes the form of a rod of semi-conducting lanthanon oxides, a *Nernst glower*. When a voltage is applied across the cold rod no current flows. On heating the rod, its resistance drops and the current is sufficient to maintain the rod at a high temperature and a low resistance. Ballast resistors have to be included to limit the current and to maintain steady temperatures of 1500—2000 °K. A somewhat lower temperature (1400°K maximum) is attained with a small version of an electric fire element, a *Globar*. This source has sometimes been preferred for longer wavelengths not so much on account of the greater energy at those wavelengths but because short wavelength energy causes embarrassment as stray (undispersed) radiation and this is less in low-temperature sources. High pressure mercury arcs have been employed for the far infra-red. No doubt the emission is essentially the same but from a hotter centre: and here stray radiation is partially eliminated by the quartz envelope which is not transparent between 4 and 50 μ.

Incandescent solid sources may obviously be used for the near infra-red, the visible and the near ultra-violet; that is, on the short wavelength side of the maximum in Fig. 9. Clearly extension into the ultra-violet requires a high temperature. The Nernst filament is only suitable for the near infra-red and visible. The tungsten filament lamp (the ordinary electric light bulb) operates at about 3000° and provides useful radiation out to 3500 Å. Projection lamps and headlight bulbs are brighter and some are designed to run at higher temperatures. Their life is thereby shortened but the quality of the visible emission is improved for photography and the ultraviolet content is greater. The ultraviolet emission beyond 3500 Å is improved by use of a silica envelope instead of a glass one. In the visible and near-infra-red the *Pointolite* provides a useful and concentrated source: it is also incandescent tungsten but it is raised to high temperature by electron bombardment from an arc maintained in an argon atmosphere.

Continuum discharge sources

Beyond the range of the tungsten lamp the ultraviolet region is catered for by discharge lamps. In particular the high-current hydrogen discharge (800—1000 V AC, passing a few amps in 1—2 mm of hydrogen, sometimes water-cooled) emits a continuum which extends from 3500—1200 Å. At wavelengths below 1950 Å the absorption by atmospheric oxygen requires that the lamp be operated in a vacuum, or in an atmosphere of nitrogen. Below 1850 Å its quartz window becomes opaque and must be replaced by fluorite; below 1450 Å a nitrogen atmosphere absorbs and must be replaced by vacuum, by helium or by hydrogen. It is evidently possible to achieve a continuous discharge spectrum with helium in the region 900—600 Å. Beyond that however discontinuous high-energy spark spectra must be used. These have been shown to emit at wavelengths as low as 4 Å and thus to overlap with the X-ray region. There is little application to chemistry here.

Another discharge lamp which was developed primarily for high-speed photography has found considerable use in the absorption spectrophotometry of transient species (see p. 308). A large condenser, charged to a high voltage, is discharged through a suitable tube in a circuit with minimum resistance and inductance. A very bright emission of very short duration results. Such *pulsed discharge tubes* are generally specially designed tubes containing argon or xenon, but it is also possible to operate some mercury discharge lamps under pulse conditions. The quality of the emission and the duration of the flash depend on the initial voltage and the capacity of the condenser. Energies of 10,000 J can be dissipated in about 10^{-4} sec for photochemical excitation and the emission is largely continuous throughout the visible and the ultraviolet. For photography of the absorption spectrum a single flash of about 1000 J and lasting some 50 μsec (from a 70 μF condenser at 4000 V) gives a sufficient exposure.

Monochromatic high frequency sources

There remain at the other end (long wavelength) of the electromagnetic spectrum the sources required for microwave and radiofrequency absorption spectroscopy. The emission from a hot body (at any attainable temperature) becomes too small to be used at wavelengths greater than $\frac{1}{2}$ mm (500 μ, corresponding to 20 cm^{-1}). On the other hand, the monochromatic sources emitting in the microwave and radio regions are capable of an output of milliwatts. Radiofrequency sources for nuclear magnetic resonance are more-or-less conventional oscillators, tuneable and designed with particular regard for stability.

Microwave generators include spark discharge, klystrons and magnetrons: of these the klystron is the most frequently used. This is essentially a resonant cavity in which electromagnetic oscillations are induced by the passage of an electron beam. In the reflex type of klystron (Fig. 10) a negatively charged

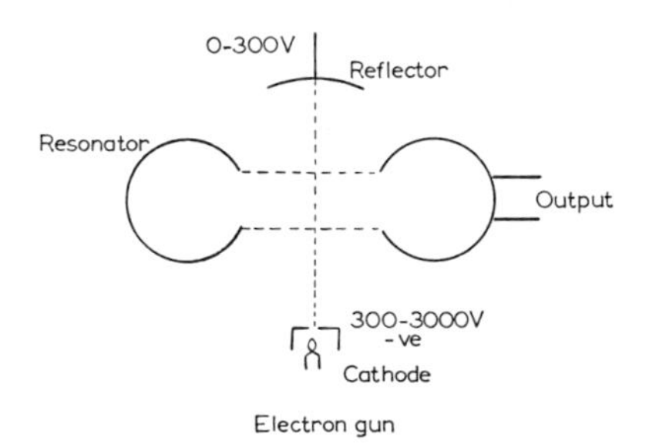

Fig. 10. Reflex klystron microwave generator (diagrammatic).

reflector returns the electron beam across the resonator. The effect of oscillations in the resonator is to bunch the electrons so that on their return after reflection they induce further oscillations in the cavity. That is, if the beam velocity and the reflector distance are property adjusted, the cavity receives more energy from the reflected beam than it delivers to the outgoing electrons and thus microwave power is generated.

The frequency of a klystron is governed essentially by the size of the cavity and klystrons are available to cover the wavelength range 5 mm—50 cm (frequencies 60,000—600 Mc/s). This range is sufficient for a study of the rotational spectra of fairly heavy molecules and for electron spin resonance (absorption at 28,000 Mc/s for a field of 10,000 gauss), but for lighter molecules it may only be possible to observe the first one or two rotational transitions below 60,000 Mc/s. The range is limited by the dimensions of the cavity: this fact, and the development of the associated equipment, has made the centimetric wave region the most studied. The wartime development of microwave techniques in radar directed attention particularly to three bands of frequency. These are designated X, K and Q, centred on 3 cm, 1.25 cm, 8 mm respectively. (9,000, 21,600 and 33,800 Mc/s).

An individual klystron may be varied in frequency by about 5 % on either side of the nominal frequency by mechanical adjustment of the cavity. This means that a range of klystrons is usually required. Small adjustments of 50—100 Mc/s are made by adjustment of the voltage on the reflector and this facility is used for sweeping the frequency for oscilloscope presentation.

For the more difficult region of wavelength below 5 mm (frequency greater than 60,000 Mc/s) polychromatic radiation from a spark discharge has been used, but the monochromatic feature of klystrons is such an advantage that development has been pressed in this direction. Harmonics in klystron oscillators, or harmonics generated from klystron fundamentals by semiconductor crystals, have extended the range to less than a millimetre (1000 μ, frequency 30,000 Mc/s, wave number 10 cm^{-1}) and thus practically to bridge the gap between centimetric waves and the far infra-red.

Spectrographs and Monochromators

In so far as all optical spectroscopy is concerned with polychromatic radiation it is necessary to separate the energy into its constituent wavelengths, to *disperse* it. This is the function of prisms, diffraction gratings or filters. Such use of a prism or grating always entails (a) a front slit illuminated by the source which is being examined, (b) a collimating spherical lens or parabolic mirror so situated that the slit is at its focus, (c) the dispersing element which receives parallel rays from the collimator and transmits or reflects different wavelengths in parallel rays but in different directions, and (d) a telescope lens or mirror which brings the rays of different wavelengths to a focus as a spectrum on a *focal plane* (Fig. 11a). There is an image

References p. 64–66

of the front slit in a different position in the focal plane for each constituent wavelength of the incident radiation. There are numerous ways in which these essential features can be laid out and Figs. 11—14 show some of them. We shall not concern ourselves here with the detailed optics or the advantages and disadvantages of each system, but merely with some of the factors which should be understood by the chemist who uses instruments based on these systems.

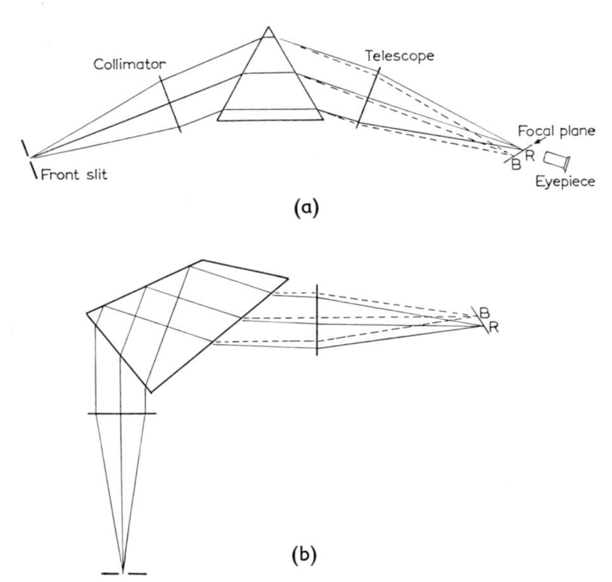

Fig. 11. (a) Simple prism spectroscope. (b) Spectroscope with Abbé-Pellin-Broca constant deviation prism. Dashed line indicates path of light of shorter wavelength.

In the first place it is possible to place a translucent screen in the focal plane so that a visible spectrum may be seen in its entirety. The range of the screen for visual use may be extended into the ultraviolet by coating the screen with a fluorescent material. More usefully a photographic plate in the focal plane allows a permanent record to be taken, as in Figs. 58 and 59, pp. 142 and 145. Such an instrument is a *spectrograph*.

If for the visible region, an eyepiece is provided for looking at a small section of the focal plane, we have a *spectroscope*. Provision has to be made for scanning the spectrum and this may be done either by moving the telescope alone, or by moving both the collimator and the telescope together to preserve minimum deviation at all wavelengths. Alternatively the prism may be rotated — as for instance in the constant-deviation spectroscope employing an Abbé-Pellin-Broca prism (Fig. 11b) — and this has the effect of moving the focal plane across the field of view of the eyepiece. The angular position of the prism table may be calibrated against wavelength.

The term *monochromator* applies to those instruments, (*e.g.*, Figs. 12(b), 12(d), 13) in which a second slit is placed in the focal plane. This slit isolates a particular wavelength (or, more correctly, a narrow band of wavelengths) of radiation which is refocussed on to a detector whose output is a measure

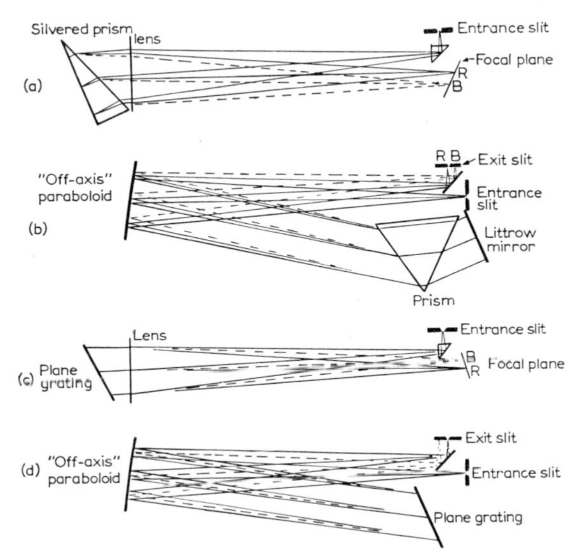

Fig. 12. Littrow optics; (a) silvered prism and lens; (b) plane mirror, prism and parabolic mirror; (c) plane reflection grating and lens; (d) plane reflection grating and parabolic mirror. Dashed line indicates path of light of shorter wavelength.

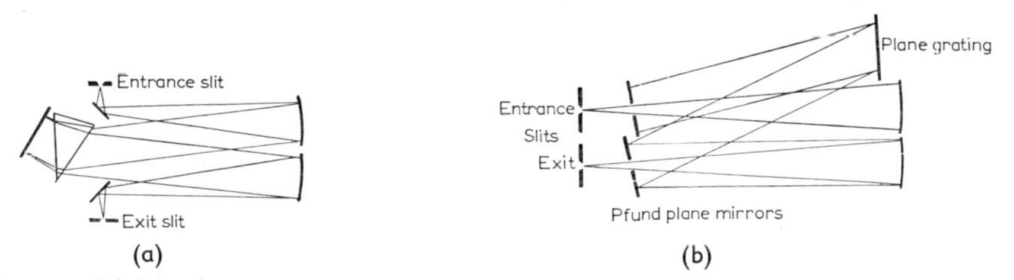

Fig. 13. (a) Pfund optics with prism and mirrors. (b) Pfund-Hardy optics with plane grating and mirrors.

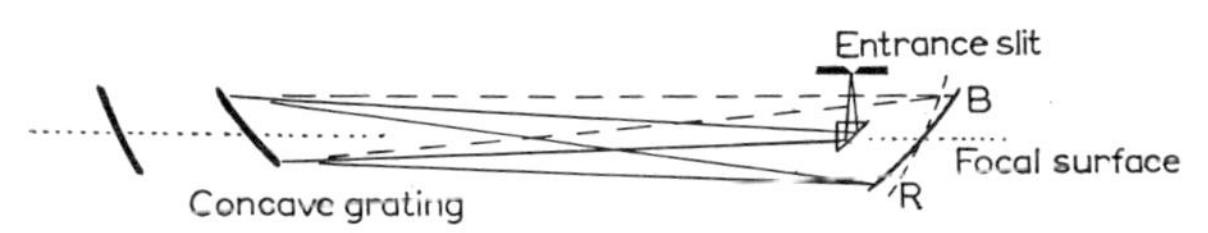

Fig. 14. Eagle mounting of a concave grating.

References p. 64–66

of the energy at the chosen wavelength. It is sometimes possible to scan the spectrum by tracking the exit slit and detector along the focal plane but it is usually much more convenient to arrange to move the dispersion element so that the focal plane moves across a fixed exit slit. This is the arrangement in most commerical spectrometers. For this purpose mirrors are preferred over lenses as focussing elements because they are achromatic. Lenses have different focal lengths at different wavelengths (*chromatic aberration*) and are thus unsuitable for monochromators of fixed focal length. Mirrors are, of course, also preferred in the infra-red because the materials which transmit in the infra-red are not very suitable for lenses.

A commercial development in the field of emission spectroscopy is the *polychromator*. Employing a grating, and considerable focal length, the instrument has a focal plane which is a yard in length between 1960 and 8000 Å. There is thus sufficient room for exit slits to be set, each with its own detector, at predetermined wavelengths. This would be of little value for scanning a spectrum but it is ideal for the rapid and simultaneous estimation of emission at known and characteristic wavelengths by up to 30 elements in a mixture. A vacuum polychromator employing a fluorite prism and four exit slits is also available to cover the range 1600—1850 Å.

Among the optical arrangements shown in Figs. 11—13 both plane gratings and prisms are shown as dispersion elements and both mirrors and lenses as means of focussing. The dispersive effects of reflection gratings and prisms result in the order of wavelengths being reversed on changing from prism to grating. The letters B and R in the figure refer to the blue (or shorter wavelength) and the red (or longer wavelength) ends of the spectrum.

Perhaps the commonest optical layout in commercial spectroscopic instruments (u.v. and visible spectrographs, u.v. and i.r. spectrometers) is the Littrow system (Fig. 12). The underlying principle is the use of reflection so that the same element (lens or mirror) serves as collimator and telescope. The focal lengths of collimator and telescope are of course identical and the front slit would lie in the telescope's focal plane save that the entrance slit is somewhat displaced from the optical axis of the lens or mirror. The focussed spectrum is displaced to the other side of the axis sufficiently to permit a small plane mirror to deflect it through 90° to a more convenient position for the exit slit. The requirement of an expensive off-axis parabolic mirror in the Littrow system is obviated by the Pfund system, (Fig. 13), where the holes in the plane mirrors permit the use of on-axis paraboloids. Rotation of the prism, or of the Littrow plane mirror or of the plane grating, enables selection of wavelength and is usually effected by means of a micrometer screw acting on a lever. For plane gratings the coupling can be arranged so that there is direct proportionality between wavelength and micrometer screw travel: when the micrometer is driven by a constant speed motor this results in a linear wavelength scan. With prisms linearity in either wavelength or wavenumber can only be accomplished by means of specially cut cams.

Prisms and Prism Materials

Prisms — and also lenses and plane windows — can be made from a variety of materials and the choice of material depends mainly upon the three factors: transmissive quality, dispersion, and cost. Of the first the relevant details are given in Table 3 with the ranges of transmission and an indication of suitability for prisms, lenses and windows. Originally tried and used as natural crystals the optical materials now used are practically entirely synthetic: controlled growth from molten salts yields large single crystals from which are cut prisms and window blanks. In the case of quartz there is some difference in the optical properties of the natural crystal and

TABLE 3

TRANSPARENCY OF VARIOUS OPTICAL MATERIALS, AND SUITABILITY FOR PRISMS, LENSES AND WINDOWS (P, L, W)

	Range	Use
Lithium fluoride	1100 Å — 9 μ	P(L)W
Calcium fluoride, fluorite	1250 Å — 12 μ	P(L)W
Quartz	1850 Å — 4.5 μ	P L W
Glass	3000 Å — 2.7 μ	P L W
Sodium chloride	2000 Å — 20 μ	P(L)W
Potassium chloride	2000 Å — 21 μ	P W
Potassium bromide	2100 Å — 30 μ	P W
Sapphire (Al_2O_3)	— 5.5 μ	W
Periklase (MgO)	— 9.5 μ	W
Barium fluoride	— 14 μ	W
Silver chloride	1 μ — 28 μ	W
Thallium bromoiodide, KRS-5	5000 Å — 40 μ	P W
Caesium iodide	2200 Å — 50 μ	P W
Quartz	50 μ —	P L W

fused silica: in particular the fused material is not doubly refracting. The transmission range for glass depends upon the composition. Flint glasses generally transmit somewhat further into the near infra-red than do crown glasses: and the limit in the ultra-violet is at about 3500 Å for these optical glasses. Borosilicate glass (Pyrex) transmits out to 3000 Å, and other special glasses have been devised to transmit well in the ultra-violet, or the infra-red.

Dispersion is the next factor in the choice of material. It is usual to use a prism at minimum deviation (where the ray traverses the prism symmetrical-

References p. 64–66

ly, Fig. 15a) and it is readily shown that the *deviation* (θ), the angle (α) of the prism, the angles of incidence and refraction (i and r) and the refractive

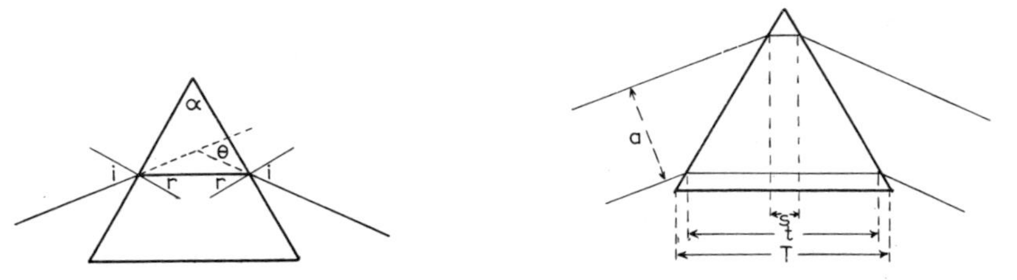

Fig. 15. Path of rays through a prism at minimum deviation.

index (n) are connected by Snell's law:

$$n = \frac{\sin i}{\sin r} = \frac{\sin\frac{\theta + \alpha}{2}}{\sin\frac{\alpha}{2}} \tag{1.17}$$

or:

$$\sin\frac{\theta}{2} + \cos\frac{\theta}{2}\tan\frac{\alpha}{2} = n\tan\frac{\alpha}{2} \tag{1.18}$$

It is clear that θ depends upon n and that the separation of adjacent wavelengths λ and $\lambda + d\lambda$ into rays at different angles θ and $\theta + d\theta$ depends upon there being a difference in refractive index, dn, between the two wavelengths. Dispersion depends upon the variation of n with λ. This is shown for different materials in Fig. 16. The *angular dispersion* $d\theta/d\lambda$ of a prism is given approximately by:

$$\frac{d\theta}{d\lambda} = \frac{dn}{d\lambda}\cdot\frac{2\tan i}{n} = \frac{dn}{d\lambda}\cdot\frac{2}{n}\tan\frac{\theta + \alpha}{2} \tag{1.19}$$

which shows more clearly that the separation of adjacent wavelengths depends upon the rate of change of n with λ; upon the *characteristic dispersion* of the material. For practical purposes what matters is the resolution of the adjacent wavelengths at the focal plane and this is governed by the *resolving power* (R) of the prism. This is the product of the angular dispersion and the aperture, a, of the collimating lens (the width of the parallel beam). It can be shown, with reference to Fig. 15(b) that:

$$R = a\cdot\frac{d\theta}{d\lambda} = (t - s)\frac{dn}{d\lambda} \tag{1.20}$$

If the whole of the prism face is filled with light, $(t - s)$ is just the thickness of the base of the prism. Thus resolution is improved by a large prism angle; but this cannot be increased indefinitely since energy losses by reflection become important — a 60° prism is the usual compromise. The resolution also depends upon the characteristic dispersion so that it can be seen from Fig. 16 (n against λ) and Fig. 17 ($dn/d\lambda$ against λ) that the dispersion of each material increases rapidly as the limit of transparency is reached.

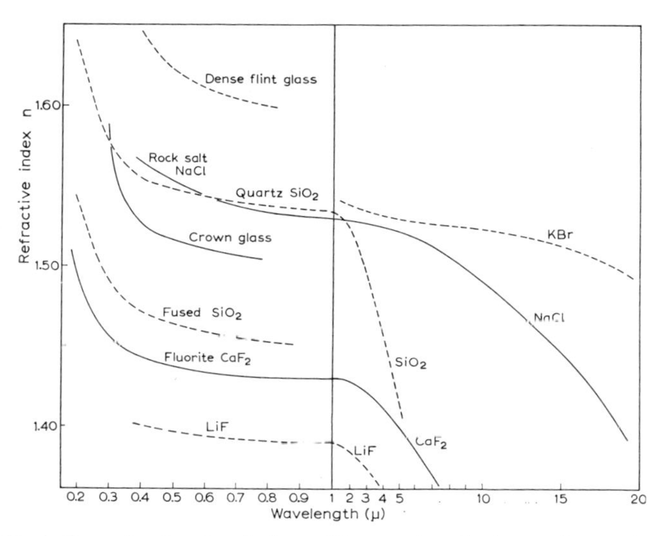

Fig. 16. Variation of refractive index with wavelength for various prism materials.

It is therefore best from the point of view of resolution to use a prism material close to its absorption edge. The preferred materials are given in Table 4. It may well be that cost prohibits the use of, say, six prisms in the infra-red region, 1—50 μ: cost is certainly an important factor where the price of a single prism may be as much as £800. Infra-red spectra which are satisfactory for the purpose of identification ("fingerprints") in the region 2.5—15 μ can be obtained with a sodium chloride prism alone: but even for identification it is sometimes desirable to have the extra resolution around 3 μ which is obtainable with a lithium fluoride prism. The present trend in commercial instrumentation is to avoid large prisms altogether and to use gratings as the main dispersive elements.

The physical strength, the ability to take an optical polish, and the reaction to atmospheric moisture are further factors in the choice of material. Silver chloride and, to a lesser extent, thallium bromiodide are unsuitable as prism materials because they are too soft to retain the shape of a large

References p. 64–66

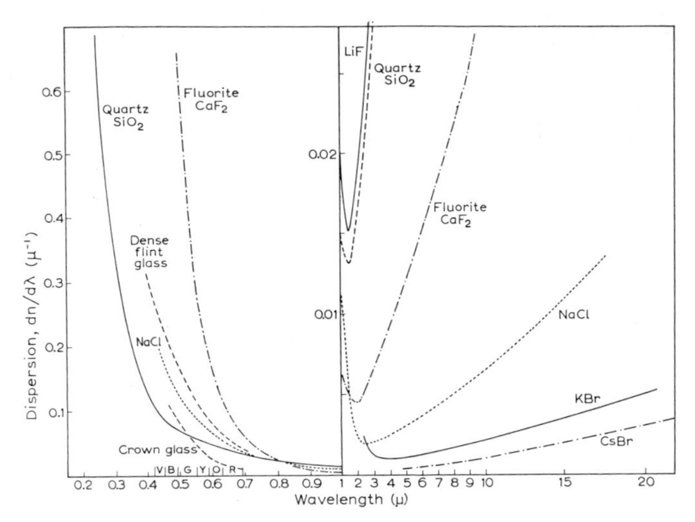

Fig. 17. Variation of characteristic dispersion with wavelength for various prism materials.

TABLE 4

RECOMMENDED PRISM MATERIALS

1200 — 2000 Å	lithium fluoride or fluorite
2000 — 4000 Å	quartz
4000 Å — 1 μ	glass
1 μ — 2.7 μ	quartz
2.7 — 5 μ	lithium fluoride
5 — 9 μ	fluorite
9 — 15 μ	sodium chloride
15 — 25 μ	potassium bromide
25 — 50 μ	caesium iodide

prism. Silver chloride, furthermore, is photosensitive. All the materials in Table 4 make good prisms but alkali chlorides, bromides and iodides are hygroscopic and the polished surfaces are very susceptible to etching in moist air. The prisms can be kept in desiccators, and the spectrometer case should be kept dry and at a temperature somewhat higher than that of the room. It is worth noting however that in the last resort it is better to risk the danger of surface etching than to drop the prism: the faces can always be repolished.

Lenses are readily available in glass and quartz so that transmission optics are feasible from 2000 Å—2.7 μ. Although lenses have been made in lithium fluoride, fluorite and sodium chloride it is usually preferable, on grounds of cost, to use mirrors for the optics of spectrometers required to operate outside the limit of transparency of quartz. It is often more convenient to use mirrors even within those limits, since mirrors are achromatic.

In passing we may note that prisms have been employed which contain a liquid enclosed by plane windows. For the windows the same transparency limits apply and possible liquids, which have some advantages over the solids in regard to dispersion, are water, monobromonaphthalene, ethyl cinnamate, and aqueous solution of barium mercuric bromide.

Gratings

Although Rowland demonstrated as early as 1882 the value and optical simplicity of spectroscopes employing concave gratings and these are still in use (*e.g.* Fig. 14), our principal concern is with plane reflection gratings. These may be optically flat metal surfaces upon which a series of regularly spaced grooves have been ruled. Alternatively they may be replicas taken from master rulings by laying down a plastic film on the master, stripping the film off when it has set, laying it down on an optically flat glass plate, and finally coating it with evaporated aluminium. Until recently good gratings were very expensive (although not so expensive as the prisms which they replace in the infra-red region) but the method of ruling developed by Merton and Sayce[7,8] has made available excellent replica gratings at about 35/- per square inch. There has been a consequent development of monochromators making use of them. The optical theory of gratings is thoroughly discussed in various books on practical spectroscopy. Here we note merely the simple results for a plane echelette grating in a Littrow mounting.

Fig. 18(a) shows the situation in which a grating with rulings, distance l apart, makes an angle α between its normal and the incident beam. Observation in a Littrow mounting is confined to the same direction as the direction of incidence and it is clear that light of wavelength λ will be diffracted in that direction when the grating is so oriented that

$$2l \sin \alpha = n\lambda \qquad (1.21)$$

where n is an integer, referred to as the *order* of diffraction. At any one angle α there will be diffracted wavelength λ_α in the first order, but also $\frac{1}{2}\lambda_\alpha$ in the second, $\frac{1}{3}\lambda_\alpha$ in the third and so on. Because of this overlapping of orders it is necessary to effect a preliminary sorting of wavelength regions. This can be done with filters (p. 29) which are transparent, say, to λ_α but not to $\frac{1}{2}\lambda_\alpha$ and shorter wavelengths. Alternatively a double monochromator is used, in which the first section uses a low resolution prism dispersion (*fore-prism*) and the second a grating. In this case the fore-prism unit must be coupled by

References p. 64–66

means of a cam to the diffraction grating in such a way that it presents to the front slit of the grating monochromator a broad band of wavelengths centred on the wavelength for which the grating is set, and maintains this relationship as the grating scans a spectrum.

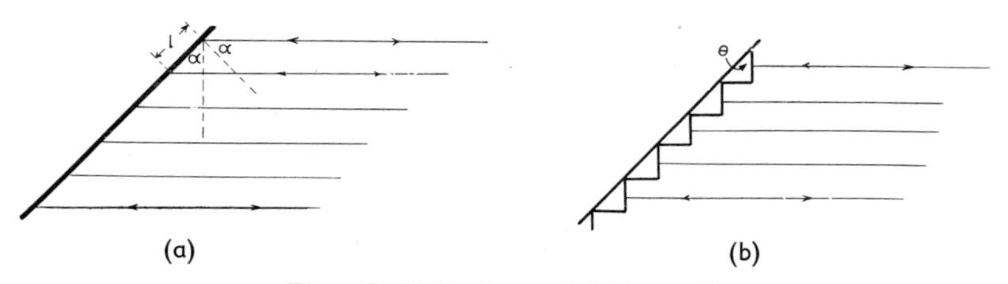

Fig. 18. Reflection echelette grating.

A reflection grating becomes an echelette grating when it is ruled in such a way that there is a series of plane facets lying at an angle θ to the surface of the grating — Fig. 18(b). The effect of this is to concentrate much of the diffracted energy around the diffraction angle $\alpha = \theta$. Because of this effect θ is called the *blaze angle*. Suppose, for example, we have a grating with 2450 lines per inch. The groove spacing is thus 0.01 mm or 10 μ. Suppose that the blaze angle is 27°, then the first-order wavelength which is diffracted at an angle $\alpha = 27°$ is 9 μ, by equation (1.21). The grating is said to be "blazed at 9 μ in the first order". Energy in the region of 9 μ will be mainly diffracted in the first order, *i.e.* in direction close to a diffraction angle of 27°. On the other hand, energy in the region of 4.5 μ will be mainly diffracted in second order since this also corresponds to a diffraction angle close to 27°. Similarly radiation of wavelength 3 μ will be mainly concentrated in the third order, and so on. Visible light at about 0.45 μ (*e.g.* the mercury blue line at 4350 Å) is diffracted in the twentieth order at 27° but, since the orders for such wavelengths are closely spaced and those near to the twentieth are diffracted at angles still close to 27°, the effect observable with such visible lines is a gradual increase in intensity of each order up to a maximum at the blaze angle and then a fall in intensity in higher orders. The appearance of many orders of short-wavelength radiation is valuable for calibration of gratings in the infra-red.

The theoretical resolving power of a grating is given by the formula

$$R = Nn \tag{1.22}$$

where N is the total number of rulings on the grating, and n is the order used. A bigger ruled area therefore makes for higher resolution (provided the whole area is illuminated) and the resolution does not depend upon the wavelength except in so far as the wavelength governs the order which it is feasible to use.

Imperfections in the ruling of gratings, or those introduced in taking replicas, lead to loss of resolving power or to the appearance of false spectral lines, or "ghosts". These need not be discussed here, but further information should be sought by those who are going to work with gratings.

Filters

There are many situations in which wavelength separation is required but without the necessity of achieving the degree of spectral purity obtainable with monochromators. It may be sufficient merely to isolate a band of wavelengths, or to ensure that only those wavelengths above, or below, a certain value are transmitted. For this purpose filters are used.

Though not strictly dispersion elements, a set of filters each transmitting a narrow band of wavelengths can serve a similar purpose. Absorptiometers (colorimeters) working in the visible region are provided with a set of coloured gelatine filters. The transmission curves of a typical set are shown in Fig. 19. Where it is desired to use for quantitative analysis an empirical

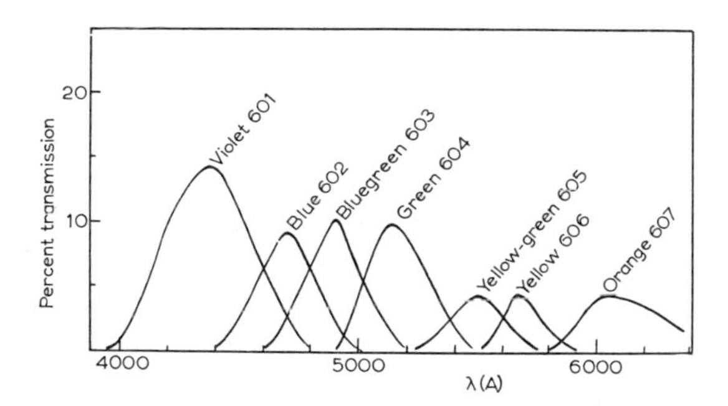

Fig. 19. Transmission curves for Ilford gelatine filters in visible region.

correlation between intensity of absorption and concentration of a coloured substance the sensitivity is increased if light of the right colour is used rather than white light. Thus for a blue solution (which is blue because it absorbs red light) it is appropriate to use a red filter. An approximate absorption spectrum of the coloured substance can be obtained by noting the percentage absorption when each filter is used in turn.

Filters made from glass of special composition are available for the visible and also for the near-ultraviolet and the near infra-red. Most of these transmit fairly broad bands, although the black glass developed for isolation of the 3660 Å band of the mercury arc spectrum has a fairly narrow transmission range. Where the purpose is to isolate a particular line in a line spectrum it may not be necessary to use a narrow-band filter if the other lines in the spectrum are not close. Sometimes the isolation is achieved by using two filters, having absorption edges just below and just above the desired wavelength. Sets of solutions and glasses for the isolation of the principal ultraviolet lines of the mercury arc spectrum are listed by Bowen[9] and by Kasha[10]. Other such filters and filter combinations can be devised as required by reference to the absorption spectra of various glasses published by the manufacturers and of various solutions published in International Critical Tables[11, 12].

The isolation of the visible mercury arc lines is required for Raman spectroscopy and it is normally convenient to use solutions in an annular space surrounding the Raman cell. The blue line (4358 Å) is most used for colourless materials and a saturated solution of aqueous sodium nitrite absorbs wavelengths below 4358 Å: praseodymium chloride absorbs from 4400—4800 Å and the rhodamine dye 5GDN-extra from 4600—5400 Å. For excitation by the violet line (4047 Å) dilute sodium nitrite solution removes the 3660 Å line and a solution of iodine in carbon tetrachloride removes longer wavelengths. Basic potassium chromate and copper sulphate solutions are used for the green line (5461 Å).

Isolation of narrow spectral regions can also be accomplished by means of interference filters and scattering filters. In the first, two transparent plates, on to which have been evaporated partially reflecting metal films, are separated by a transparent dielectric. Multiple reflections, occurring within the cell, lead to interference and only a narrow band of wavelengths is transmitted, the central wavelength being governed by the dimensions and the nature of the dielectric. The width is 100—200 Å and the maximum transmission about 50 %. Available for the near ultraviolet and visible they are now being developed for the near infra-red. The scattering, or *Christiansen*, filter is essentially a transparent dielectric (a liquid, or air) in which is suspended a transparent powder. At those wavelengths at which the refractive indices of powder and dielectric are different the beam is scattered in all directions, but light is transmitted at the wavelength at which the refractive indices are equal. For example quartz chips suspended in mixtures of cyclohexane and decahydronaphthalene give bands about 100 Å in width in the region 2500—2800 Å. The maximum transmission is 30 % and the band centre is sensitive to the composition of the liquid mixture and to temperature. Such filters have been used in all regions. See, for example, Kayser and Sawyer (Ref. 64 and Ref. 49, p. 297).

In grating spectrometers, particularly in the infra-red, there is a need to eliminate unwanted higher orders; that is, to filter out shorter wavelength radiation which will appear in higher orders at the same angle as the wanted wavelength. For this purpose one can use the F-centre filters of potassium chloride or bromide produced by controlled treatment of the crystals with X-rays and heat. These can be made so that they are opaque out to 1, 2, 3 or 5 μ and transparent at longer wavelengths. Other longwave-pass filters are made from lead sulphide, selenide and telluride. Prism instruments for the far infra-red suffer from stray (undispersed) radiation predominantly of wavelengths around the peak of the Nernst filament at 1—2 μ: this can be considerably reduced by use of an F-centre filter, or by roughening one of the plane mirrors near to the detector so that short wavelength radiation following that path is scattered rather than reflected on to the detector. A plane grating of fine enough ruling can be similarly used for selective reflection.

Another device to overcome the problem of stray light, in instruments in which the light beam is chopped, is to use a chopper of a material opaque at the wanted wavelength and transparent to the main components of the stray light. Thus a quartz chopper interrupts radiation from 4—50 μ; but that from 1—4 μ passes unmodulated and is therefore not noticed by an amplifier tuned to the frequency of chopping.

Detectors

Detectors include such selective devices as the eye, the photographic plate, photoelectric cells, crystal detectors and radio-receivers; and the non-selective thermal devices, the bolometer and the thermocouple.

The eye has a limited range but within that range (7000—4000 Å roughly) is at once the most sensitive and one of the most misleading of detectors. It can discriminate colour and can therefore give a reliable indication of the approximate wavelength of monochromatic light, but it can give the same colour sensation for certain mixtures of other wavelengths. It is notoriously misleading in the estimaton of intensity except in judging equal intensities of two adjacent areas of monochromatic light of the same wavelength. The spectral sensitivity also depends upon the intensity. Normal *photopic* vision employing the cones on the retina has maximum sensitivity on the yellow-green border: for low intensities the retinal rods are used (*scotopic* vision) with a maximum sensitivity in the green (Fig. 20).

Photographic detection

Photography is a valuable means of detection for all wavelengths shorter than 1.3 μ; *i.e.*, for the ultraviolet, visible and near-infra-red. A wide variety

References p. 64–66

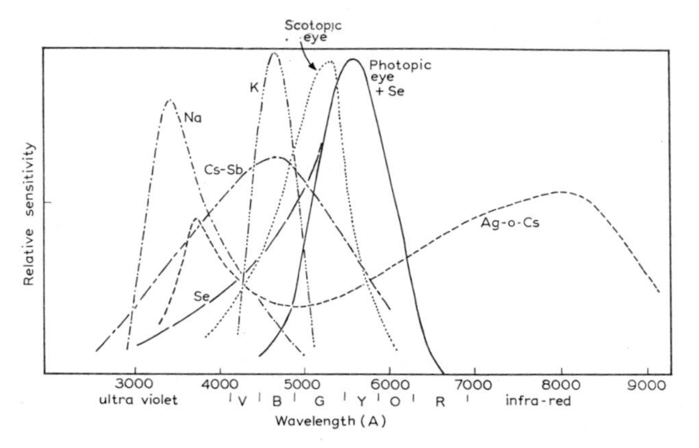

Fig. 20. Sensitivity of various detectors: the eye, various photocathode coatings (Na, K, Cs–Sb, Ag–O–Cs), and the selenium barrier-layer cell. The relative sensitivity scale permits comparison of sensitivity of any one detector at various wavelengths but does not allow comparison between different detectors at one wavelength.

of emulsions has been developed for various spectral regions. Sensitivity (speed) and the resolving power (graininess) are also variable and can be selected to suit the problem. Sensitivity varies with wavelength and density of blackening varies more-or-less as the logarithm of the exposure time. The accurate estimation of intensity requires care and experience: a good account is given in Harrison, Lord and Loofburow[50]. Photography has the great advantage over all other modes of detection that the plate can be placed in the focal plane and record the complete spectrum in one exposure. If the principal concern is the measurement of frequency (or wavelength) of spectral lines the method is an obvious choice.

Photocells

Photoemissive cells are vacuum or gas-filled tubes in which the cathode is coated with a photoemissive film such as caesium and acts as light receiver. Quanta of more than a characteristic threshold frequency have sufficient energy to release electrons from the cathode coating to be collected on the anode (Fig. 21a). The current which flows is a measure of the number of quanta, or of the intensity of light falling on the cell. The sensitivity of these cells varies with wavelength in a manner which depends upon the nature of the coating, upon the gas filling, and upon the nature of the envelope (Fig. 21).

The detection and measurement of very low levels of intensity has been made possible in the last decade by photoemissive cells in which the original

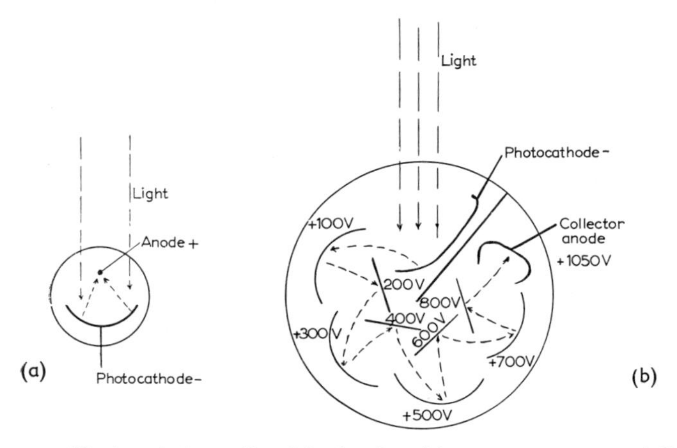

Fig. 21. Photoemissive cells; (a) simple, (b) 9-stage photomultiplier.

photoelectrons are attracted to secondary electrodes at which they release some five times their number for attraction to the next secondary. In the nine-stage *photomultiplier* shown in Fig. 21(b) the gain is therefore 5^9, *i.e.* of the order 10^6. The spectral sensitivity depends upon the photocathode coating as in the simple cells, and cells are available for the visible, ultraviolet, and the near-infra-red as far as 1 μ. It should be remembered, in using spectrometers in which photomultipliers are used, that the cells can be damaged by overloading in the final stages of multiplication. They are designed for intensities so low that even the high gain is insufficient to cause overloading: they should not be exposed (when the anode voltage is on) to intensities sufficient for operation of a simple one-stage cell. The high gain of photomultipliers has made it possible to record Raman lines directly, rather than by photography with lengthy exposures. The response time of photomultipliers is somewhat less than the simple cells, but all types of photoemissive cells are sufficiently fast to permit interruption of the light at frequencies up to 2000 c/s without reducing the sensitivity below 80 % of the low frequency value. The AC signal (at the modulation frequency) can then be amplified very readily.

A number of cells depend for their action upon the related phenomena of *photoconduction* and *photo-voltaic* effect: selenium, thallium sulphide, lead sulphide, and lead selenide are materials exhibiting these effects which have been successfully used in photo-cells. Selenium is used widely in photovoltaic (barrier layer) cells and responds in the ultraviolet and visible (Fig. 20). The cell develops a potential when exposed to light and its response is linear with intensity over a ten-fold range of intensity.

Thallium sulphide ("thalofide") was the first photoconductive material

References p. 64–66

to be used for infra-red detection. More recently lead sulphide and lead selenide have extended the range of available photocells for the infra-red to 5 μ; or to nearly 8 μ when the lead selenide cell is cooled to liquid air temperature (Fig. 22). The absolute sensitivity of these cells is considerably

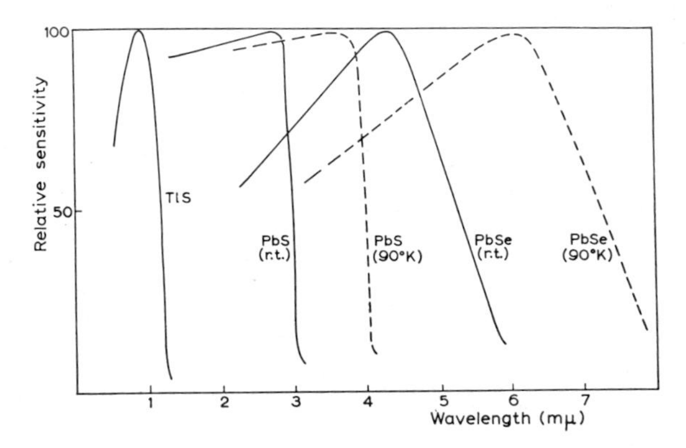

Fig. 22. Variation of sensitivity of three photoconductive detectors. The sensitivity of each is relative to its own maximum.

greater than that of the thermal detectors: 10—100 times at the peak of sensitivity. Their response is very rapid (time constant 10^{-4} sec) and they are used with chopped radiation, generally at 800 c/s: the AC signal is amplified by an amplifier tuned to that frequency.

Thermal detectors

The thermal detectors are *thermocouples* and *bolometers,* the former generating a small voltage, the latter changing resistance, as a consequence of a small rise in temperature on absorbing radiative energy. A variety of thermocouple elements may be used; bolometer elements may be metallic strips (Fe, Ni, Au) or semi-conducting thermistors. The spectral response depends only upon the "blackness" of the receiving area. For absolute radiometry in the u.v. or visible it is possible to use bulky thermocouples, or thermopiles. For the relative measurements required in spectroscopy thermocouples and bolometers need to be small enough to respond to interruption of the radiation at frequencies 10—20 c/s. They need to have time constants less than 5×10^{-2} secs and AC amplification is then possible. The development of such fast thermal detectors was probably the principal factor in the growth of the infra-red spectrometer as a universal laboratory tool in the last two decades: previously very sensitive galvanometers and temperamental optical amplifiers had to be used.

Microwave detectors

Thermal detectors can also be used for microwave power, but almost exclusively crystal rectifiers are employed. A fine wire in contact with semi-conducting silicon or germanium constitutes a rectifier whose capacity is nevertheless small enough to permit the detector to be used at very high frequencies. A technique often employed, which is analogous to chopping the radiation in optical spectroscopy, is to subject the microwave absorption cell to an alternating potential field. For polar molecules (which are in any case the only ones which absorb on rotation) the absorption frequencies are shifted by an electric field (Stark effect). The potential field is made to alternate at a convenient radio-frequency (say 100 kc/s) in square-wave form with one half-cycle at zero potential. The molecules then experience no field for half a cycle, and a constant field V_s during the other half (Stark modulation). Microwave radiation at frequencies corresponding to $V = 0$ and to $V = V_s$ is absorbed in each alternate half cycle. The crystal rectifier thus experiences radiofrequency modulation of microwave power when the klystron source is swept through the two microwave frequencies ν_0 and ν_s (Fig. 23). The radiofrequency signals are readily amplified and presented

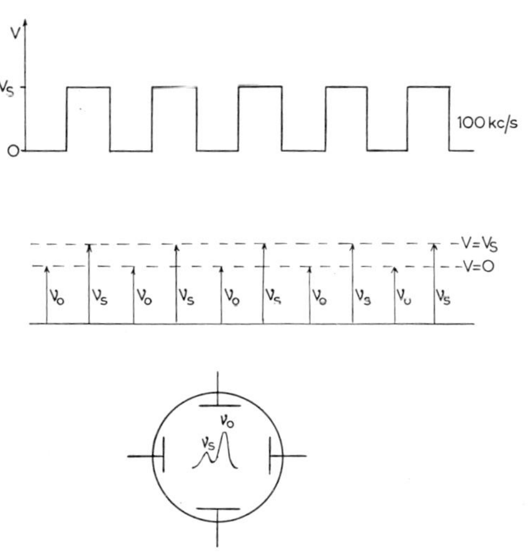

Fig. 23. Stark modulation of microwave absorption, ν_0 is the absorption frequency at $V = 0$, ν_s the absorption frequency at $V = V_s$.

to the oscilloscope. Both undisplaced lines (ν_0) and Stark components (ν_s) appear, the Stark components being identified by their movement when the amplitude of the Stark modulation (V_s) is varied).

References p. 64–66

Modulation of radiation falling on any detector has considerable advantages. The principal one is the ease of amplification of the resulting AC signal (although the low chopping frequency for thermal detectors raises special problems). Furthermore the amplifier may be sharply tuned to the frequency of modulation and thus only notices radiation which has been so modulated. The effect of stray radiation in optical spectroscopy is much reduced and spectrometers need not be confined to dark rooms or protected from 50 c/s artificial light. Detector noise (which can be considerable in crystal rectifiers and photomultipliers) is not amplified and its relative effect is therefore much reduced.

Presentation

The final step in the measurement of spectra is the drawing of the graph of energy against wavelength. The information obtained by the detector has to be converted to diagrams such as Fig. 1 (a–f).

Visual detection of a spectrum only permits frequencies to be recorded. A line may be noted at a wavelength setting of, say, 5085 Å and this may be reproduced on paper by drawing a line vertically on a horizontal wavelength axis. Only a rough indication of intensity can be given by labelling the line strong, medium or weak (and these will always be relative terms) and some fuller information may be given by labelling the line sharp, or diffuse. The shape of a band is even less satisfactorily recorded visually. The method is, however, excellent for measurement of the wavelengths at which there are fairly sharp changes in energy and these are as well presented in tabular form as graphically.

Photographs

Photographic detection (atomic line spectra, Raman spectra) immediately presents a form of graph of intensity against wavelength. Fig. 1(a) is an example. Wavelength appears as abscissa: intensity is recorded not as ordinate, but in a third dimension, as density of blackening of the photographic emulsion. This may be sufficient if the principal concern is with accurate wavelengths. If intensity is also important it can be plotted as ordinate with the aid of a *microphotometer* (more correctly a *microdensitometer*). In Fig. 24 a thin beam of light of constant intensity is focussed on to the photographic plate and the transmitted light is measured by the photocell. The amplified output of the photocell is made to actuate the recorder pen. The recorder paper is moved forward as the photoplate is tracked through the light beam. Synchronous drive of these movements ensures that the paper movement is accurately related to distance along the plate, and thus to wavelength. In this way Fig. 1(a) is converted to Fig. 1(b). The resultant graph now presents a measure of intensity as ordinate. Conversion of this to actual intensity requires either a knowledge of the relationship between density and intensity at different wavelengths or empirical comparison with plates exposed under standard conditions.

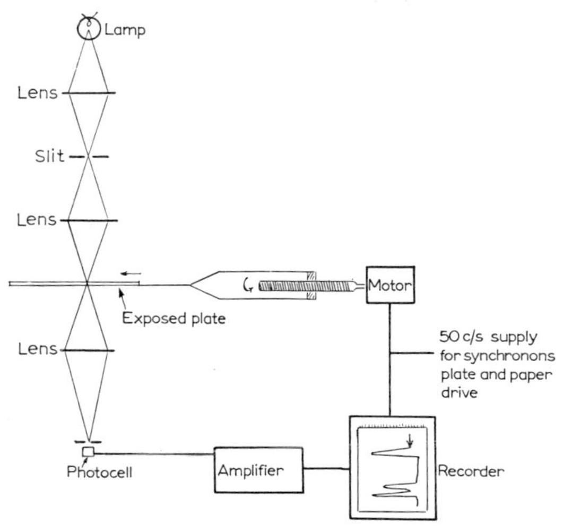

Fig. 24. Recording microdensitometer.

It may be noted that photographic records of emission appear on the developed plate as dark lines or bands on a light background. Absorption lines or bands appear bright on a dark background. While it is possible to reverse either form (positive from negative) they are best reproduced as observed, to avoid confusion.

Direct recording

All other forms of detector give an electrical output which can be measured as a galvanometer deflection. This can be plotted point-by-point against wavelength, as it is changed manually, but time and effort are saved by automatic recording. The detector output is amplified and fed to a recording potentiometer or recording milliammeter. If the wavelength setting of the monochromator is driven in synchrony with the recorder paper, the result is a graph of a measure of energy against a measure of wavelength (*e.g.* Fig. 1(c) and (d)). This is the form in which emission, Raman, and fluorescence spectra are conveniently presented. The scanning mechanism on the monochromator may include a cam designed to give a linear connection between distance on paper and either wavelength or frequency; but this is a matter of wavelength calibration (p. 42). Relating detector output to

References p. 64–66

absolute intensity is a matter of intensity calibration (p. 47).

A cathode ray oscilloscope offers an alternative mode of presentation if the scanning mechanism of the monochromator is fast enough to cover a useful range of wavelength, followed by immediate return to the starting point, in a time short enough for the persistence of image on the oscilloscope screen (less than 1 sec). This is regular practice in microwave absorption, in nuclear magnetic resonance and in electron spin resonance, where the wavelength range is scanned at 50 c/s or faster. CRO presentation has also been done in the infra-red over a wavelength range sufficient to include one or two bands. For a permanent record, the CRO picture can be photographed and if the intensity at a particular wavelength changes with time this can be recorded with moving film. It is also found convenient in spin resonance spectroscopy to record the derivative of the absorption curve rather than the absorption itself. For this purpose the magnetic field is modulated at 50 c/s over a very small range of magnetic field while the main field is slowly varied (say 5 gauss/minute). The resulting signal has a 50 c/s component which is proportional to the tangent of the absorption curve at the value of the main field. A phase-sensitive detector picks out the 50 c/s component whose amplitude is then presented on a pen recorder. The recorder paper movement and the main field variation are synchronized so that the record is of the derivative against the magnetic field.

Comparative recording

When, as in optical absorptiometry, absolute intensity is not required but rather a comparison between incident intensity and absorbed intensity, the spectrum is recorded as a graph against wavelength either of percentage absorbed or of percentage transmitted. The two curves (I_0 and I) of Fig. 1(d) are to be presented in the form of Fig. 1(e). This conversion can be done by recording both curves of Fig. 1(d) and carrying through a point-by-point comparison. Point-by-point measurement of I and I_0 without recording is also possible and not much more tedious. Double beam arrangements were devised to enable the instrument to make the comparison and present the result directly as a graph, such as Fig. 1(e). In the double beam system described on p. 9 a variable attenuator in the blank (or I_0) beam maintains equality between the two beams so that its position is a measure of the fraction absorbed. Direct mechanical connection to a pen moving vertically on paper which moves horizontally as the wavelength is changed leads to the required graph. Alternatively the I_0-beam has no attenuator but a discriminator permits the AC signal received by the detector to be divided into parts due to the two beams. The amplified signals are then compared on a recording potentiometer wired for ratio recording.

Standard presentation

In so far as absorption spectra are sufficiently characteristic "fingerprints" it is desirable that they should be presented in a standard way for ready comparison. This is the case with infra-red spectra where it is fairly generally agreed that spectra shall be plotted as percentage absorption against a linear wave number scale with wave number increasing from right to left (wavelength increasing non-linearly from left to right). Frequencies of microwave absorptions are also very characteristic, but they are better tabulated. So also are the characteristic chemical shifts of nuclear magnetic resonance, (p. 255) and the wavelengths of emission lines characteristic of elements (p. 44).

Resolution

Resolution refers to the ability to separate lines of slightly differing wavelength. In optical spectroscopy a criterion of resolution can be given and the resolving power of a spectrometer can be calculated. Because it is necessary to collect a cone of light from the entrance slit it turns out that the image on the focal plane, even for an infinitely narrow slit illuminated with truly monochromatic light (λ), is not a geometrical image but a diffraction pattern. The pattern comprises a central peak centred on the position in the focal plane corresponding to λ, with fringes at distances along the focal plane which depend upon the numerical aperture of the spectrometer. Fig. 25

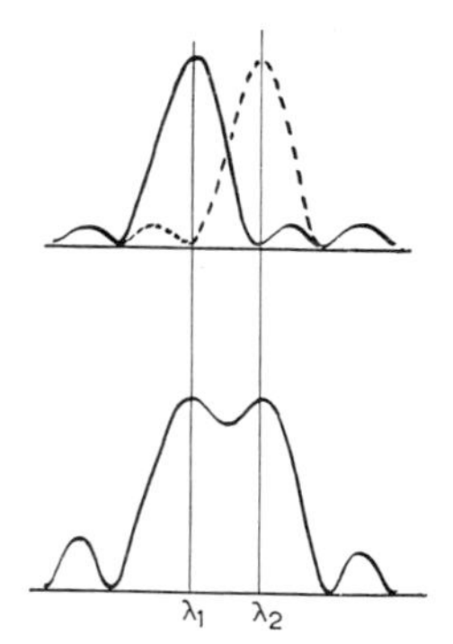

Fig. 25. Rayleigh's criterion for resolution.

shows two such patterns for two wavelengths λ and $\lambda + \delta\lambda$ with the centre for λ coinciding with the minimum for $\lambda + \delta\lambda$. The summation of these two patterns is shown underneath and it is clear that for any smaller difference between λ and $\lambda + \delta\lambda$ it would not be possible to distinguish two "lines". This is Lord Rayleigh's criterion for resolution, that the smallest difference

References p. 64–66

$\delta\lambda$, which an instrument can resolve is that which corresponds to coincidence in position at the focal plane of the central peak for $\lambda + \delta\lambda$ and the first minimum for λ. The smaller $\delta\lambda$ the better the resolution: hence the dimensionless number $\lambda/\delta\lambda$ is a measure of resolution. It is, in fact, the *resolving power* R of the instrument, depending upon the dispersion element as discussed on p. 24 (prisms) and p. 28 (gratings). Values of 1,000 to 60,000 correspond to medium resolution in commercial instruments.

In practice the slit is not infinitely narrow. It has to be opened sufficiently to allow enough energy to fall on the detector. The energy is proportional to the slit-width and in a monochromator, where entrance and exit slits are generally controlled together with equal width, the energy is proportional to the square of the slit width. The effect of opening the entrance slit is to broaden the image on the focal plane and thus to increase the minimum wavelength difference for distinguishing two lines (Fig. 26 a, b, c). The

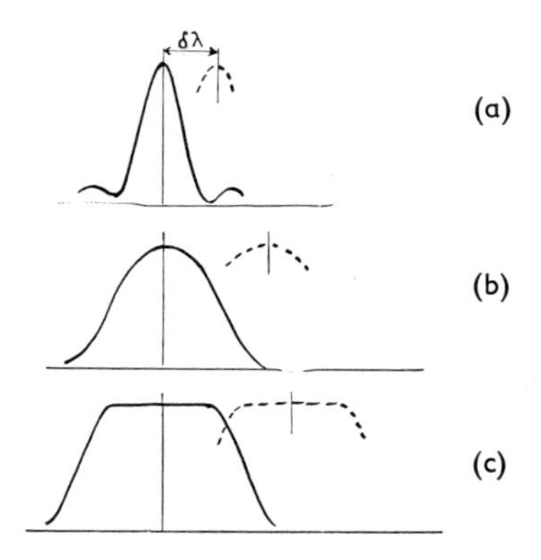

Fig. 26. Increasing the entrance slit width decreases the resolution.

effect of opening the exit slit to a width greater than the distance along the focal plane corresponding to $\delta\lambda$ (or of using a photographic emulsion whose grain size is greater than that distance) is to observe patterns due to λ and $\lambda + \delta\lambda$ as one and thus to lose resolution. In so far as actual resolution depends upon the slits it is useful to define an *effective slit width* or *slit range*, which is the actual slit width translated into a wavelength (or frequency) range. For this purpose, the linear dispersion of the spectrometer is required. Thus, for a prism spectrometer:

$$\delta\lambda = \delta x \cdot \frac{a}{fR} = \delta x \cdot \frac{a}{fT\,\mathrm{d}n/\mathrm{d}\lambda} \qquad (1.23)$$

and for a grating spectrometer:

$$\delta\lambda = \delta x \frac{a}{fR} = \delta x \frac{a}{fn'N} \tag{1.24}$$

where $\delta\lambda$ is the wavelength range corresponding to a distance δx in the focal plane, a is the width of the parallel beam, f is the focal length of the telescope, and R is the resolving power. T is the base thickness of the prism of characteristic dispersion $dn/d\lambda$ (assuming the prism is filled with light). N is the number of grating rulings illustrated and n' is the diffraction order used (see equations (1.20) and (1.22)).

A number of other instrumental factors operate to limit resolution and they include optical imperfections, the way in which the slit is illuminated, the grade and contrast of photographic emulsions. In recording spectrometers there may be apparent loss in resolution if the rate of scanning is too high. Thus, if the slits are narrowed for improved resolution, the gain of the amplifier has to be increased and consequently the damping adjusted to smooth out the increased noise. The reduced speed of response demands a slower scanning speed for faithful recording.

Molecular spectra in the infra-red and visible may involve bands or large numbers of lines which are too closely spaced for resolution by instruments of medium resolving power. The contour of the band is then recorded. On the other hand no spectral line is truly monochromatic and various factors contribute to the broadening of lines and bands. The image of a line of finite width is the sum of a range of overlapping slit images of varying intensity. If the natural width of a line or band is the comparable with the actual resolution of the instrument the result is a broadening of the line and a reduction of peak height. The natural band width needs to be more than ten times the resolution before this effect becomes negligible: or, to put it another way, the result of decreasing the slit width is to increase the peak height and diminish the width of the recorded band. However, beyond a certain point, when the slit range is about 1/10th of the band width, further increase of resolution is without effect. This has an important bearing on intensity measurement (p. 50) and also means that for a large number of applications of spectroscopy in chemistry available resolution is now adequate, since for a variety of reasons, bands are broad.

Resolution in microwave and spin resonance spectroscopy is of another order. Generally the lines are very narrow and instruments are capable of resolving lines separated by less than 0.1 Mc/s. This, represents a resolving power of the order of 500,000: even then the limit is in the width rather than in the instrumental capacity. In nuclear spin resonance experiments peaks have been separated at 2 c/s in 40 Mc/s, a resolving power of 2×10^7.

References p. 64–66

Wavelength or Frequency Calibration

An optical spectrograph is usually supplied by the makers with an indication of the wavelength scale on the focal plane. A monochromator may have anything from an accurately marked wavelength drum to one which is marked merely in arbitrary numbers. A rough indication of frequency in a microwave spectrometer can be obtained from the length at which resonance occurs in an adjustable resonant cavity. In spin resonance experiments the magnetic field strength is roughly calculable from the design of the magnet and its magnetic sweep coils. For many purposes these more-or-less approximate indications may be adequate: the manufacturer's calibration will often be sufficient. On the other hand it may become necessary either to check the calibration or to carry out a complete calibration and for this purpose standard wavelengths are required.

The ultimate wavelength standard is the wavelength of the red line in the atomic spectrum of cadmium, defined as 6438.4696 Å in dry air at 15°C and 760 mm pressure. A very recently adopted new standard of length defines one metre as equal to 1650763.73 wavelengths in vacuum of a particular well recognised line in the discharge spectrum of the krypton isotope, ^{86}Kr (in an Engelhard standard lamp at − 210°C). The line in question therefore has standard wavelength of 6057.80211 Å in vacuum and 6056.1252_5 Å in air at 15°C and 760 mm pressure. Accurate comparison with this by interferometric methods gives secondary standards in the very rich arc spectrum of iron and copper and in the discharge spectra of neon and krypton. They cover wavelengths from 2447 to 7032 Å and the range is being extended in both directions. These spectra and others derived from them for a large number of elements are catalogued in various tables. For further discussion of accurate wavelength determination see Sawyer (Ref. 49, Chap. 9).

Optical spectrometers

For calibration of any spectrograph in the visible and ultraviolet it is convenient to photograph the arc spectrum of iron or copper on the same photographic plate. Unknown wavelengths are then obtained by interpolation between the two nearest standards. Linear interpolation is possible over a small range even for prism instruments; over a wider range for grating spectrographs.

Spectroscopes (visual detection) or monochromators (with all detectors which produce a galvanometer deflection, or a paper record) generally require less accurate calibration. A standard should suffice whose wavelength is known to an accuracy of one part in $10R$, where R is the resolving power of the instrument. This means, usually, that an accuracy of 0.1 Å is

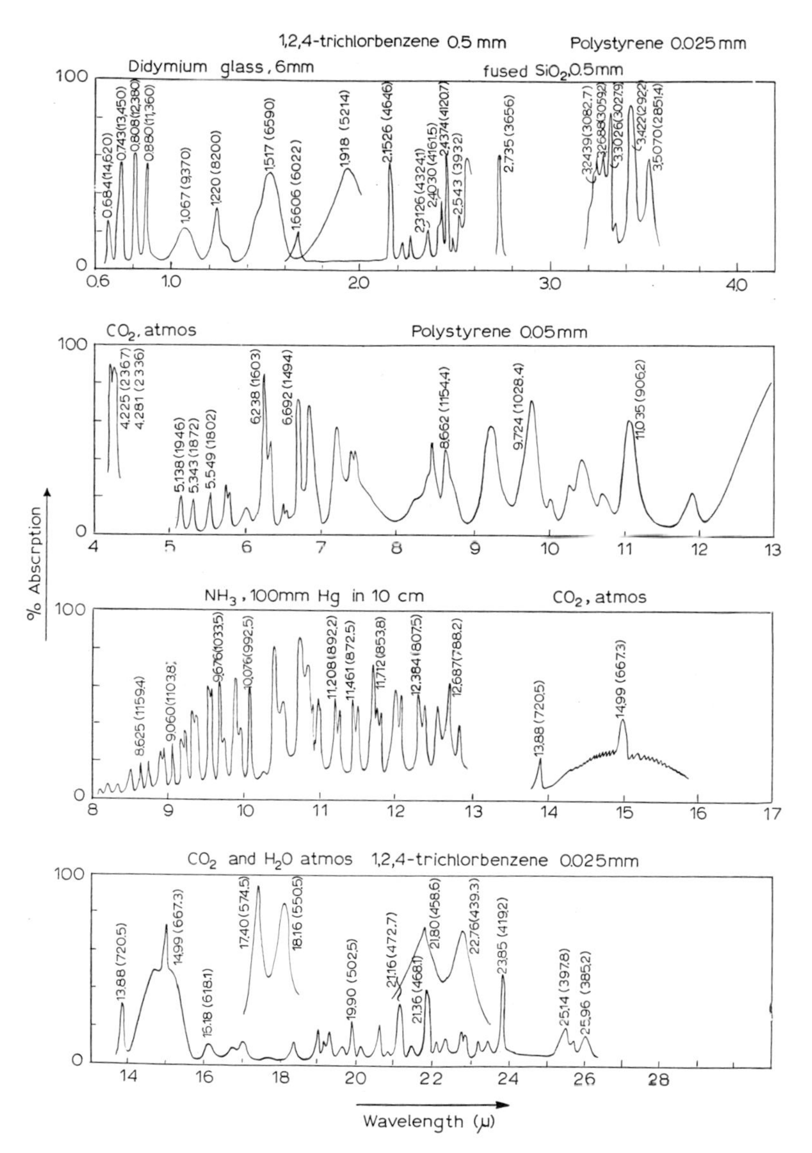

Fig. 27. Some standard absorption spectra in the infra-red suitable for calibration of medium resolution instruments. Wavelengths (and wavenumber) in vacuum are indicated.

References p. 64–66

TABLE 5

WAVELENGTHS (IN AIR) AND CORRESPONDING WAVENUMBER (IN VACUUM) OF MERCURY LINES USEFUL FOR CALIBRATION OF SPECTROMETERS

	λ_{air} (Å)	ω_{vac} (cm^{-1})	λ_{air} (μ)	ω_{vac} (cm^{-1})
	2536.52	39,412.3	0.7729	—
	2967.28	33,691.1	1.01398	9859.4
	3021.50		1.12866	8857.7
	3125.66	31,984.0	1.35703	7367.0
	3131.56	31,923.8	1.36728	7311.0
	3131.84	31,920.9	1.39506	7166.3
	3650.15	27,388.3	1.52952	6536.2
	3654.83	27,353.3	1.69202	5908.5
	3663.28	27,290.2	1.69419	5900.9
	4046.56	24,705.4	1.70727	5855.7
	4077.83	24,515.9	1.71099	5843.0
blue	4358.34	22,938.1	1.81307	5514.0
	4916.04	20,335.9	1.97009	5074.5
green	5460.74	18,307.5	2.24929	4444.6
	5769.69	17,327.1	2.32542	4299.1
yellow	5789.69	17,267.3		
	5790.66	17,264.4		

adequate in the u.v. and visible, and 0.001 μ in the infra-red. Practically any line spectrum is known to this accuracy: the mercury arc provides a very convenient range of such lines up to 2.3 μ: they are listed in Table 5. In the infra-red region standard absorption spectra have to be used. Fig. 27 shows absorption spectra of polystyrene, atmospheric water vapour and carbon dioxide, ammonia, and liquid 1, 2, 4-trichlorbenzene with some wavelengths marked. Wavelengths in vacuum (p. 5) are quoted since they are required for direct conversion to wave number. Further calibration data can be found in the literature (Ref. 13 (1.5—24 μ); Ref. 14 (3—24 μ); Ref. 15 (0.6—2.6 μ); Ref. 16).

The standards are used in the following way. If the monochromator has a wavelength drum which is marked in wavelength, it is sufficient to compare each standard wavelength, λ_s, with the instrument reading, λ_0, at which the standard line appears and to plot $\delta\lambda = \lambda_s - \lambda_0$ against λ_0. Should there be any systematic error, a smooth curve drawn through these points can be used as a correction curve. In a recording instrument there are many ways in which the calibration can shift: from time to time it becomes necessary either to readjust the monochromator or to prepare a new correction curve. One source of error is pen alignment on the recorder. It is usual for a second pen on the recorder to make fiducial marks on the edge of the paper corresponding to certain positions of the wavelength drive. It need hardly be stressed that λ_0 should be read off from these marks in preparing a correction curve.

If the monochromator has a wavelength adjustment marked only in arbitrary numbers, or in distance of travel of a micrometer screw, the task of calibration is more troublesome. In the ultraviolet and visible there is probably a sufficient number of spectral lines to make it hardly necessary to use a dispersion formula to interpolate between standards: if an interpolation is required for a prism instrument, the Hartmann formula

$$\lambda = \lambda' + \frac{C}{T' \pm T} \qquad (1.25)$$

may be found adequate over a small range. λ', C and T' are constants and T is the scale or drum reading. The alternative $\pm$ sign accounts for the possibility of λ decreasing or increasing with increase in T. In the infra-red the calibration standards are so far apart as to require a good interpolation formula. McKinney and Friedel (Ref. 17; see also Refs. 18, 19) have shown that the formula

$$T = T_0 + \frac{B}{(\omega^2 - \omega_0^2)} \qquad (1.26)$$

leads to a satisfactory linear interpolation if ω_0 is chosen from one of the Restrahlen wave numbers for the material of the prism, B and T_0 being constants. It appears, however, that this formula fails in the region where $dn/d\lambda$ for the material goes through a minimum and in that region an additional term $A\omega^n$ can be found to improve the interpolation. An alternative method of interpolation is to use interference fringes from a thin cell (see p. 55).

Grating monochromators can be constructed so that there should be accurate linearity between T and λ_{air}. Use of λ_{vacuum} changes the slope by about 1 part in 4000 but should not introduce any noticeable curvature until the resolving power exceeds 100,000. Although there are no useful mercury lines beyond 2.3 μ, grating settings can be calibrated in the infra-red with the visible mercury lines in high diffraction orders. The problem of identifying the orders requires that one absorption band be measured in the first order but its accuracy need only be sufficient to locate a region and to make it possible to decide to which order an observed line is to be assigned.

Microwave and radio frequencies

A similar location problem arises in microwave absorption and for this purpose a cavity wavemeter is used. The attainable accuracy is about one part in 3000, or about 10 Mc/s, which is sufficient to distinguish between the harmonics in the more accurate frequency standards. These are quartz crystal oscillators which can be checked as necessary against broadcast radio transmissions of frequencies standardized against the astronomical clock; or the radio transmission may be used directly. In either case the frequency (100 kc/s; 2.5, 5 or 10 Mc/s) has to be multiplied to produce microwave frequencies which are accurate harmonics of the received frequency. Interpolation between the harmonics (separated by 30 Mc/s) is effected by mixing them with the klystron output to produce a beat frequency between 0 and 30 Mc/s. The beat is fed to a radio-receiver, which can be tuned over this range and calibrated against a low-frequency oscillator covering the same range, and thence to the same oscilloscope as that on which the microwave absorption line is displayed. For a given setting of the receiver there is one klystron frequency (within the 30 Mc/s range) which beats with the standard, and it is at that frequency that a sharp pulse is produced on the oscilloscope screen. The receiver is tuned so that the pulse coincides with the absorption line. Suppose, for example, that coincidence is found when the receiver is tuned to 20 Mc/s, the standard frequency harmonics being 27,000 or 27,030 Mc/s, it follows that the possible klystron frequencies could be 26,980, 27,010, 27,020, 27,050 Mc/s. A pulse arising from the first and second would move across the screen to lower microwave frequency as the receiver frequency was increased; from the third and fourth it would move to higher microwave frequency. This test, coupled with a wavemeter indication good to 10 Mc/s, enables the correct assignment to be made and once this is done the frequency value can readily be given to an accuracy of better than 1 part in 10^6 (30 kc/s at 30,000 Mc/s).

Such precise determination of absolute frequency (or magnetic field in spin resonance measurements) is not generally called for in chemical applications. An internal or substitution standard is normally sufficient. Electron spin resonance is much concerned with the measurement of free radical concentrations and this requires the area of the resonance "line". The chemical significance of nuclear magnetic resonance spectra in solids lies in the line shape: in liquids and gases, it is to be found in the relative rather than the absolute position of lines. Hence the relationship has to be established between interval of magnetic field and distance on the X-axis of the oscilloscope screen even when the frequency is not the major goal. The proton resonance signal in liquid water has been very accurately measured and it is normally used as the standard. In nuclear magnetic resonance studies this is simply done either by reference to water already present in the sample, or by another probe coil containing water. In electron spin resonance a separate probe coil containing water is fed from a variable radio-frequency oscillator (calibrated against a quartz oscillator as standard) and the proton resonance line is displayed independently on the screen of a double beam oscilloscope, which also displays the electron resonance line. Since the magnetic field is being swept through a range sufficient to cover the absorption line, the proton-resonance frequency can be altered so as to correspond to two magnetic field values close to each end of the sweep. The proton resonance and the field strength are connected by the formula

$$\frac{H}{\text{(gauss)}} = 2.3487 \times 10^{-4} \frac{\nu}{\text{(c/s)}} \tag{1.27}$$

and then the positions on the screen are marked with appropriate values of the field. Intermediate values can be interpolated.

Intensity Measurement

Although absolute intensities of emission are of considerable importance in quantitative analysis by means of arc and flame spectra, the greater chemical interest lies in the determination of relative intensities in absorption. Whereas measurement of absolute intensities is critically dependent upon the spectral response of the detector, in the estimation of relative intensities this factor no longer matters except in so far as it determines the region in which the instrument can function at all. There are nevertheless other factors which need to be borne in mind.

Beer's law

Much of quantitative analysis by absorptiometry relies upon the law, commonly known as Beer's law, which is in fact a combination of a law involving thickness of sample, propounded by Bougner and later by Lambert, and one involving concentration put forward by Beer. It states that, for a parallel beam of light passing through a homogeneous medium of thickness l in which the absorbing species is present in molar concentration c, the intensity of transmitted light I and of incident light I_0 are in the ratio:

$$\frac{I}{I_0} = \mathrm{e}^{-kcl} \tag{1.28}$$

By general agreement the law is usually expressed in the form

$$D = \log_{10}\left(\frac{I_0}{I}\right) = \log_{10}\left(\frac{1}{1-a}\right) = \varepsilon cl \tag{1.29}$$

where D is the *optical density* (thus defined) and ε is the *molar extinction coefficient*. a is the fraction of light absorbed and is included because many recording spectrophotometers plot a against wavelength. If the units are c mole l^{-1} and l cm, the units of ε are mole^{-1} l cm^{-1}. Practical formulae are required for calculating cl (mole l^{-1} cm) for samples which are either liquid, in solution, gaseous, or solid dispersed evenly over a known area. Table 6 gives those formulae in terms of relevant quantities.

There is good theoretical ground for supposing that Beer's law should hold

TABLE 6

MOLAR DECADIC EXTINCTION COEFFICIENT $\varepsilon = D/cl$

Sample		cl (mole l^{-1} cm)
Liquid:	density ρ g cm^{-3}; molecular weight M; thickness l' mm	$100 \dfrac{\rho l'}{M}$
Solution:	concentration c molar; thickness l' mm	$0.10\, cl'$
Gas:	pressure p mm Hg; thickness l cm; temperature T°K	$0.016 \dfrac{pl}{T}$
	at temperature 300°K	$5.34 \times 10^{-5} p$
Solid:	m mg dispersed evenly over ½″ circular disc: molecular weight M	$0.788 \dfrac{m}{M}$

rigorously for monochromatic light and for sufficiently narrow slits. In practice neither of these conditions is satisfied and there arise the questions of the actual form of the dependence upon cl and the best means of assessing ε, or related quantities.

In the first place, ε (and therefore also I) is evidently a function of λ or ω, for this is the nature of an absorption spectrum. Moreover a line, or a band, has finite width. The derivation of the exponential law connecting the fraction of light transmitted and the product cl shows that if the light is not monochromatic but comprises a band of wavelengths the law holds strictly only if ε is constant over the band. In practice this means that Beer's law holds to a good approximation when ε does not vary greatly over the portion of band selected by the slit. It would not be expected to hold when the slits accept a whole band, or, even a portion of what appears to be a continuous spectrum but is in reality a number of unresolved lines. In that case there are reasons (see Refs. 20, 21 and Ref. 54, p. 41) for expecting the absorption to be proportional to the square-root of cl. When the slits are narrow compared with the band width Beer's law generally holds provided the conditions are standardized. An apparent value of ε is obtained appropriate to the conditions. Because bands in the ultraviolet and visible, especially of species in solution, are generally broad compared with the available slit ranges the values of ε are close to limiting values and are comparable from instrument to instrument. How far this is true can be judged from tests carried out with a standard chromate solution measured in a number of laboratories (Refs. 22-24). There still remain some hazards in exact determination of ε mainly arising from the mis-alignment of the sample, from cell errors, from scattering, from non-parallel radiation, from stray light. These sources of error are discussed elsewhere (Refs. 25, 26; Ref. 55, p. 261).

References p. 64–66

Additivity of optical density

If there is present more than one species which absorbs at the given wavelength setting and each obeys Beer's law, then, of course, the total optical density is the sum of the individual optical densities

$$D = \varepsilon_1 c_1 l + \varepsilon_2 c_2 l + \varepsilon_3 c_3 l + \ldots \qquad (1.30)$$

i.e. it is optical density which is additive, not the percentage absorption. Fig. 28 illustrates this in two ways. In (a) are shown four bands such as

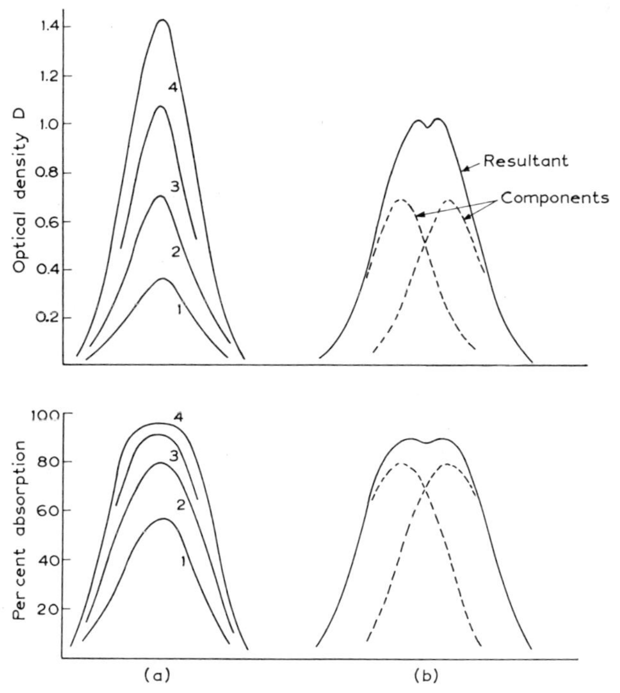

Fig. 28. Optical density is additive but percentage absorption is not: (a) one absorbing species at different concentrations; (b) two absorbing components of a mixture.

would arise from a single absorbing species at four different values of cl, in the ratio 1 : 2 : 3 : 4. The optical densitie sare in the same ratio but it is evident that when the curves are plotted as percentage absorbed against wavelength there is no additivity: note in particular the flattening of the highest curve. It can be seen that, for analytical purposes, estimation of D, and hence of concentration, becomes inaccurate when the absorption exceeds 80 %. It is preferable to measure D between about 0.2 and 0.7 (or 30 % and 80 % absorption). In (b) is drawn the resultant curve when two

References p. 64–66

similar bands overlap. Plotted against optical density the resultant curve is the sum of the component curves. This is evidently not the case when percentage absorption is plotted. It is clear that, in any case, overall optical density is not directly proportional to the concentration of any one species if other bands overlap at the wavelength used: and in such a case Beer's law would not appear to hold until corrections had been made for overlap by subtracting the optical densities contributed by the neighbouring bands.

Integrated intensity

Although it is often satisfactory to use D values determined from a part of a band — usually the peak value — for analytical purposes the method does not give an absolute measure of intensity. Furthermore it may not even lead to results which are comparable between spectrometers, especially in the infra-red where the slit range on instruments in general use is often not much smaller than the band width. The significant quantity for an absolute measure of intensity is the area under the absorption curve (of ε against ω). Since, in estimating this from observed data, slit effects must be taken into account the results should be more closely comparable between instruments.

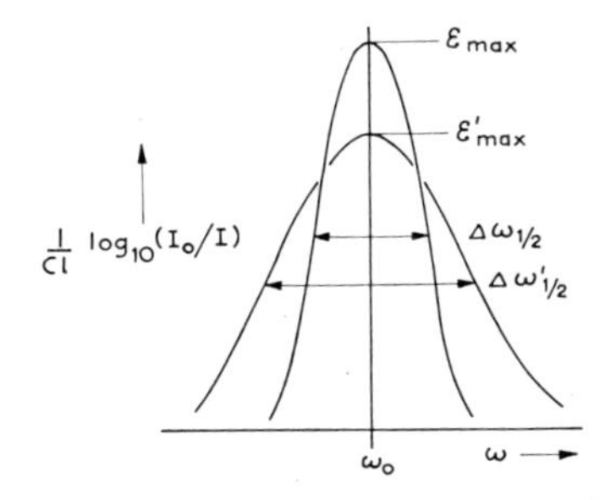

Fig. 29. True and apparent absorption bands. ε'_{max} and $\Delta\omega'_{1/2}$ refer to the apparent, or observed, band.

The problem is essentially as follows. In Fig. 29, the narrow band records the true variation of ε with ω. It is characterized by a value of $\varepsilon = \varepsilon_{max}$ at the wave number ω_0 and by a natural width at half the peak height of $\Delta\omega_{\frac{1}{2}}$. The *integrated absorption intensity*:

$$A = 2.303 \int_0^\infty \varepsilon \, d\omega = \frac{1}{cl} \int_0^\infty \ln\left(\frac{I_0}{I}\right) d\omega \qquad (1.31)$$

Now, given a spectrometer with sufficient resolution, it may be possible to draw the true absorption curve: the test is whether further narrowing of the slits leads to a change in the measured absorption. If the resolution is insufficient, the recorded curve is, as we have seen, a broader curve with lower peak value. If the apparent values are denoted by primes, $\varepsilon'_{max} < \varepsilon_{max}$, $\Delta\omega'_{\frac{1}{2}} > \Delta\omega_{\frac{1}{2}}$ and the *apparent integrated intensity*

$$B = 2.303 \int_0^\infty \varepsilon' d\omega \tag{1.32}$$

is somewhat less than A. The logical procedure to get from B to A would be to measure B at various slit ranges S (which can be computed for the instrument by equations (1.23) or (1.24)) and to extrapolate to zero S. This however is unsatisfactory since B approaches A as a limiting value and in order to detect the limit it is necessary to go to values of S which the instrument is supposed to be incapable of reaching. A form of extrapolation can be worked out if the true absorption curve is assumed to conform to a particular function. Ramsay[27] has done this for a Lorentz function[28,29]:

$$\ln\left(\frac{I_0}{I}\right) = 2.303\, D = \frac{a}{(\omega - \omega_0)^2 + b^2} \tag{1.33}$$

where $a/b^2 = 2.303 D_{max}$ and $2b = \Delta\omega_{\frac{1}{2}}$, and has provided tables from which, given $S/\Delta\omega'_{\frac{1}{2}}$ and D'_{max}, the ratios $D_{max}/D'_{max} = \varepsilon_{max}/\varepsilon'_{max}$ and $\Delta\omega'_{\frac{1}{2}}/\Delta\omega_{\frac{1}{2}}$ can be found, and hence D_{max} and $\Delta\omega_{\frac{1}{2}}$. As a measure of the true intrinsic absorption intensity A can now be readily obtained by simple integration of the Lorentz curve: *i.e.*

$$A = 2.303 \cdot \tfrac{1}{2}\pi \cdot D_{max} \Delta\omega_{\frac{1}{2}}/cl = 2.303 \cdot \tfrac{1}{2}\pi\, \varepsilon_{max} \cdot \Delta\omega_{\frac{1}{2}} \tag{1.34}$$

and for a linear measure of concentration the corrected D_{max} and $\Delta\omega_{\frac{1}{2}}$ should conform to

$$D_{max} \cdot \Delta\omega_{\frac{1}{2}} = \text{constant} \times cl \tag{1.35}$$

Further details of Ramsay's corrections and of other extrapolation procedures are given by Jones and Sandorfy (Ref. 55, p. 271).

Raman scattering intensity

In Raman spectroscopy the intensity of a line is not exponentially related, but directly proportional, to the concentration of the species giving rise to the line. The constant of proportionality can only be obtained by calibration of the instrument under the same operating conditions: either peak height or integrated area may be used as a measure of intensity and the concentration is expressed as molarity (although the use of volume fraction has also been suggested). The measurement of absolute intensity, or even of standard intensities for comparison between instruments, is a more difficult matter which we shall not pursue. Carbon tetrachloride is a generally accepted standard, but there remain various instrumental factors to be taken into account.

References p. 64–66

Microwave absorption intensity

The measurement of intensity in microwave absorption in gases involves principles which can be compared with those in optical spectrophotometry. The measured quanttiy is γ, the fraction of energy absorbed per unit length of sample. The line has width, which arises predominantly from molecular collisions (pressure broadening) and thus is a function of ν. For medium pressures (which in microwave work means something less than 0.5 mm Hg) the Lorentz curve discussed above gives a good representation of line shape. Available resolution is however generally sufficient to enable the true line shape to be traced. The integrated intensity

$$\int_0^\infty \gamma \, d\nu = \text{constant} \times n \tag{1.36}$$

and is thus of interest in yielding information about the properties of the molecule (through the constant) if n the number of molecules per cc is known. If the constant can be calculated from the molecule, n can be found. At pressures at which the line has the Lorentz shape:

$$\int_0^\infty \gamma \, d\nu = \tfrac{1}{2}\pi\gamma_{max}\Delta\nu_{\frac{1}{2}} = \text{constant} \times n \tag{1.37}$$

where γ_{max} is the peak absorption. $\Delta\nu_{\frac{1}{2}}$ is the width of the line at half height, and is proportional to total pressure, P. Thus, $\gamma_{max} = kn/P$: the peak absorption is directly proportional to the concentration of the component in a mixture and is thus independent of total pressure. For a pure compound n/P is constant at constant temperature and γ_{max} is independent of P over a wide range of pressure For accurate assessment of concentration, it is necessary to take into account the dependence of $\Delta\nu_{\frac{1}{2}}$ on the composition of the gas: that is, that gases are not equally effective in line broadening by collision. But in any case the calibration of measurement of γ_{max} is most conveniently done by reference to γ_{max} for a gas for which absolute values are known.

Microwave absorption intensity is also measured in electron spin resonance as a measure of the number of unpaired electrons in the sample. The area of the absorption curve is measured directly or, if a Lorentz curve is appropriate, is obtained from the product of peak height and half-band width. The required number of free electrons can then be calculated from the instrumental factors, but more conveniently, the area measurement is calibrated against the area of absorption for a standard substance, 1,1-diphenyl-2-picryl hydrazyl, which has 1.53×10^{21} free electrons per gram.

It should be noted that in all the intensity measurements mentioned which call for an estimate of band or "line" area the wavelength or frequency calibration (horizontal axis) is involved as well as the vertical response of the instrument.

Sample Treatment

In the previous section we made no more than passing reference to the intensities of emission spectra, taking the view that the problems involved were those of a rather specialized analytical technique whereas absorptiometry held more direct interest for the chemist. The reason for this lies in the fact that absorptiometry has to do mainly with molecular species and not with elementary analysis. For the same reason, we are now concerned with the treatment of samples for absorption studies. Preparation for emission spectroscopy is discussed in the special monographs[49,50].

Absorption cells

Gases, liquids and solutions need containers, and, furthermore, the containers need windows transparent to the wavelengths of interest. Suitable window materials for various regions are listed in Table 3, p. 23. The windows must be plane and have parallel faces and must be parallel with each other. Cells required to be transparent in the visible and ultra-violet are made of glass or quartz and the windows are fused on to the body of the cell of the same material. For gases a length (say from 5—20 cm) of wide-bore tubing has fused-on end windows and a side-arm and tap to permit evacuation and filling. For liquids and solutions fused cells of thickness 0.5, 1, 2 and 4 cm are usually required and they are supplied with open tops and lids, or with closed tops and filling ports. Where liquids or solutions have been subjected to chemical treatment in a vacuum system a reaction flask may be fitted with a side arm carrying a fused-on cell with parallel windows. The liquid is transferred to the cell, still under vacuum or a controlled atmosphere, when the measurement is to be made. Cells thinner than 0.5 cm may be needed for intensely absorbing materials and, although fused cells can be made, there is some virtue in having demountable cells for ease of cleaning.

Demountable cells are necessary for cells used in the infra-red because the common window materials, sodium chloride and potassium bromide, cannot be fused to the body of the cell. Furthermore the windows need to be repolished from time to time. This is especially true of work with reactive gases, such as the volatile fluorides. Attack of the windows or deposits arising from decomposition, or sometimes from hydrolysis, can give rise to spurious absorption which can only be checked by measurements on the empty cell.

References p. 64–66

Gas cells, again 5—20 cm, may be of glass or metal and have flat-flange ends to which polished window discs are attached by a suitable grease or by a gasket seal with appriopriate clamping. Cells for solutions usually need to be of thickness 0.1—1 mm and for pure liquids from capillary thickness to 0.1 mm. Here the two windows are separated by a washer of appropriate thickness and the cell may be filled either as it is being made up or, in a semi-permanent cell, filling tubes are provided and the liquid is injected from a syringe.

Of the other window materials in Table 3, fluorite, sapphire, periclase, barium fluoride, silver chloride and thallium bromiodide are insoluble in water and remove one impediment to the measurement of infra-red spectra of aqueous solutions. The other difficulty with aqueous solutions is that water is so intense an absorber in the infra-red that capillary thicknesses have to be used to get any reasonable regions of transmission. In consequence, the solution has to be fairly concentrated in order to achieve measurable absorption by the solute.

Measurement and adjustment of path length

For satisfactory quantitative work the thickness of the sample must be accurately known. In the larger cells used for gases and for ultra-violet spectroscopy of solutions the measurement of thickness presents no difficulty. On the other hand, in cells of thickness less than a millimetre, accuracy of direct measurement is low. Micrometer measurements of the thickness of spacers used is accurate enough in itself, but the spacers are usually made of a soft metal such as lead in order to provide an effective seal. The thickness of a cell of these dimensions, which can be kept made up for a series of spectra, is measured with considerable accuracy by means of the interference effects which are obtained when an apparent absorption spectrum is taken of the empty cell (Fig. 30). The thickness, t cm, of the cell is given in terms of the wave number difference $\delta\omega$ (cm^{-1}) between one peak and the mth peak from it by the formula:

$$2t = m/\delta\omega \qquad (1.38)$$

Variable cells are available for all regions of optical spectroscopy and they have several advantages. The thickness is varied by an accurate micrometer screw and therefore when the distance between the inside faces of the plates has once been determined (by an interference method, for example) other thicknesses may be set to within the accuracy of such screws, say 0.003 mm. This represents an accuracy of about 1 % in the thicknesses commonly used for solutions in the infra-red. The bands which are measured

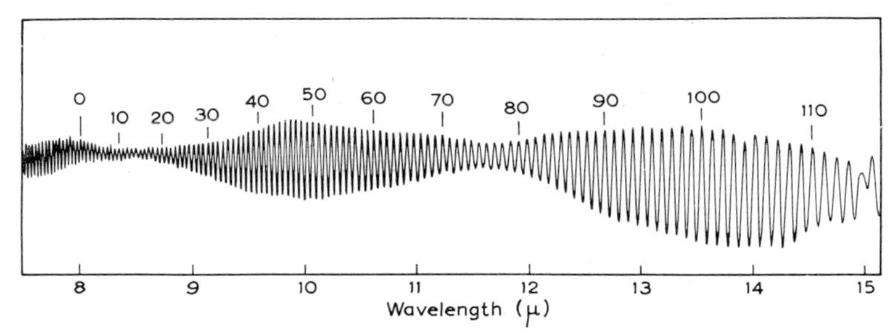

Fig. 30. Interference pattern for measurement of cell thickness. In this case 100 fringes between 8μ (1250 cm^{-1}) and 13.55μ (738 cm^{-1}) gives $t = 987\mu$.

in pure liquids are usually so intense as to need thicknesses 0.01 to 0.05 mm and the accuracy is then not likely to be much better than 10 %.

Solvent compensation

Continuous variation of cell thickness is a useful facility in double beam operation for elimination of the contribution of the solvent to the spectrum of a solution. The variable cell contains pure solvent and is placed in the reference beam, where its thickness is adjusted until only the solute spectrum is observed. It should be noted that for concentrated solutions the two thicknesses will not be equal, since the concentration of the solvent will be less in the solution than in the pure liquid. Care should be taken in the interpretation of spectra in the regions of strong solvent absorption. Although it is an obvious fact when solution and solvent are separately measured that solute bands are completely obscured at wavelengths at which the solvent absorption is complete it is not always so obvious in double beam operation. No amount of balancing will alter the fact that there is no energy coming through in either beam. On the other hand, the appearance of a solute band can be markedly improved by balancing when an overlapping solvent absorption does not exceed, say, 80 %.

Other factors

In that all absorptiometry is essentially a comparison of two intensities it is desirable that as far as possible the measured differences should be confined to the actual absorption of the sample and the reference. The adjustment of the thickness to balance the solvent absorption is just one factor. For dilute solutions cells of equal thickness are called for. The cells should also be matched as well as possible in regard to absorption by the window material and to losses by reflection. Mis-alignment of cells in the beam is another possible source of error: the windows should be placed normal to the beam.

References p. 64–66

A simple but common fault is not to fill the cell sufficiently to cover all the light which is to enter the spectrometer. In thick cells which permit a length of horizontal liquid surface this error may be aggravated by internal reflection from that surface. Needless to say the liquids should be free of suspended particles which may cause loss by scattering. This is more important in the ultraviolet and visible than in the infra-red: we shall see (pp. 60 and 63) that it can be all important in Raman and fluorescence spectroscopy. Solvent purification is necessary for accurate absorptiometry of a solution: again this matters more in the ultraviolet than in the infra-red because very small traces of impurity can give rise to intense absorption in some regions of the ultraviolet, whereas in the infra-red it is the bands due to the solvent itself which are the principle limitation to its use.

Solvents

There are a number of factors which influence the choice of a solvent for absorption spectroscopy. Solubility is an obvious one: in as much as extinction coefficients for electronic transitions tend to be higher than for vibrational, ultraviolet spectra generally require much smaller concentrations than infra-red and Raman. Consequently solubility is a less limiting factor for ultraviolet spectra. On the other hand the higher intrinsic intensity of ultraviolet transitions makes solvent impurities much more troublesome. For this reason the chemical suppliers now market solvents specially purified for ultraviolet absorptiometry.

Account must also be taken of possible interaction of solvent and solute: physical perturbations of the sort discussed on p. 282 may change the solute spectrum; chemical perturbations may change the identity of the solvent. A simple example of the latter is proton exchange: it is easy enough to forget that an organic compound containing a labile H atom will promptly exchange it for a D atom on solution in D_2O. As to physical effects solutes are generally less likely to be affected by non-polar solvents but, for the same reason, are less likely to be soluble.

The final consideration is the absorption spectrum of the solvent. The spectra of some commonly used solvents are given in Fig. 31. It will depend upon the intensity of the solute spectrum what value of cl will give a reasonably measurable value of optical density. Solubility puts an upper limit on the concentration c and hence a lower limit on the cell thickness l (and these may be further limited by the particular application of the absorption measurements). Clearly so far as solvent absorption is concerned the smaller the cell thickness the better. Absorption by the solvent can be allowed for as discussed above.

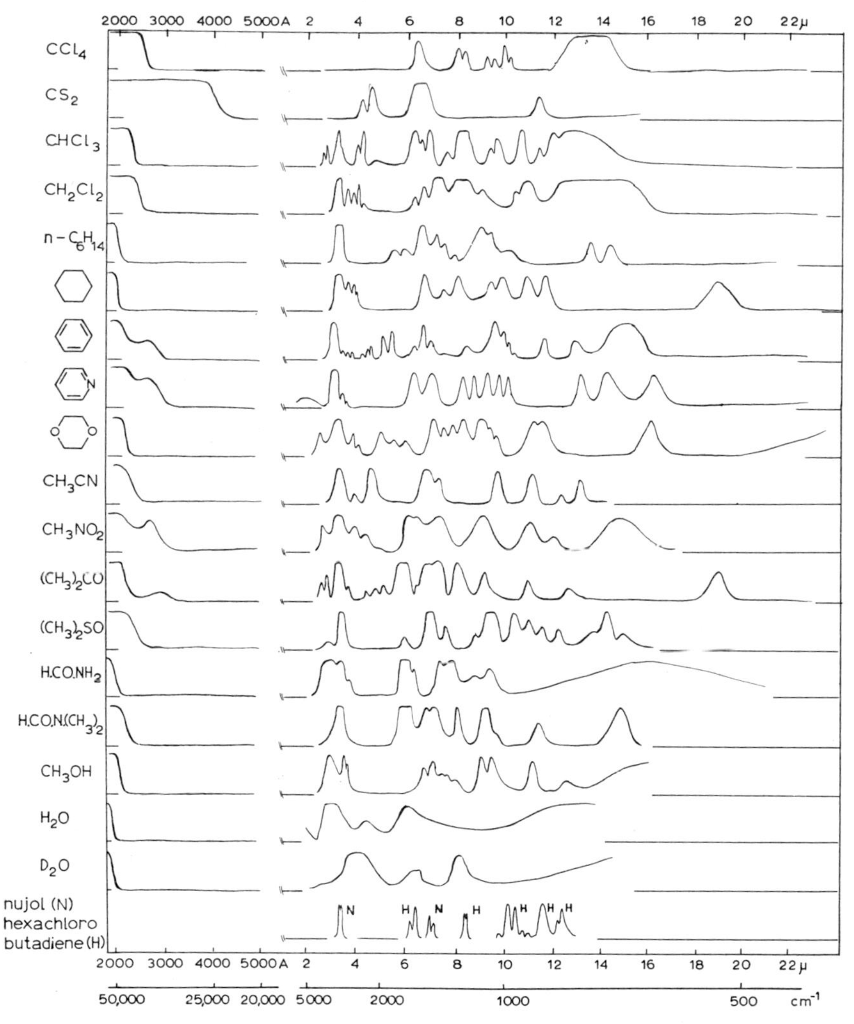

Fig. 31. Ultraviolet and infra-red absorption bands of various solvents. They are drawn with only sufficient detail to assist in choice of solvent, and refer to 1 cm in the ultra-violet and 0.2 mm in the infra-red.

The absorption of water (and D_2O) in the infra-red is so intense that very thin cells have to be used and consequently high concentrations of solute are necessary to make *cl* sufficiently large. On that account little work has been done in this region with aqueous solutions: nevertheless useful information can be obtained from solutions of the order of 0.1 to 1 *M*. Solvent balancing

References p. 64–66

is easier with hard polished barium fluoride as cell windows than with soft silver chloride.

Raman spectroscopy of solutions is subject to similar considerations. The solvent must not absorb the primary light, nor must its own Raman spectrum obscure that of the solute. In view of the weak nature of the Raman effect in any case solutions need to be of concentration greater than about 0.1 *M*.

Solids

Measurement of absorption in solids is easy when the material can be got as a thin enough (say 0.01—0.1 mm) transparant film. When the solid is only available as a powder it is more difficult to obtain its spectrum. Infra-red spectra of solid powders have been obtained by making a paste or mull of the powder with liquid paraffin (Nujol), hexachlorobutadiene or a fluorinated oil. The aim is to reduce the scattering and the mull is placed between two windows and measured as a liquid. An alternative technique, which has been much used in the last few years and which avoids interference from absorption by the mulling agent, is to grind a few milligrams of the powder with about 200 mg of dry potassium bromide and to press this into a $\frac{1}{2}''$ diameter circular disc, under vacuum, at 30—40 tons per square inch. The resulting disc is usually a clear transparent window from which good absorption spectra of the solid can be obtained. There are cases in which this treatment has been shown to modify the spectrum. In the absence of this effect the technique can be used for quantitative analysis: either a microbalance must be used to measure 2 mg with accuracy or a larger quantity is weighed and mixed throughly with a larger weighed quantity of potassium bromide, 200 mg aliquots being taken from the mixture. See Table 6 for calculation of extinction coefficients.

Temperature variation

Some mention should be made of cells to be used at temperatures other than room temperature. Cells have been designed to take solid and liquid samples to low temperatures and to obtain as solids materials which are gaseous at room temperature. To measure spectra of liquids which are gaseous at room temperature is possible if the thickness of sample can be fairly large and especially if the region of study permits quartz or glass as window material: for thin samples in the infra-red the measurement is difficult. Cold cells[30–33] generally have to have some provision such as a vacuum jacket and jets of dry air for preventing condensation on windows. High temperature cells[34,35] of various sorts generally employ electrical heating. Absorption studies of solution equilibria may not require high temperature, but they do need constant temperature[36–39]: instrument manufacturers have only recently begun to supply thermostatted cell holders.

Quantities required

Some indication has been given in this account of the sizes of sample which are required. It cannot be more than an indication since extinction coefficients vary so widely. For vibrational spectra in the infra-red ε may have any value between 1 and 1500: ultraviolet and visible spectra similarly, but charge-transfer spectra in this region may give values as high as 100,000. In Table 6 (p. 48) formulae for ε are given in terms of the measured optical density, and various concentrations and thickness terms in measurements in gases, liquids, solutions and solids (KBr disc). If the probable extinction coefficient is known then these formulae can be used to ascertain the optimum conditions for absorption: say 50 % absorption or $D = 0.3$.

For the majority of substances it is reasonable to say that normal spectroscopic equipment will obtain absorption spectra from about 10 ml of gas (about 0.5 millimole), 0.1 ml of liquid, 0.5 ml of 0.1 M solutions or 2 mg of solid in the infra-red; and 5 ml of 0.01 M solution in visible and ultraviolet instruments. Normally there will be sufficient material available for these modest requirements (and samples are recoverable, even from potassium bromide discs) but smaller samples can be dealt with. Reflecting microscopes have been used as micro-illuminators so that the image of the slit at the focus of illumination is small enough that as little as 0.1 mg of liquid covers all the light which enters the front slit.

Microwave absorption

The cell in which microwave absorption is measured is commonly a length of the waveguide run, separated off from the rest by gasket-sealed mica windows. Provision is made for admission of gaseous samples and for pumping. Since the pressure of sample used is only about 10^{-2} mm it should be clear that good evacuation and outgassing is called for. Since it is desirable, for a number of reasons, to keep the sample at a low temperature the absorption cell is surrounded by a trough of powdered solid carbon dioxide provided the vapour pressure of the sample exceeds 10^{-2} mm at that temperature (− 80°C). On the other hand it is possible, for less volatile samples, to heat the absorption cell.

Spin resonance experiments use glass for the sample tube. For nuclear magnetic resonance the tube containing about 1 ml of the sample liquid or solid is surrounded by the probe coil. This has to fit inside a Dewar vessel which must nevertheless be thin enough to go between the pole pieces of the magnet. The size and shape of the sample are of some consequence on account of inhomogeneity in the magnetic field. Spherical containers have been used as well as spinning sample tubes. For electron spin resonance the sample is placed in a tube in a cavity resonator between the poles of the magnet. Reasonable care in positioning is needed because of the geometry of the resonator and the homogeneity of the field. Naturally, since electronic spin resonance detects and measures free unpaired electrons, oxygen has to be removed.

Raman Spectra

There are some pecularities in the practice of Raman and fluorescence spectroscopy which warrant separate sections for their consideration. The spectra are still measured with the aid of prism or grating spectrographs: so far as Raman spectra are concerned, a suitable spectrograph is one of wide aperture and (usually) medium dispersion. Absence of stray light is an important requirement, best met by use of a double monochromator.

The observation of a Raman spectrum entails illumination of the sample with monochromatic light and observing the light scattered at right angles to the incident radiation. Fig. 32 shows a typical arrangement of a mercury

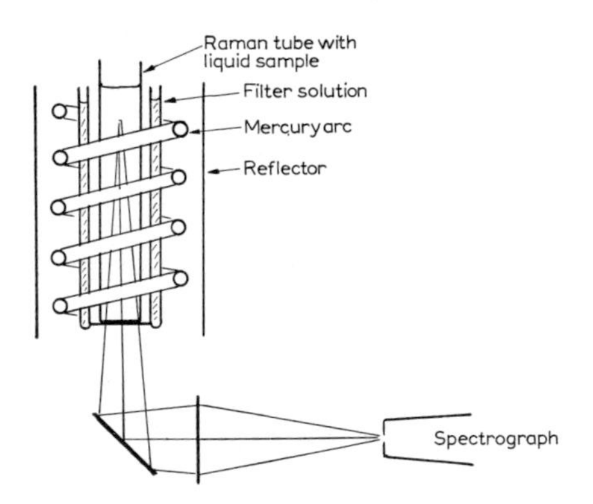

Fig. 32. Observation of Raman spectra.

vapour lamp as the source of energy in a monochromatic "light furnace", irradiating the sample horizontally. Scattered light is gathered through a plane window at the base of the Raman tube and is focussed into the front slit of the spectrograph. Any such arrangement presents the problem of how to illuminate the spectrograph when the source is extended along the optical axis. What conditions have to be met to ensure that the maximum amount of scattered light is accepted by the spectrograph? It is usually to adjust the optics to fulfil the "Nielson conditions"; *i.e.* so that the image of the slit is focussed at the top of the Raman tube and the collimator lens at the bottom[40]. This adjustment is much assisted by the use of a small mercury lamp temporarily placed in the focal plane of the instrument.

The measured spectrum contains the very intense exciting line (the Rayleigh line) and, displaced from it, a series of comparatively weak lines whose distance from the main line is converted to frequency or wave number and is called the Raman shift (Fig. 33). So weak are the Raman lines that

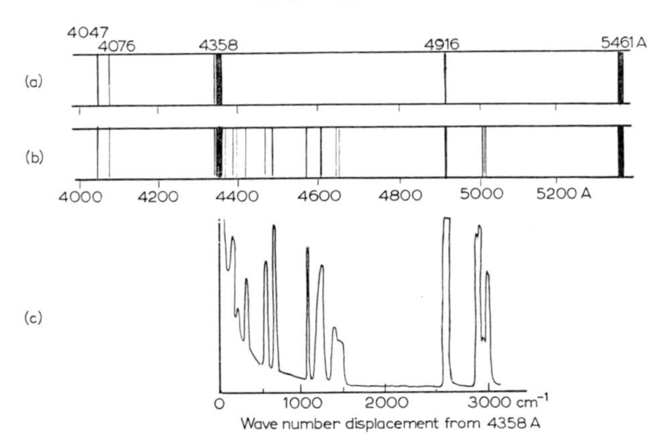

Fig. 33. Typical Raman spectrum. (a) Mercury arc spectrum, photographic; (b) spectrum of scattered light, showing Rayleigh scattering (undisplaced lines) and Raman scattering (displacement of 4358A line); (c) microdensitometer trace or photoelectric record of part of (b).

long exposures are required for their photographic detection. Photomultipliers are used for direct measurement. In either event the Rayleigh line is very heavy or over-exposed and this is aggravated if scattered light is not kept to a minimum. A broadened undisplaced line results from scattering by fine particles and may obscure small Raman shifts. For this reason it is necessary for good quality spectra either to filter the liquids and solutions, or to distil them into the Raman tubes, in order that the Rayleigh line shall not be further intensified. Fluorescent materials should also be excluded.

The choice of the wavelength of the exciting line is governed by the following considerations. In the mercury arc spectrum (Table 5, p. 44) the lines which are useful for Raman excitation are the ultraviolet lines 2537 and 3650 Å, the violet 4047 Å, the blue 4358 Å and the green 5461 Å. Helium gives a useful red line at 5875.6 Å and cadmium a red line at 6438 Å[41].

Filters for the isolation of these lines have been mentioned (p. 29). The intensity of the Raman effect increases in proportion to λ^{-4}: it is therefore 16 times more intense at 2500 Å than at 5000 Å. On the other hand the resolution required for a separation of 5 cm^{-1} at 2500 Å, is 0.25 Å and at 5000 Å is 1 Å: this is readily attainable in prism and grating spectrographs. Some samples may be photochemically sensitive, in which case longer wavelength excitation is preferred. Samples should also be transparent to the exciting line and thus for coloured solutions only one line may be appropriate. Where it can be used the mercury 4358 Å line has the advantage that the interval to longer wavelength corresponding to Raman shifts up to 4000 cm^{-1} contains only one other arc line, 4916 Å. Raman shifts arising from 4047 Å may overlap but that line and its accompanying Raman shifts can be reduced in intensity by filters if the overlap is troublesome. It may be worth noting that the whole of the region from, say 50—5000 cm^{-1}, is accessible to one optical arrangement. Absorptiometric measurements over that region require at least four prisms in one spectrometer and a coarse grating in another.

The size of liquid sample which is generally required is between 10 and 100 ml but cells down to 1 ml or less can be used without too much loss of energy. Smaller samples than that can only be measured with considerable care in setting up and with long exposure. Solutions in normal cells need to be more concentrated than 0.1 M, at any rate for satisfactory photoelectric measurement of line intensity. Recent advances in design of Raman sources and of cells in which multiple reflection makes better use of the scattered light have made possible [42]the measurement of high quality Raman spectra of gases: the technique is not, however, a routine laboratory procedure. This qualification also applies to Raman spectra of solids.

If liquids are to be measured at low temperature the Raman vessel can be surrounded by a cooling jacket through which it is convenient to circulate cold air evaporated from liquid air at a controllable rate[43]. A jet of dry air directed at the bottom window of the cell prevents condensation of atmospheric moisture. For solutions, in which chemical equilibria are fairly sensitive to temperature, water from a thermostat bath should be circulated through a jacket surrounding the vessel. Another jacket is also required for the filter solutions used to isolate the particular line in the mercury spectrum.

Polarization

A valuable feature of Raman spectra from the point of view of structure determination is the behaviour to polarized light. A simple practical device (for excitation between 4200 and 6000 Å) is to surround the cell with tubes of Polaroid, first with the plane of polarization vertical and then horizontal[44]. The ratio of the two intensities, $\rho = I_v/I_h$ of a line measured under these conditions is the *depolarization factor*. The theoretical maximum for ρ is 6/7 when the line is said to be depolarized. Polarized lines have $\rho < 6/7$. This method does not give exact values of the ratio and it is not easy to measure it exactly. In consequence it is not easy to label a line as polarized or depolarized if ρ is close to 6/7.

Fluorescence

As in Raman spectroscopy the fluorescence effect is one which is excited by incident radiation and is observed in a direction perpendicular to the incident beam, in order to avoid interference from it. Apart from this superficial similarity, however, the phenomena are quite different. Fluorescence is the re-emission of light at various frequencies after absorption of the exciting light. Generally a fluorescence spectrum of a molecule is an approximate mirror-image of the absorption spectrum in the ultraviolet (Fig. 34) the fluorescence being at longer wavelengths (lower frequencies) than

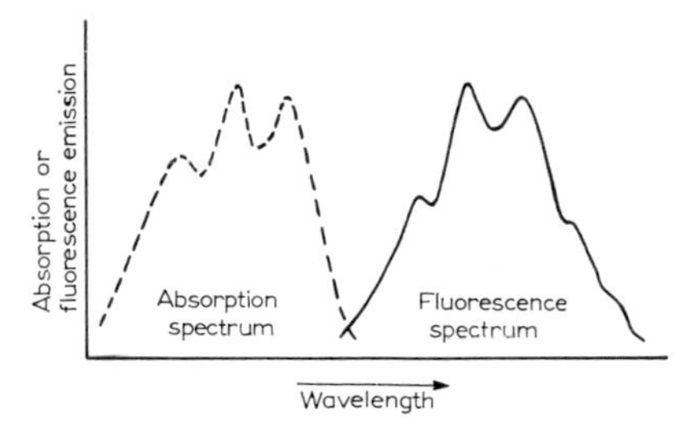

Fig. 34. Fluorescence — comparison of absorption and fluorescence spectra.

the absorption. The entire fluorescence spectrum is generated at the same absolute frequencies by light of any wavelength in the absorption region. It is therefore possible to examine the fluorescence spectrum with a monochromator after excitation by a chosen mercury arc line, suitably filtered. The intensity of fluorescence is considerable and with the use of photomultipliers a fluorescence spectrum may be recorded[45,46] of such materials as anthracene, anthranilic acid, umbelliferone, at concentrations as low as 1 part in 10^9.

The intensity of fluorescence permits the excitation arrangement to be less complex than in Raman spectroscopy, since there is not the need to conserve all available energy. A cell of square cross section, with all sides optically plane and in a compartment suitably baffled to eliminate stray reflections, is illuminated with the filtered light from a mercury arc, a hydrogen lamp or

References p. 64–66

a tungsten filament lamp. The exciting light does not need to be monochromatic. The fluorescence is focussed on to the entrance slit of the spectrometer (Fig. 35) which is of sufficient resolving power if it suffices for the absorption spectrum. The solution should be free from scattering particles and solvents should be free from fluorescence. The material of the cell itself should be free from fluorescence. Some fused quartz does show a weak fluorescence from ultraviolet excitation. Photochemical decomposition is a

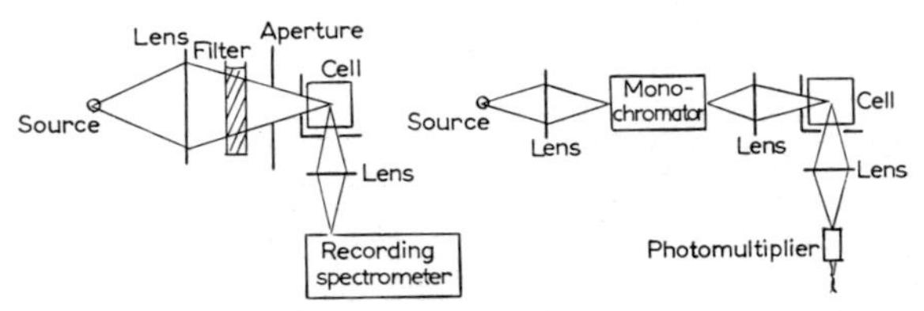

Fig. 35. Arrangements for measuring (a) fluorescence spectra, and (b) excitation spectra.

possible inconvenience, as indeed it is in the measurement of absorption. Quenching of fluorescence (self-quenching) can be overcome by dilution of the solution: quenching by oxygen is eliminated by bubbling nitrogen through the solution.

An alternative way of observing fluorescence is shown in Fig. 35(b) in which all the fluorescence is observed and its change in intensity recorded as the wavelength of the exciting light is changed. In principle the intensity of the exciting light should not be changed, only the wavelength. In fact both the intensity of the exciting light and the sensitivity of the fluorescence change with wavelength and corrections for these have to be made after calibration against known standards. The resultant spectrum is the *excitation spectrum* and it bears a close similarity to the absorption spectrum. By this means the absorption spectra of substances which fluoresce can be measured at concentrations as low as 10^{-8} M and without separation from other materials provided they do not also fluoresce[47,48].

REFERENCES

1 D. H. RANK, *Advances in Spectroscopy*, Vol. I, Interscience, 1959, p. 79.

2 *Intern. Critical Tables*, Vol. VII, McGraw-Hill, 1930, p. 2, and similar sources.

3 J. J. HEIGL, B. F. DUDENBOSTEL, J. F. BLACK and J. A. WILSON, *Anal. Chem.*, 22 (1950) 154.

4 H. L. WELSH, E. J. STANSBURY, J. ROMANKO and T. FELDMAN, *J. Opt. Soc. Am.*, 45 (1955) 338.

5 L. A. WOODWARD and D. N. WATERS, *J. Sci. Instr.*, 34 (1957) 222.

6 N. S. HAM and A. WALSH, *Spectrochim. Acta*, 12 (1958) 88.

7 G. D. DEW and L. A. SAYCE, *Proc. Roy. Soc.* (*London*), A 207 (1951) 278.

8 R. G. N. HALL and L. A. SAYCE, *Proc. Roy. Soc.* (*London*), A 215 (1952) 536.

[9] E. J. BOWEN, *Chemical Aspects of Light*, Oxford Univ. Press, 1946, Appendices I and II.
[10] M. KASHA, *J. Opt. Soc. Am.*, 38 (1948) 929.
[11] E. P. CARR, M. L. SHERRILL, V. HENRI, *Intern. Critical Tables*, Vol. V, McGraw-Hill, 1929.
[12] W. C. GOLMES, *Intern. Critical Tables*, Vol. VII, McGraw-Hill, 19aa.
[13] A. R. DOWNIE, M. C. MAGOON, T. PURCELL and B. CRAWFORD, *J. Opt. Soc. Am.*, 43 (1953) 941.
[14] E. K. PLYLER, L. R. BLAINE and M. NOWAK, *J. Research Natl. Bur. Standards*, 58 (1957) 195.
[15] N. ACQUISTA and E. K. PLYLER, *J. Research Natl. Bur. Standards*, 49 (1952) 13.
[16] L. H. JONES, *J. Chem. Phys.*, 27 (1957) 1229; *ibid.*, 24 (1956) 1250.
[17] D. S. MCKINNEY and R. A. FRIEDEL, *J. Opt. Soc. Am.*, 38 (1948) 222.
[18] W. GUY and J. H. TOWLER, *J. Sci. Instr.*, 28 (1951) 103.
[19] G. W. BETHKE, *J. Opt. Soc. Am.*, 46 (1956) 560, gives an alternative interpolation formula.
[20] J. STRONG, *J. Franklin Inst.*, 32 (1941) 1.
[21] F. C. STRONG, *Anal. Chem.*, 24 (1952) 338.
[22] G. W. HAUPT, *J. Opt. Soc. Am.*, 42 (1952) 441; *J. Research Natl. Bur. Standards*, 48 (1952) 414.
[23] N. T. GRIDGMAN, *Photoelec. Spectrometry Group Bull.*, 4 (1951) 67.
[24] J. A. A. KETELAAR, J. FAHRENFORT, C. HAAS and G. A. BRINKMAN, *Photoelec. Spectrometry Group Bull.*, 8 (1955) 176.
[25] E. R. BLOUT, C. R BIRD, D. S. GREY, *J. Opt. Soc. Am.*, 40 (1950) 304.
[26] A. E. MARTIN, *Trans. Faraday Soc.*, 47 (1951) 1182.
[27] D. A. RAMSAY, *J. Am. Chem. Soc.*, 74 (1952) 72.
[28] H. A. LORENTZ, *Proc. Acad. Sci. Amsterdam*, 8 (1906) 591.
[29] J. J. FOX and A. E. MARTIN, *Proc. Roy. Soc. (London)*, A 167 (1938) 257.
[30] J. POWLING and H. J. BERNSTEIN, *J. Am. Chem. Soc.*, 73 (1951) 1815.
[31] E. L. WAGNER and D. F. HORNIG, *J. Chem. Phys.*, 18 (1950) 296.
[32] A. WALSH and J. B. WILLIS, *J. Chem. Phys.*, 18 (1950) 552.
[33] R. C. LORD, R. S. MCDONALD and F. A. MILLER, *J. Opt. Soc. Am.*, 42 (1952) 149.
[34] R. E. RICHARDS and H. W. THOMPSON, *Trans. Faraday Soc.*, 41 (1945) 183.
[35] L. BROWN and P. HOLLIDAY, *J. Sci. Instr.*, 28 (1951) 27.
[36] H. CAMPBELL and J. A. SIMPSON, *Chem. & Ind. (London)*, (1953) 887.
[37] J. K. GRANT, *Chem. & Ind. (London)*, (1955) 942.
[38] N. MACLEOD, *Chem. & Ind. (London)*, (1960) 342.
[39] D. N. GLEW and R. E. ROBERTSON, *J. Sci. Instr.*, 33 (1956) 27.
[40] J. R. NIELSEN, *J. Opt. Soc. Am.*, 20 (1930) 701; *ibid.*, 37 (1947) 494.
[41] H. STAMMREICH and R. FORNERIS, *J. Chem. Phys.*, 22 (1954) 1624.
[42] B. P. STOICHEFF, *Advances in Spectroscopy*, Vol. I, Interscience, 1959, p. 91.
[43] J. A. ROLFE and L. A. WOODWARD, *Trans. Faraday Soc.*, 50 (1954) 1030.
[44] D. H. RANK and R. E. KAGARISE, *J. Opt. Soc. Am.*, 40 (1950) 84.
[45] R. L. BOWMAN, P. A. CAULFIELD and S. UDENFRIEND, *Science*, 122 (1955) 32.
[46] S. UDENFRIEND, *J. Pharmacol. Exptl. Therap.*, 120 (1957) 26.
[47] C. A. PARKER and W. J. BARNES, *Analyst*, 82 (1957) 606.
[48] C. A. PARKER, *Analyst*, 84 (1959) 446.

The techniques of optical spectroscopy are covered fairly extensively and considerable attention is given to the relevant principles of optics, to the photographic process and intensity determination, to the determination of wavelength and the identification of atomic spectra, in

[49] R. A. SAWYER, *Experimental Spectroscopy*, 2nd ed., Chapman and Hall, 1951.
[50] G. R. HARRISON, R. C. LORD and J. R. LOUFBUROW, *Practical Spectroscopy*, Blackie, 1948.

A useful reference for discussion of the general principles of physical optics is

[51] F. E. JOHNSTON and H. E. WHITE, *Fundamentals of Optics*, 3rd ed., McGraw-Hill, 1957.

A detailed account of the operation of some instruments manufactured by Hilger and Watts, Ltd. is given by

[52] C. CANDLER, *Practical Spectroscopy*, Hilger and Watts, 1949.

Strongly recommended for practical details and for a good selection of examples of application is

[53] G. F. LOTHIAN, *Absorption Spectrophotometry*, 2nd ed., Hilger and Watts, 1958.

For the infra-red there is the monograph

[54] R. A. SMITH, F. E. JONES and R. P. CHASMAR, *The Detection and Measurement of Infra-red Radiation*, Oxford Univ. Press, 1957,
which examines closely the behaviour and design of detectors, the emission of various sources, the transparency and dispersion of prism and window materials, the use of filters and the optical layout of various monochromators.

The only experimental techniques covered in

[55] W. WEST (Ed.), *Chemical Applications of Spectroscopy*, Interscience, 1956,
are in the chapter by R. N. JONES and C. SANDORFY on infra-red and Raman spectrometry.

[56] A. D. CROSS, *Introduction to Practical Infra-red Spectroscopy*, Butterworth, 1960,
devotes 22 pages to techniques and includes a useful comparison of commercial infra-red instruments.

[57] J. H. HIBBEN, *The Raman Effect and its Chemical Application*, Reinhold, 1939,
gives a brief account of earlier experimental techniques (Kohlrausch's book gives practically none).

The techniques (including circuitry) of microwave and radiofrequency spectroscopy are fully covered by the following books

[58] C. H. TOWNES and A. W. SCHAWLOW, *Microwave Spectroscopy*, McGraw-Hill, 1955.
[59] W. GORDY, W. V. SMITH and R. F. TRAMBARULO, *Microwave Spectroscopy*, Wiley, 1953.
[60] D. J. E. INGRAM, *Spectroscopy at Radio and Microwave Frequencies*, Butterworth, 1955.
[61] D. J. E. INGRAM, *Spectroscopy of Free Radicals*, Butterworth, 1959.
[62] E. R. ANDREW, *Nuclear Magnetic Resonance*, Cambridge Univ. Press, 1955.
[63] J. A. POPLE, W. G. SCHNEIDER and H. J. BERNSTEIN, *High Resolution Nuclear Magnetic Resonance*, McGraw-Hill, 1959.

There remain the monumental:

[64] H. KAYSER, *Handbuch der Spektroskopie*, Hirzel, 1905—33, and various volumes of
[65] WIEN-HARMS, *Handbuch der Experimentalphysik*, Akad. Verlag, Leipzig, 1926.

This list is not intended to be exhaustive, but merely adequate. Much practical detail is naturally only to be found in the original papers which describe particular applications.

Papers devoted specifically to technique are to be found in
Journal of the Optical Society of America,
The Review of Scientific Instruments, and
The Journal of Scientific Instruments.

Chapter 2

CHARACTERIZATION OF STATES: ELECTRONIC STATES

The central theme of spectroscopy, which is stated at the beginning and restated with variations at many places in this book, is that all spectra arise from transitions between energy levels. We shall see, however, that the levels have properties, of which the characteristic energy is the one which manifests itself when the atom or molecule changes from one level to another. The levels represent the *characteristic states* of the atom or molecule, and the primary task of the spectroscopist is to identify the characteristic states from observation of the transitions between them. The properties of the characteristic state are, in turn, closely related to the identity of atoms or molecules, to the atomic or molecular structure, and to the energetics of chemical processes: information on these is, for the chemist, the purpose of the exercise.

It has been said that to attempt to derive information about atomic and molecular structure from spectra is like trying to reconstruct a grand piano from the noise it makes when thrown downstairs. Undoubtedly the magnitude of the task was of that order. Even if the units of construction, nuclei and electrons, were taken for granted the present understanding of the way they are put together owes a great deal to spectroscopy. Planck showed that the spectral distribution of black-body radiation (Fig. 9, p. 16) could only be understood if it were recognised that the emission of radiative energy occurred only in *quanta* of energy, the energy content of which was constant for a given frequency but varied in proportion to the frequency of the radiation. The energy of the quantum is given by

$$\varepsilon = h\nu = hc\omega = hc/\lambda \tag{2.1}$$

where, as we have seen on p. 7, h is Planck's constant, ν is the frequency, c is the velocity of light, and ω and λ are the vacuum values of the wave number and the wavelength respectively. Evidently Planck made this proposition in 1900 with some reluctance but there seemed no other way of accounting for the falling off of energy (Fig. 9) at short wavelengths, or high frequencies, than to say that the emitter was the less likely to muster the energy required for a quantum the larger the quantum became. Five years

References p. 127–128

later Einstein suggested that the photoelectric effect implied not only discontinuity in the process of emission of light from the black body but quantization of radiation itself*. Bohr then recognised (1913) that this, taken in conjunction with the most striking feature of atomic line spectra — that emission of energy from atoms occurred only at certain frequencies — implied that the atoms can exist only in certain discrete stationary states (quantum states or characteristic states) having characteristic energies E_1, E_2, E_3, . . . such that $E_1 - E_2 = \varepsilon_{12}$, $E_1 - E_3 = \varepsilon_{13}$ . . . or generally

$$E_i - E_j = \varepsilon_{ij} = h\nu_{ij}, \qquad E_i > E_j \tag{2.2}$$

A loss of energy from the higher level i to the lower level j is accompanied by the *emission* of a quantum of frequency ν_{ij}. An increase of energy from the lower to the higher level is the result of *absorption* of a quantum of frequency ν_{ij}.

It is common to express the energies of the levels in the units of frequency or wave number. They are then referred to as *spectral terms, e.g.*

$$\frac{T_i}{(\mathrm{sec}^{-1})} = \frac{1}{h}\frac{E_i}{(\mathrm{erg})} \quad \text{or} \quad \frac{G_i}{(\mathrm{cm}^{-1})} = \frac{1}{hc}\frac{E_i}{(\mathrm{erg})} \tag{2.3}$$

so that the frequencies or wave numbers associated with transitions are:

$$\nu_{ij} = T_i - T_j \quad \text{or} \quad \omega_{ij} = G_i - G_j \tag{2.4}$$

In fact, of course, spectral terms were so called before Bohr perceived their meaning. They were the direct outcome of the observation, made by Ritz in 1908 and known as the *combination principle,* that the frequencies of lines throughout the atomic spectrum of a particular atom could be expressed as sums and differences of two other lines in the same spectrum. For example, corresponding to a line at frequency ν_{32} there could be found two other lines in the same spectrum at frequency ν_{31} and ν_{21} such that

$$\nu_{32} = \nu_{31} - \nu_{21} \tag{2.5}$$

An alternative statement of the principle is that, for the spectrum of a given element, there can be found a set of numbers, or spectral terms (T_1, T_2, T_3, . . .) such that all frequencies in the spectrum can be expressed as the difference of two terms. Thus:

$$\nu_{32} = T_3 - T_2, \qquad \nu_{31} = T_3 - T_1 \quad \text{and} \quad \nu_{21} = T_2 - T_1 \tag{2.6}$$

which is equivalent to (2.5). On the other hand (2.6) also enables a large number of spectral lines to be represented by a smaller number of terms and

* The arguments for the quantization of radiation need not concern us. Radiation is only observed by its interaction with matter and the spectroscopist is really only concerned with the states of matter before and after interaction. It is convenient thus to separate matter and radiation and consequently to regard the radiant energy as being transferred in quanta or *photons*.

suggests a diagrammatic representation — a *term scheme* — in which the spectral terms are shown as horizontal lines placed at distances proportional to the term values (Fig. 36). Distances between the lines are then propor-

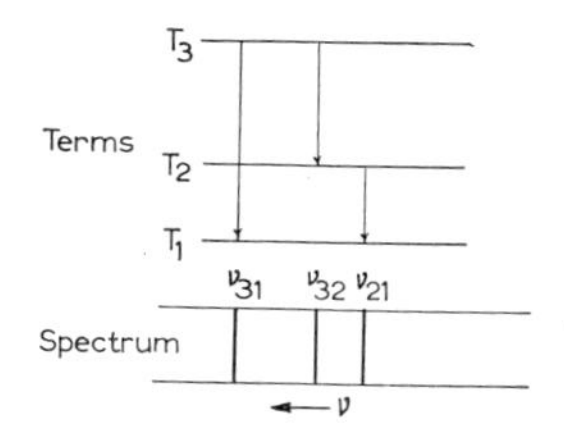

Fig. 36. A term scheme.

tional to the differences between spectral terms. It remained to Bohr to show that the spectral terms were direct measures of the energies of the characteristic states of the atom.

To construct a set of spectral terms from the observed spectral frequencies is not as easy a task as might appear for at least two reasons. The first is that a spectral line does not obviously carry with it the subscript label, which we have used above, to tell us which pair of spectral terms it belongs to: it is merely a number. Without the aid of some theoretical guidance, only trial and error will reveal another pair of lines with which it combines as in (2.5). Secondly it transpires that although all observed frequencies conform to the Ritz combination principle, it is not found that all possible combinations can be observed. Indeed many possible combinations do not occur. Fortunately it is not now necessary to compile such a term scheme without theoretical assistance. Discernible regularities, which are the basis of the theory, appear in the term values or in the series of frequencies.

The first such regularity to be perceived was found by Balmer in 1885 in the hydrogen atomic spectrum and others in the same spectrum were discovered later. This spectrum is particularly simple since each series belongs to one simple set of spectral terms, $T_1, T_2, T_3 \ldots$, such that

$$T_n = R/n^2 \tag{2.7}$$

where R is the Rydberg constant.

We have seen how Bohr recognised the existence of characteristic states and identified spectral terms with the energies of these states. He went further and devised a mechanical model which succeeded in predicting the R/n^2 formula for the hydrogen atom and a correct value for the Rydberg constant. The model used Newtonian mechanics and Coulomb's law. It rejected the idea that an accelerated electric charge must continuously emit radiation and substituted an arbitrary condition for non-radiating stationary states.

References p. 127–128

That the particular quantum rule was one which could not be extended to more complex systems with complete success does not matter (except that the simple derivation of the formulae for the hydrogen atom can still be accurately reproduced by students who may not have appreciated the more fundamental aspects of Bohr's contribution). What was important was the demonstration that, in dealing with atoms and molecules, some part of classical mechanics was inappropriate. Bohr's initial success with the hydrogen atom showed this even though his quantum rule was a compromise which later had to be rejected in favour of a more fundamental modification of mechanics from which quantization appeared as a natural consequence of the basic postulates.

The new mechanics arose in two different ways. The first most clearly owes its origins to spectroscopy but is more difficult mathematically and conceptually. Heisenberg discovered in 1925 how mechanics could be altered in such a way as to yield the observed combination of frequencies. The idea was immediately developed with the aid of matrix calculus by Born and Jordan into the so-called *matrix mechanics*, the only quantities upon which the theory was based being the frequencies and intensities of radiation associated with transitions between the various states of the atom. On the other hand, de Broglie's theoretical investigations (1925) of the consequences of associating waves with massive particles (as waves and photons are associated in the behaviour of light) coupled with the practically simultaneous discovery of electron diffraction, led Schrödinger to develop *wave mechanics*, and to represent physical quantities by differential coefficients (operators). The equivalence of matrix mechanics and wave mechanics was soon recognised and the name *quantum mechanics* is now generally used.

It is not the purpose of this book to describe in detail the course of the development of quantum mechanics or the methods of solving the equations which arise in the quantum mechanical description of particular systems. Such descriptions can be found in the books listed at the end of this chapter (p. 127). On the other hand it is important to see that the task of constructing an energy level diagram from the spectroscopic data was lightened by the perception of regularities in the simplest of term schemes, by modification of the empirical formulae to accommodate more complex schemes and subsequently by the development of a mechanics adequate in principle to interpret all the spectral data. In providing such an interpretation quantum mechanics introduces parameters of structure and configuration which are of interest to the chemist. The problem therefore becomes one of finding the proper values of those parameters to correspond with the observable spectroscopic transitions and their intensities.

Experiment and theory in the period 1900—1930 advanced very much hand-in-hand. Atomic and molecular spectra provided an abundance of data requiring interpretation: the theoretical advances both stimulated further experiment and were themselves induced by the need to interpret the results. The scope of the theory has now, of course, extended far beyond spectroscopy: the whole of theoretical physics is based upon its mathematical methods. So far as chemical spectroscopy is concerned we may take the basic interpretation as valid (the validity ultimately rests on the success of the interpretation) and consider how we may use it to derive chemical information from spectroscopic data. Therefore we describe first some of the basic concepts of quantum mechanics. We shall then, in this chapter consider particularly those systems for which a reasonably detailed interpretation of *frequency values* can be given in terms of specific arrangements of electrons and nuclei, stating without derivation the quantum mechanical results we need.

Quantum Mechanical Basis

It is possible to imagine a system of nuclei and electrons, point masses and charges, in a particular configuration at a particular time. This image is based on common experience of familiar things. Newtonian mechanics gives expression to the experience in equations of motion which imply that at a particular time the system has a particular configuration, always calculable in principle and sometimes in practice. In other words, an imagined configuration at a given time either occurs or it does not. Quantum mechanics drops this yes-or-no attitude to the question of configuration and instead enquires into the probability that an imagined configuration will occur.

The basis of ordinary Newtonian mechanics is the conservation of kinetic and potential energy. A particle of mass m moves with potential energy $V(xyz)$ and kinetic energy T and the sum of the two may be written:

$$\begin{aligned} H &= T + V \\ &= \frac{1}{2m}(p_x^2 + p_y^2 + p_z^2) + V(xyz) \end{aligned} \tag{2.8}$$

The Hamiltonian H can be defined for any system as a more general function of position coordinates and momenta. In conservative systems, it is constant and expresses the energy of the system in terms of its coordinates and momenta. For systems of ordinary particles acting under forces which can be derived from a potential the Hamiltonian takes the form given in (2.8). The law of conservation of energy is then simply expressed in the *Newtonian dynamical law*:

$$H = E \tag{2.9}$$

References p. 127–128

where E is a constant of the motion. For a given system there is a corresponding Hamiltonian (2.8) and the dynamical problem becomes one of finding how the configuration of the particle — the coordinates x, y, z — varies with time in such a way as to satisfy the equation (2.9). Such functions are solutions of the equation of motion. The configurations they describe are the ones that occur: others do not.

Fundamental postulates

A valid approach to quantum mechanics is to set down as basic postulates the rules for modifying the Newtonian law. We require now an equation whose solution gives not the configuration itself but the probability that the configuration occurs. We will write this solution as $\Psi(xyzt)$. The modifying rules are such as to convert total energy, momentum and coordinates into corresponding *quantum mechanical operators*, or differential coefficients, which operate upon Ψ yielding a differential equation. Thus:

for H write $-\dfrac{\hbar}{i}\dfrac{\partial}{\partial t}$

for p_x write $\dfrac{\hbar}{i}\dfrac{\partial}{\partial x}$, and similarly for p_y and p_z

for x write *multiply by* x, and similarly for y and z.

The quantity $\hbar$ (crossed h) is frequently used in quantum mechanics for $h/2\pi$: it is the natural unit of angular momentum. $i = \sqrt{-1}$. It is common practice to continue using the symbol H for the Hamiltonian operator representing energy. In its quantum mechanical significance therefore

$$H \equiv -\frac{\hbar}{i}\frac{\partial}{\partial t} \tag{2.10}$$

The operators are to operate upon Ψ. With the aid of the modifying rules equation (2.8) of classical mechanics thus becomes the differential equation:

$$H\Psi = -\frac{\hbar}{i}\frac{\partial \Psi}{\partial t} = -\frac{\hbar^2}{2m}\left(\frac{\partial^2}{\partial x^2} + \frac{\partial^2}{\partial y^2} + \frac{\partial^2}{\partial z^2}\right)\Psi + V(xyz)\Psi \tag{2.11}$$

Equation 2.11 is the *dynamical law* of quantum mechanics.

The function Ψ may be complex (*i.e.* of form $f + ig$) in which case Ψ^* represents its conjugate $(f - ig)$. Ψ is not itself the probability of occurrence of any configuration. The product $\Psi\Psi^*(xyzt)$ (a real quantity, $f^2 + g^2$) gives the required probability that the particle will have the configuration x, y, z at time t.

Characteristic states

We are interested in those configurations — *stationary states* — for which the probability of their occurrence remains unchanged in time. More realistically we may say that interest is centred on those states that remain unchanged for a time which is long compared with the time required to characterize them by spectroscopic observation of transitions between states. Such states are those for which the energy E has precise value. They are solutions of the equation

$$H\Psi = E\Psi \tag{2.12}$$

in which H is the Hamiltonian operator (2.10) and E is a multiplying number. Note the formal similarity to (2.9).

When it is required that $\Psi\Psi^*$ shall be independent of time the differential equation (2.12) has solutions of the form:

$$\begin{aligned}\Psi(xyzt) &= \psi(xyz)\mathrm{e}^{-i(E/\hbar)t}\\ \Psi^*(xyzt) &= \psi^*(xyz)\mathrm{e}^{+i(E/\hbar)t}\end{aligned} \tag{2.13}$$

in which ψ is independent of time. Such a solution permits (2.13) to be rewritten as

$$H\psi = E\psi \tag{2.14}$$

or, using (2.10), as:

$$\frac{\hbar^2}{2m}\left(\frac{\partial^2}{\partial x^2} + \frac{\partial^2}{\partial y^2} + \frac{\partial^2}{\partial z^2}\right)\psi + (E - V)\psi = 0 \tag{2.15}$$

or:

$$\nabla^2\psi + \frac{2m}{\hbar^2}(E - V)\psi = 0 \tag{2.16}$$

These are forms of the quantum mechanical equation of motion generally known as the *Schrödinger equation*. It can of course, be generalized for a system of particles and its solutions depend upon the nature of the system as expressed in the potential function $V(xyz)$.

If $\Psi\Psi^*$ is the general configuration probability, it can be seen from (2.13) that $\psi\psi^*$ is the configuration probability in the steady state. $\psi\psi^*(xyz)$ for a single particle gives the probability that the particle is to be found at position (xyz). It follows from this interpretation that $\psi(xyz)$ must have only one value at any position xyz, must not be infinite at any position xyz, must not change suddenly at any position xyz, and must become zero when x, y or z is infinite. In other words ψ must be everywhere single-valued, finite and continuous and must vanish at infinity.

The effect of imposing these boundary conditions is that there are only certain solutions $\psi_1, \psi_2, \psi_3 \ldots \psi_n$ which satisfy them, and with each solution

References p. 127–128

there is a corresponding value E_1, E_2, $E_3 \ldots E_n$ of the parameter E. The permitted solutions ψ are called variously *characteristic functions, proper functions, eigenfunktionen* (in German), or *eigenfunctions* (in anglicized German). Because of the form of the Schrödinger equation and because of its origins in de Broglie's theory the ψ_n are also called *wave functions*. The corresponding E_n are the *characteristic* or *proper values*, the *eigenwerte*, or the *eigenvalues*. E has the dimensions of energy and the E_n are the eigenvalues of energy: in other words, the characteristic numbers E_n are the energy levels of the particle.

Energy appears in quantum mechanics as having a peculiar significance because of its appearance in the dynamical law; because of the way in which it stands in relation to time (energy operator). Momentum, p_x, stands in a similar relation to the position coordinate, x, and it may be mentioned here that it is possible to set up a Schrödinger equation in momentum, proper solutions of which arise only for certain eigenvalues of momentum.

We thus see how, in principle, quantum mechanics predicts the existence of characteristic states — quantum states — in which the energy assumes a definite value with certainty. The nth state has energy E_n and is characterized by the function of coordinates $\psi(xyz)$, giving the *probability amplitude* at all points xyz. The function $\psi\psi^*$ (ψ^2, if ψ is not complex) gives the *probability density* at all points xyz. Since the probability of finding the system somewhere in space must be unity,

$$\int_{-\infty}^{\infty} \psi\psi^* \, d\tau = 1 \tag{2.17}$$

Degeneracy

If there are a number, m, of eigenfunctions ψ_n, ψ_{n+1}, $\psi_{n+2} \ldots \psi_{n+m}$ all having the same eigenvalue E_n then the energy level E_n is said to possess *degeneracy*: it is *m-fold degenerate*. If a perturbation (experimental or theoretical) applied to the system gives different eigenvalues for each of the m characteristic states, the perturbation is said to resolve or split the degeneracy.

Uncertainty principle

Inherent in the quantum mechanical description of states in terms of probability there are limitations to the accuracy with which certain quantities can be simultaneously measured, or meaningfully specified. The pairs of quantities, momentum and its corresponding position coordinate, or energy and time, are subject to such limitation. We have seen that the probability of finding a particle in a specified position is, in general less than unity. Unit probability implies certainty in the specification of position. Quantum mechanics asserts that to demand observation of position with certainty one must accept complete uncertainty in the corresponding momentum: this, in terms of the wavelength associated with the particle, implies complete uncertainty as to the wavelength.

The *uncertainty principle* has been formulated by Heisenberg as follows. Suppose momentum p and its corresponding coordinate q can be measured simultaneously with uncertainties Δp and Δq, such that zero values of Δp and Δq correspond to complete certainty. Then

$$\Delta p \cdot \Delta q \geqq \tfrac{1}{2}\hbar \qquad (2.18)$$

Similarly for energy and time

$$\Delta E \cdot \Delta t \geqq \tfrac{1}{2}\hbar \qquad (2.19)$$

The relationship (2.19) particularly shows the spectroscopic significance of the stationary state. It implies that to specify energy exactly ($\Delta E = 0$) requires that the system remains in the same state for ever ($\Delta t = \infty$). Conversely it implies that the energy of a short-lived quantum state (finite Δt) is subject to some uncertainty in energy ($\Delta E \neq 0$). In consequence spectral lines arising from transitions from such levels are not sharp but are broadened by the spread of frequencies. Quantities which are together subject to uncertainty relations such as (2.18) and (2.19) are referred to as *complementary* and the operators representing them are non-commutative. It is worth noting that the two products in (2.18) and (2.19) have the dimensions of action (erg sec) as does $\hbar$ itself.

General Interpretation of Spectra

Atomic spectra arise from transitions between electronic energy levels. Although transitions at all wavelengths from the radio region to hard X-rays occur in atomic spectra, the major region of activity is the visible and the ultraviolet, indicating energy levels lying 10,000—100,000 cm^{-1} apart. It is reasonable to suppose that the considerable activity of molecules in this region also implies electronic transitions.

Molecules have other possible motions besides electronic and in accordance with this they also show strong spectral activity in the infra-red and, for gaseous molecules, in the far infra-red and microwave regions also. A satisfactory agreement with the observed spectra is obtained when the infra-red spectra are interpreted as arising from transitions between the energy levels of quantized molecular vibration (lying some 100—4,000 cm^{-1} apart) and the longer wavelengths as due to transitions between rotational energy levels (lying 0.1—10 cm^{-1} apart). This interpretation must now be examined more closely.

References p. 127–128

Separation of motion in molecules

Although it is possible to imagine molecules as vibrating and rotating and to regard these motions as independent, it is strictly not a permissible quantum mechanical description. The characteristic functions describe the probability distributions for the complete system and to assume separation of electronic and nuclear motions amounts to assuming that the complete functions can be written as a product of two functions, one for electrons and one for nuclei. Thus:

$$\psi_{v\varepsilon}(q_n q_e) = \psi_\varepsilon(q_e) \cdot \psi_{v\varepsilon}(q_n) \qquad (2.20)$$

in which the subscripts v and ε refer to quantum numbers for nuclear and electronic motion and the q_n and q_e stand for the coordinates of all the nuclei and all the electrons. The Schrödinger equation in $\psi_{v\varepsilon}(q_n q_e)$ becomes two equations; one in $\psi_\varepsilon(q_e)$ from which electronic states are to be obtained for assumed nuclear configurations, and one in $\psi_{v\varepsilon}(q_n)$ in which a set of nuclear configurations is found for each electronic quantum state denoted by quantum number, ε. That is to say, the $\psi_\varepsilon(q_e)$ describe the behaviour of the electrons moving in a field of fixed nuclei with potential energy V_e, V_e varies with internuclear distance and thus the ψ_ε and E_ε also depend upon the internuclear distance. There is, in general, a different E_ε for each electronic state. If to each E_ε we add the Coulomb interaction of nuclei, also dependent upon internuclear distances, we have the potential $E_\varepsilon + V_o$ under which the nuclei move. The curve of $E_\varepsilon + V_o$ against the single internuclear distance in a diatomic molecule is the potential curve for the molecule (see p. 125) and there is a different potential curve for each electronic state. The $\psi_{v\varepsilon}(q_0)$ describe the nuclear motion under that potential. The validity of this so-called *Born-Oppenheimer approximation* depends upon the ratio of the masses of electrons and nuclei and thus upon the relative spacing of the energy levels. Mathematically the Schrödinger equation only separates into two equations if all the terms like $\partial\psi_\varepsilon/\partial q_n$ can be neglected; that is, if ψ_ε varies sufficiently slowly with change in nuclear configuration.

We shall see (p. 318) that the intensities of electronic transitions agree with the idea, in classical terms, that the heavy nuclei have no time to move appreciably during the interval of an electronic transition. The times for electronic transitions and nuclear vibrations, measured by the inverse of their radiation frequencies, are 1.3×10^{-15} sec for a supposed electronic transition at 4000 Å and 3.3×10^{-14} sec for a typical vibration at 10 μ. There are many unexamined assumptions in such an argument and it may be preferable to assert that the separation of states into electronic and nuclear parts, and indeed of nuclear motion into vibration and rotation, rests upon the energy differences. So long as there is a considerable difference in the spacing of levels we may write for a molecule

$$E = E_{\text{electronic}} + E_{\text{vibration}} + E_{\text{rotation}}$$

and accordingly split a quantum into the imagined parts

$$h\nu = \Delta E = \Delta E_{\text{electronic}} + \Delta E_{\text{vibration}} + \Delta E_{\text{rotation}}$$
$$= h\nu_{\text{e}} + h\nu_{\text{v}} + h\nu_{\text{r}} \qquad (2.21)$$

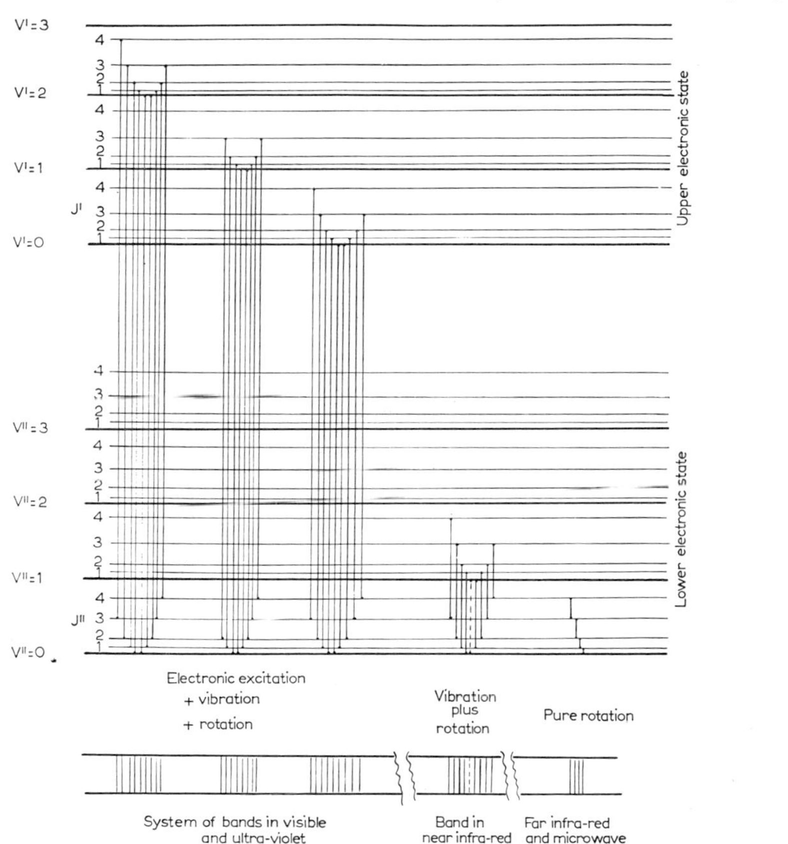

Fig. 37. General interpretation of molecular spectra. The scale of the diagram is distorted; the separation of the vibrational levels is too large compared with that of the two electronic states; the separation of the rotational levels is much too large compared with that of the vibrational levels.

With this separation it becomes possible to account for the various parts by solution of independent Schrödinger equations, leading to rotational, vibrational and electronic energy levels.

A typical energy level diagram for a gaseous molecule is shown in Fig. 37.

References p. 127–128

Two electronic states are shown in each of which the first few vibrational levels are indicated, v' and v'' are the vibrational quantum numbers in the upper and lower electronic states respectively. With each vibrational state in either electronic level there is an associated set of rotational levels, quantum number J. The relative spacing of the rotational levels is in fact much smaller than is shown in the figure. The transition depicted by the line on the extreme left of the diagram is between (a) the $J = 3$ level of the lowest vibrational state in the ground electronic state and (b) the $J = 4$ level of the $v' = 2$ vibrational state in the upper electronic state. The quantum for this transition is made up as follows:

$$\begin{aligned} \Delta E &= E(v' = 0, J' = 0) - E(v'' = 0, J'' = 0) \\ &\quad + E(v' = 2, J' = 0) - E(v'' = 0, J'' = 0) \\ &\quad + E(v' = 2, J' = 4) - E(v'' = 0, J'' = 3) \\ &= \Delta E_e + \Delta E_{v(0,2)} + \Delta E_{r(3,4)} \end{aligned}$$

In the first approximation it is assumed that the spacing of the vibrational levels is the same in each electronic state and that the same set of rotational levels stands on each vibrational level in each electronic state. Since the electronic state undoubtedly affects the bonding in the molecule both the vibrational levels and rotational levels may be expected in fact to be different in the two electronic levels and, to a lesser extent, the rotational levels may be expected to differ in different vibrational states. Such considerations blur the distinctions between ΔE_e, ΔE_v and ΔE_r but they are best treated as refinements after the consequences of the initial approximation have been worked out.

Band spectra

Transitions between rotational levels without change in vibrational or electronic energy, give lines in the far infra-red and microwave region. Rotational changes of the same magnitude always accompany a change of vibrational energy. For example, the quantum for the transition ($v = 0$, $J = 0 \rightarrow v = 1$, $J = 0$) in hydrogen chloride corresponds to a wave number value, $\omega_0 = 2886$ cm^{-1}; and for the transition ($v = 0$, $J = 0 \rightarrow v = 1$, $J = 1$) $= 2906$ cm^{-1}. The difference, 20 cm^{-1}, corresponds to the rotational change ($J = 0 \rightarrow 1$) in the vibrational state, $v = 1$. There are in fact a number of such lines which appear on either side of ω_0, constituting the *rotational structure* of the vibrational transition ω_0. The whole collection of lines constitutes a *band* and the position of the band is governed by the magnitude of the vibrational transition. Vibrational levels are so spaced that these bands appear in the infra-red.

Electronic transitions offer the possibility of accompanying vibrational and rotational changes. For any one electronic transition there may be a number of vibrational changes, each with its rotational structure. There thus arises a system of bands, whose general position is governed by the magnitude of the electronic change, and therefore lies in the visible or ultra-violet.

The band system drawn in Fig. 37 is an ideal. Observed band systems are more complex; bands are more numerous and often overlap. For diatomic molecules an instrument of medium resolving power will normally resolve the system into bands (*gross structure*). Higher resolution is required to resolve the *fine structure, i.e.* the rotational structure of each band. Resolving power may similarly limit the apparent shape of a band in the infra-red. Some molecules are light enough for the rotational structure to be resolved by the spectrometers which are now in common use in analytical and research laboratories. With increasing molecular weight the lines in a band come close to the centre, and to each other, and resolved structure gives way to a band envelope (see p. 185).

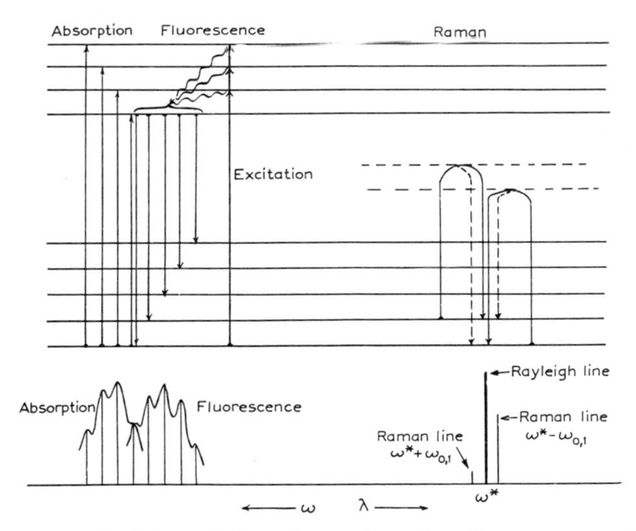

Fig. 38. To compare the interpretation of u.v. absorption, fluorescence, and the Raman effect in molecules.

Fluorescence and Raman spectra

It is now possible to extend the interpretation to fluorescence and Raman spectra. In Fig. 38 vibrational levels are shown for two electronic states of a molecule and the transitions on the left of the diagram symbolize the u.v. absorption of the molecule. The molecule is raised to any vibrational level

References p. 127–128

in the upper electronic state by absorption of light. It is then reduced to the lowest ($v' = 0$) vibrational level of that state by radiationless transfer of energy, *i.e.* by collisions. Re-emission of light then accompanies transitions to various vibrational levels in the ground electronic state: this is fluorescence. To the extent that the distribution of vibrational levels is similar in the two states, the fluorescence spectrum appears as an approximate mirror image of the absorption spectrum (as in Fig. 34, p. 63). It should be emphasized that fluorescence is a process of absorption and re-emission: the fluorescence spectrum appears at a characteristic wavelength and does not shift with the wavelength of the exciting light. Atomic fluorescence is, naturally, more restricted than molecular fluorescence since there are not the accomodating vibrational levels to permit excitation by different wavelengths. There are only the sharp electronic levels so that fluorescence can only be excited by light (or X-rays) of the same wavelength as is to be re-emitted.

Raman spectra, on the other hand, do not arise from such a process but by scattering. In Fig. 38 the dotted level is not to be interpreted as an energy level belonging to the molecule but as a representation of the energy of the exciting quantum, of wave number ω^*. Added to the energy of the $v = 0$ state either it is scattered without change (return to $v = 0$) or it is reduced by a vibrational quantum, being scattered as $\omega^* - \omega_{0,1}$ and leaving the molecule in the state $v = 1$. Added to the energy of the $v = 1$ level, either it is again scattered as ω^* leaving the molecule in the $v = 1$ level or it is scattered as $\omega^* + \omega_{0,1}$ leaving the molecule in the $v = 0$ level. Lines $\omega^* - \omega_{0,1}$ arising from different vibrational modes with different $\omega_{0,1}$ are the so-called *Stokes* lines. They are more intense than the lines at $\omega^* + \omega_{0,1}$ — the *anti-Stokes* lines — since the latter arise only from molecules which are already in the excited vibrational state. In principle the exciting radiation can be of any frequency greater than the vibrational frequency; in practice, useful visible and ultraviolet lines are provided by the mercury arc. The Raman lines differ from fluorescence in that they move with ω^*, maintaining a constant difference from ω^*. Raman shifts are not confined to vibrational changes. They have also been observed for rotational changes. In this case, however, the Raman lines are very much closer to the Rayleigh line at ω^*.

Effect of magnetic and electric fields

Spectra also arise from transitions between levels, normally degenerate, whose degeneracy has been resolved by a magnetic field. Such is the case with odd, or unpaired, electrons in an atom or molecule. Fields of the order of 10,000 gauss can be produced in the laboratory and the resulting separation is such that the quantum for transition between the separated levels has a frequency of about 28,000 Mc/s. *Electron spin resonance* is observed in this way in the microwave region. Even smaller energy level separations are produced by the interaction of a magnetic field with the magnetic moment of atomic nuclei. Nevertheless radiation of the appropriate frequency is absorbed to effect transition between the separated levels. *Nuclear spin resonance* is thus observed at radio frequencies (10—50 Mc/s).

Magnetic fields and electric fields also have perturbing effects on optical spectra. A single electronic transition to a degenerate level goes over into a number of slightly different transitions, if the levels are split by an external magnetic field. This is the *Zeeman effect*: the perturbation may be less than 28 in 500,000 kMc/s, or 0.3 Å in 6000 Å, and so requires good resolution for its observation. An internal magnetic field arising from the nucleus has a similar effect on the electronic levels of an atom. The effect is smaller, being of the order of 0.01 Å in 6000 Å. It is one cause of the *hyperfine structure* of atomic lines, which is often only apparent under the high resolution obtained with interferometry.

Perturbation of electronic levels by electric fields gives rise to the *Stark effect*. Such an effect is also observed in molecular rotation spectra, where the levels of a molecule with a permanent dipole moment are partially resolved by an electric field. Although in principle it could also appear in the rotational fine structure of infra-red bands of polar gaseous molecules, the resolution required is much too high.

Population of levels

An important feature of the general interpretation of spectra is the relative population of energy levels. The intensity of a transition depends both on the intrinsic probability that such a transition will occur and upon the number of molecules which are in the state corresponding to the starting point of the transition. We shall return to this in Chapter 5, but it is worth noting here that the number of molecules, n_i, in an upper state i relative to the number, n_j, in a lower state j is given by the Boltzmann distribution law

$$\frac{n_i}{n_j} = \frac{g_i}{g_j}\, e^{-\Delta E/kT} \qquad (2.22)$$

where ΔE is the energy difference between the two states and g_i and g_j are the degeneracies or statistical weights of the two states. Clearly n_i is negligible with respect to n_j when ΔE is much greater than kT, and only becomes appreciable as ΔE approaches kT. From Table 2 (p. 8) it can be seen that even at 1000°K very few atoms are thermally excited to upper electronic states, but at that temperature many molecules are excited to upper vibrational states. Even at room temperature those $v = 1$ vibrational levels which lie some 200—500 cm^{-1} above the $v = 0$ level are fairly well populated.

Rotational levels, separated by say 1 cm^{-1}, are populated up to high J values even at low temperatures. In absorption therefore only those transitions which arise from the ground electronic state are observed; the principal vibration bands arise from the $v'' = 0$ state; but the rotation spectrum may contain lines whose transitions originate from high J values. This partly explains the choice of transitions from among many other possibilities which are shown in Fig. 34.

If atoms or molecules are thermally or electrically excited to high energy levels then, of course, other transitions may occur and give emission spectra containing lines and bands not observed in absorption. Atomic spectra are more usually observed in emission, but it is a useful confirmation of the identification of series to observe by absorption (called the principal series in the alkali-metal spectra) that one which originates from the electronic ground state.

Spectral Activity: Selection Rules

If all possible lines were drawn between all levels in any real term scheme they would represent a large number of transitions that are not observed. Even if the considerations of the previous section were used to eliminate many of these possibilities there would still remain many that are not observed. There are clearly further limitations to the observation of spectra, or to the ability of the atom or molecule to change from one state to another, and these limitations are referred to as *selection rules*. Their theoretical explanation is based on considerations of the symmetry of the system in the upper and lower states (or the symmetries of the characteristic functions describing those states).

The assertion that a transition does occur is the same as saying that its intensity is not zero. If a transition occurs between two terms, the terms are said to *combine*. A selection rule therefore defines the conditions that two terms may combine, or the conditions that the intensity of a transition between the two levels shall be other than zero: such a transition is said to be *allowed*.

So far as absorption and emission of radiation are concerned the interaction with radiation depends upon a redistribution of electric charge between upper and lower states. In fact the transition must involve a change in dipole moment if it is to interact with the oscillating electric vector of the radiation. If the transition between states involves no change in any of the three components of electric moment the transition is *forbidden* in absorption or emission; it is *inactive*. A transition which does involve such a change is allowed.

The above rule has more detailed consequences but it can be seen immediately that the following molecular states do not combine:

(i) vibrational levels which retain either a centre of symmetry or spherical symmetry.

(ii) all rotational levels of molecules with either a centre of symmetry or with spherical symmetry; *i.e.* without a permanent dipole.

For example, the vibration of a homonuclear diatomic molecule such as hydrogen is not active in the infra-red because stretching does not give rise to an oscillating dipole, whereas the vibrational stretching of a heteronuclear diatomic molecule such as hydrogen chloride does. Rotational transitions of HCl are also active in absorption and emission, whereas those for H_2 are not, since HCl has a permanent dipole and therefore its rotation leads to an oscillating electric moment. We give examples of the operation of the rule for polyatomic molecules later.

The Raman effect does not involve absorption and re-emission but is a scattering phenomenon. Interaction in this case depends upon a change not of polarization but of polarizability. This may lead to vibrational levels participating in Raman activity which are inactive in the infra-red, and vice-versa. The stretching of the hydrogen molecule, for example, alters the polarizability and the vibration is therefore Raman-active. The polarizability of the H_2 molecule is greater along the axis of the molecule than perpendicular to it. Therefore rotation about an axis perpendicular to the molecular axis produces an oscillatory change of polarizability and, in consequence, the rotation of hydrogen is also Raman-active. Rotation of spherically symmetrical molecules however is not active in the Raman effect.

The above selection rules, although basic to the interpretation of spectra, evidently do not contribute much towards reducing the number of imaginable transitions. The more detailed rules, which have the same origin as the gross rules just discussed, are those which reduce the number of allowed transitions down to manageable proportions. They will be introduced as required in each of the relevant sections of this chapter and the next.

It should also be noted that many transitions, labelled forbidden, may nevertheless appear if they involve a change either in magnetic dipole moment or in quadrupole moment which can interact with the radiation field. The intensity of such transitions is expected to be considerably smaller than that of allowed electric dipole transitions. Conversely it may, and often does, happen that allowed transitions are not observed because the intensity is too small. What with forbidden transitions which occur and allowed transitions which do not it might be thought that the value of selection rules was marginal. This is not so: the detailed interpretation of spectra is entirely dependent upon the selection rules despite this handicap.

References p. 127–128

Electronic States of Hydrogen and Alkali Metal

The proper interpretation of atomic spectra had profound chemical significance because it fully confirmed the supposed connection between the periodic classification of the elements and the electronic structure of their atoms. In that the energetics of chemical binding, both covalent and electrovalent, depend upon the energies of the atoms and their ions, the spectroscopic determination of those quantities, and in particular of ionisation potentials, is also a valuable contribution to theoretical chemistry. Hence the determination of the energy level diagram is important for an understanding of the chemical behaviour of an element. Although this task is largely completed we must describe it because it is important to know the origin of such quantities as can be found tabulated in many chemistry textbooks and also because it provides the background against which to describe spectral observations which are not as yet so fully interpreted.

Atomic spectra, of course, have continuing chemical application in spectrochemical analysis but we shall not further discuss that aspect. The electronic states of molecules more complex than diatomic are only imperfectly understood. Such spectroscopic information as we have, however, provides data against which theories of valency may be tested.

Hydrogen atom

The remarkable simplicity of the hydrogen atom spectrum is illustrated in Fig. 39. The spectral lines can be seen as a number of series each converging to a limit, represented by a dotted line. Above the spectrum the energy level diagram shows the transitions corresponding to each line. Each level is labelled with its term value $T_1, T_2, \ldots$. If the Balmer series is taken first as defining the levels $T_2, T_3, T_4, \ldots$ because it comprises $\omega_{3,2} = T_3 - T_2$, $\omega_{4,2} = T_4 - T_2 \ldots, \omega_{n,2} = T_n - T_2$, then the simplicity of the term diagram is seen to arise from the experimental fact that all members of the Lyman series can be accurately described in terms of these levels with the addition of another level T_1. By combining the first member of the Lyman series (at 82,259.56 cm^{-1}) with the Balmer series the remaining members of the Lyman series are obtained, as is demonstrated in Table 7. The Lyman series comprises $\omega_{2,1} = T_2 - T_1$, $\omega_{3,1} = T_3 - T_1, \ldots, \omega_{n,1} = T_n - T_1$ and the observation:

$$\omega_{n,2} + \omega_{2,1} = \omega_{n,1} \tag{2.23}$$

is consistent with

$$(T_n - T_2) + (T_2 - T_1) = (T_n - T_1) \tag{2.24}$$

This is just an example of Ritz' combination principle (p. 68).

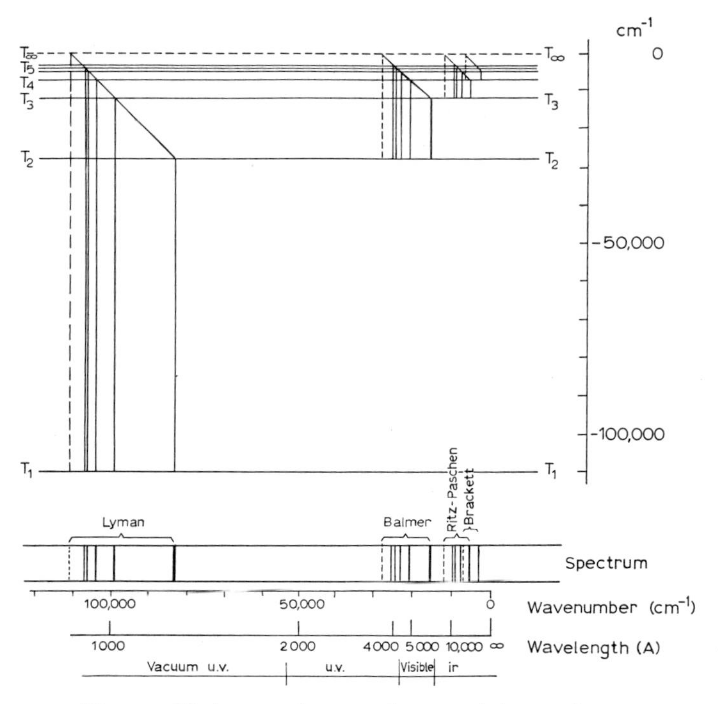

Fig. 39. Hydrogen atom spectrum and term scheme.

TABLE 7

VERIFICATION OF RITZ COMBINATION PRINCIPLE

Balmer series observed		Lyman observed		Lyman series calculated	Lyman series observed
15,228.5	+	82,259.56	=	97,488.1	97.491.5
20,564.7	+	82,259.56	=	102,824.3	102,823.5
23,032.5	+	82,259.56	=	105,291.1	105,289.7
24,373.0	+	82,259.56	=	106,632.6	106.630.2
		(all ω in cm^{-1})			

No additional levels are required for the Ritz-Paschen series (series term, T_3), the Brackett series (series term, T_4) or the Pfund series (series term, T_5): all are accurately predicted by the terms obtained from the Balmer series.

Balmer made a further discovery which, expressed in wave-number values, amounts to the statement:

$$T_n = -R_H/n^2 \tag{2.25}$$

where the presently accepted value of $R_H = 109{,}677.581\ cm^{-1}$. ($R_H$ is the

References p. 127–128

Rydberg constant. Energy is denoted on Fig. 40 as increasing upwards, and the term limit T_∞ is arbitrarily chosen as the zero of energy). $R_H = T_\infty - T_1$ and reference to Fig. 36 shows that R_H is the value of the limit of the Balmer series. Bohr showed, and quantum mechanics confirms, that:

$$E_n = -\frac{2\pi^2 \mu e^4}{h^2} \cdot \frac{1}{n^2} \quad \text{(erg)} \tag{2.26}$$

or:

$$T_n = -\frac{2\pi^2 \mu e^4}{h^3 c} \cdot \frac{1}{n^2} \quad (\text{cm}^{-1}) \tag{2.27}$$

and hence:

$$R_H = 2\pi^2 \mu e^4 / h^3 c \qquad (\text{cm}^{-1}) \tag{2.28}$$

where $\mu = mM/M + m$, and m and M are respectively the masses of the electron and proton. Inserting the appropriate values of m, M, e, h and c gave Bohr a value of R_H in very close agreement with experiment. To quote a calculated value of R_H today would, in a sense, be cheating for the simple reason that R_H is one of the most accurately known of physical constants and therefore present-day values of the other constants of which it is composed are adjusted to agree with it, assuming the Bohr formula to be correct. The validity of (2.28) now rests upon the equation being consistent with other formulae connecting the fundamental physical constants.

The energetic interpretation of T_1 and T_∞ is that T_1 is the ground state of the hydrogen atom, the one in which it normally resides unless excited, and $T_\infty - T_1 = R_H$ is the minimum energy required to remove the electron entirely, *i.e.* to ionise the hydrogen atom. R_H is thus the *ionisation potential*, so called because the energy is normally cited in electron-volts, where 1 eV is the energy acquired by an electron in acceleration across a potential drop of 1 V. 1 eV $= 1.60203 \times 10^{-12}$ erg. The quantum of this energy has a wave-number 8067.5 cm^{-1}. Consequently, $R_H = 13.595$ eV.

The quantum mechanical treatment of the hydrogen atom also gives the characteristic functions ψ_n to which the eigenvalues E_n correspond. Wave functions which refer to the motion of an electron are called *orbitals* and the term is very often extended to mean the shape of the electron distribution given by $\psi\psi^*$. We shall not describe these functions in detail[1,19,20] but note some features of them which are relevant to later discussion.

Any eigenfunction is identified by three integers n, l and m. For a given *principal quantum number*, n, the *orbital quantum number*, l, may take the values 0, 1, 2, . . ., $n - 1$ (n values). For each value of l, the *magnetic quantum number* may take the values $l, l-1, l-2, \ldots, 1, 0, -1, \ldots, -l$ ($2l + 1$ values). Thus the first three values of n permit the following sets of values (n, l, m):

$n = 1$: (1, 0, 0)
$n = 2$: (2, 0, 0); (2, 1, 1) (2,1, 0) (2, 1, −1)
$n = 3$: (3, 0, 0); (3, 1, 1) (3, 1, 0) (3, 1, −1); (3, 2, 2) (3, 2, 1) (3, 2, 0) (3, 2, −1) (3, 2, −2).

Each different set of numbers identifies a different eigenfunction or orbital and there are clearly n^2 sets for each value of n. Since the energy, for the hydrogen atom, depends only on n there are n^2 functions having the same energy, E_n. We should observe here that this degeneracy (n^2-fold) of the E_n level is inadequate for a satisfactory account of the periodic table. We shall see on p. 112 that $2n^2$ orbitals are required for each value of n and that the factor 2 was happily arranged in the postulate of electron spin.

The general way in which $\psi\psi^*$ depends upon the quantum numbers is as follows. $\psi\psi^*$ determines the probability density of the electron or, more loosely, the size and shape of the electron-cloud. The overall size increases as n increases. The shape is governed by l so that when $l = 0$ the cloud is spherical; when $l = 1$, the cloud is dumbbell-shaped with maximum electron density at points equidistant along an axis through the nucleus; when $l = 2$, the cloud is the shape of a quatrefoil with maximum electron density at four points equidistant from the nucleus along two mutually perpendicular axes[1] (see also p. 268, Fig. 120). Though they are labelled s ($l = 0$), p ($l = 1$), d ($l = 2$) for a reason connected with the interpretation of alkali metal spectra (p. 88) they are not spectroscopically observable (not, at any rate, in electronic transitions).

The orientation of the directed shapes with respect to some chosen direction is governed by m. There being no alternative orientations of a sphere, $m = 0$ when $l = 0$. But there are three possible orientations of the p-shape ($m = +1$, 0 or -1) and five possible orientations of the d-shape ($m = 2$, 1, 0, -1 or -2). It should perhaps be emphasized that this distinction of shape and direction is, in a sense, an imagined one. They are possible mathematical solutions but within the limits of the theory there is no means of distinguishing between them in the unperturbed atom. By application of a magnetic field a direction may be defined (*Zeeman effect*) and the shapes may be separated on going over to the alkali metals where other electrons influence the energy of the electron under consideration. The electron distributions are even then not observed but only their different energies.

For the hydrogen-like atoms He^+, Li^{2+}, Be^{3+}, etc. comprising a central nucleus of charge Z units and one electron, the predicted energy levels become

$$T_n = -\frac{2\pi^2\mu e^4}{h^3 c} \cdot \frac{Z^2}{n^2} = -\frac{RZ^2}{n^2} \tag{2.29}$$

where R differs slightly from R_H because of the difference in the reduced

References p. 127–128

mass μ. For example, $R_{He} = 109{,}722.263$ cm^{-1}. The same simplicity as is observed in the hydrogen spectrum should consequently appear in the He-II, Be-III, and B-IV but the corresponding series will be displaced to higher frequency or shorter wavelength by virtue of the factor Z^2. Lines are indeed found in these spectra precisely at the calculated positions. The second ionization potential of helium is slightly greater than 4×13.955 eV and the third ionisation potential of lithium is 0.072 eV greater than 9×13.595 eV.

Alkali metals

In accordance with the chemical idea of one *valency* electron, the spectra of lithium, sodium and potassium are interpreted in terms of one so-called *optical* electron. The remainder of the electrons — two for lithium, ten for sodium, eighteen for potassium — are assumed to form with the nucleus a spherically symmetrical atomic core, with overall unit positive charge. It is however immediately obvious from Fig. 40 that the lithium atom spectrum

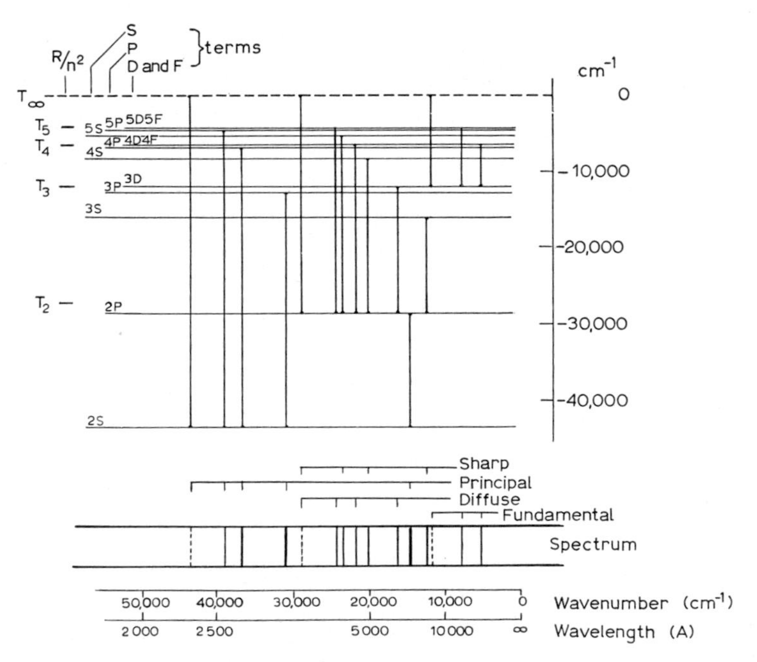

Fig. 40. Lithium atom spectrum and term scheme.

is considerably more complex than the hydrogen atom spectrum and the construction of an energy level diagram is correspondingly more difficult.

From the appearance of the spectrum it is possible to pick out series of lines; the other alkali metal spectra are similar in gross structure. In Fig. 40 the lines are labelled as belonging to:

P, the principal series; *S*, the sharp series, *D*, the diffuse series; and *F*, the Bergmann or fundamental series.

The names do not mean very much but the principal series is so called because it also appears in absorption. Its wave-number values enable a first set of levels to be drawn in: these are labelled 2*S*, 2*P*, 3*P*, 4*P* on the left of the diagram. The fact that the principal series appears in absorption identifies the 2*S* level as the ground state and the convergence of the series to its limit fixes the T_∞ level relative to 2*S*.

How are the other three series to be related to these levels? The sharp and diffuse series both converge to the same limit whose position corresponds precisely to $T_\infty - 2P$. It is reasonable therefore to draw another set of levels (3*S*, 4*S*, 5*S* . . . in the diagram) to account for the sharp series and a further set (3*D*, 4*D*, . . .) to account for the diffuse series. In lithium the fundamental series can, with reasonable accuracy, be accounted for also by transition between 3*D* and the remaining *D* levels but in the other alkali metals this combination does not reproduce the fundamental lines with sufficient accuracy. The convergence limit certainly coincides with $T_\infty - 3D$ but if the series is to have 3*D* as its common level a fourth set of terms must be added (4*F*, 5*F* . . .) such that $4F - 3D$ gives the first member of the Bergmann or fundamental series. In lithium the *D* and *F* terms coincide.

We have thus established four series of terms which are sufficient to give an account of the gross structure of the lithium atomic spectrum. The four series are given by:

principal series	$\omega = nP - 2S$ *	$n = 2, 3, 4 \dots$
sharp series	$\omega = nS - 2P$	$n = 3, 4, 5 \dots$
diffuse series	$\omega = nD - 2P$	$n = 3, 4, 5 \dots$
fundamental series	$\omega = nF - 3D$	$n = 4, 5, 6 \dots$

Each series thus has its own running term, which depends upon a quantum number *n*. The fixed terms 2*P* and 3*D* for the sharp, diffuse, and fundamental series, occur as the first running terms in the principal and diffuse series. The fixed term for the principal series, however, does not occur in any other way and we must give a reason for labelling it *S*. A similar set of four series is found for each of the alkali metals.

* It is more usual to see this written 2*S*-*nP* since energy level diagrams are often depicted with energy *increasing* downwards. It is consistent with our practice in this book of showing the lower states as having lower energy to write *n*-2*PS* in order to obtain a positive wave-number.

The sets of terms converge in a manner similar to the $T_n = R/n^2$ terms for hydrogen. The positions of T_2, T_3, T_4 and T_5 are indicated in Fig. 37. Indeed the D and F terms in lithium can be fairly accurately represented by R/n^2 (and approximately so in the other alkali spectra) and it should be noted that they do not start below $n = 3$ and $n = 4$ respectively. A good representation of the terms can in fact be given by the formulae:

$$nS = -\frac{R}{(n - \sigma_s)^2}; \quad nP = -\frac{R}{(n - \sigma_p)^2};$$
$$nD = -\frac{R}{(n - \sigma_d)^2}; \quad nF = -\frac{R}{(n - \sigma_f)^2} \tag{2.30}$$

The numbers $(n - \sigma)$ are the experimental quantities and the σ values, called the *quantum defects*, clearly depend upon the number chosen for the integer n. σ_d and σ_f are so small as to make the choice of n clear. The values of n which we assign in order to evaluate σ_s and σ_p are different from the arbitrary numbers by which experimental spectroscopists originally labelled the terms, but their theoretical justification will appear later. The lowest values of n and the corresponding σ-numbers appropriate to the four sets of terms for four alkali metals are therefore as follows:

	lowest quantum number n				quantum defect			
	S	P	D	F	σ_s	σ_p	σ_d	σ_f
Li	2	2	3	4	0.41	0.04	0	0
Na	3	3	3	4	1.37	0.88	0.01	0.001
K	4	4	3	4	2.33	1.77	0.15	0.01
Rb	5	5	4	4	3.20	2.72	1.23	0.01

It will be seen that positive values of σ imply that all the terms numbered n lie below the corresponding hydrogen term R/n^2. The fixed term for the principal series can now be identified as an S term since its value is found to conform to the formula for S terms.

The close similarity of the formulae (2.30) to the simple RZ^2/n^2 terms for hydrogen-like atoms led to the suggestion that the levels are essentially one-electron hydrogen-like levels in which the degeneracy is split by the perturbing effect of the atomic core. An alternative representation of the levels could be given by a formula

$$n(S, P, D, F) = -\frac{RZ^2_{\text{eff}}}{n^2} \tag{2.31}$$

The effective charge Z_{eff} then has the following values for the lowest values of n:

	Z_{eff}				Z
	S	P	D	F	
Li	1.26	1.02	1.00	1.00	3
Na	1.84	1.41	1.01	1.00	11
K	2.40	1.79	1.05	1.01	19
Rb	2.78	2.19	1.44	1.01	37

Now it can be seen that Z_{eff} does not even approximate to the full nuclear charge Z, but that this charge is screened by the inner electrons of the atomic cores. When the screening is complete, $Z_{eff} = 1$, since the core contains $Z - 1$ electrons, and then each nth term $= R/n^2$ and $\sigma = 0$.

According to the quantum mechanical description the shape of the distribution of one electron moving in the central coulomb field of a unit positive charge is governed by the quantum number l. The higher the value of l the greater the probability of finding the electron some distance away from the centre and the closer the atomic core approximates to the point charge, $Z_{eff} = 1$. As l decreases the electron spends more time near the nucleus and the screening by the inner electrons is no longer complete. An increase in Z_{eff}, and a value of $\sigma > 0$, reflects this state of affairs. Thus we associate the S states with $l = 0$, P states with $l = 1$, D states with $l = 2$ and F states with $l = 3$. The fact that D and F states do not appear for n less than 3 and 4 respectively is in agreement with the theoretical restriction on l that it cannot exceed $n - 1$.

The existence in the alkali metal spectra of only four series illustrates a selection rule which can also be derived theoretically. Inspection of the four combination equations for the four series shows that S terms only combine with P; P with S and D; D with P and F. Thus it is observed that transitions occur only between states in which l differs by one and not between states of the same l nor between states in which l differs by more than one. The selection rule is that $\Delta l = \pm 1$.

Closer inspection of the lithium atom spectrum under high resolution, or of the sodium spectrum under medium resolution, reveals that the lines we have been discussing are in fact pairs of lines. The familiar sodium yellow line is a doublet, wave-lengths 5889.95 and 5895.92 Å: the wave-number separation is 17.2 cm^{-1}. Double lines indicate double levels and analysis of all the alkali metal spectra shows that the P, D and F terms are all double (though D and F are only noticeably so in rubidium and caesium) but not the S terms. Thus the three pairs of lines corresponding to the first three transitions in the sharp series, *viz.* $5S - 4P$, $6S - 4P$, $7S - 4P$, are, for potassium, each separated by 94 cm^{-1}. This would be the case if the S levels were single and the common $4P$ level double. In the principal series of potassium, however, where the common term is the single $4S$, the separation

References p. 127–128

of the doublets, and therefore also of the corresponding P levels, vary as follows: $4P - 4S$, 94 cm^{-1}; $5P - 4S$, 19 cm^{-1}; $6P - 4S$, 8 cm^{-1}. Fig. 41 shows this part of the energy level diagram for potassium.

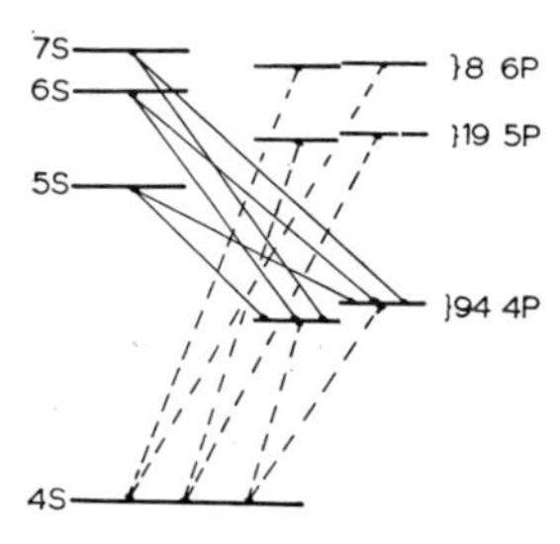

Fig. 41. Part of the potassium energy level diagram showing the doublet P levels and corresponding doublet transitions. The full lines correspond to transitions originating in the same doublet level (equal separation): the dashed lines involve different doublet levels (different separations).

Angular Momentum

It is convenient now to extend the physical interpretation of the quantum number l and other matters arising before discussing further the spectra which initially raised the question.

Space quantization

Reverting to the quantum mechanics of the hydrogen atom we find that l determines not only the shape of the electron distribution but also the value of the angular momentum associated with that distribution. It was perhaps easier to visualize the angular momentum of an electron moving in definite Bohr orbits than that of an electron cloud in which circular motion, in the classical sense, is not envisaged. Nevertheless, in quantum mechanics angular momentum is calculable and, like energy, has definite values associated with particular states. For the hydrogen atom the angular momentum has the values

$$M = \sqrt{l(l+1)}\,\hbar, \text{ where } l = 0, 1, 2, \ldots n-1, \text{ and } \hbar = \frac{h}{2\pi} \qquad (2.32)$$

Furthermore the component of this momentum in an arbitrarily chosen direction is also quantized, and may take the values

$$M_z = m\hbar, \text{ where } m = 0, \pm 1 \ldots \pm l, \text{ and } \hbar = \frac{h}{2\pi} \qquad (2.33)$$

The significance of the quantization of the component is that not only is the value of the total angular momentum restricted but also its direction. In Fig. 42 the angular momentum vector of magnitude $\sqrt{2}\,\hbar$ (for $l = 1$) is shown as making three possible angles (45°, 90° and 135°) with a chosen axis so that the component along that axis is $+\hbar$, 0 or $-\hbar$.

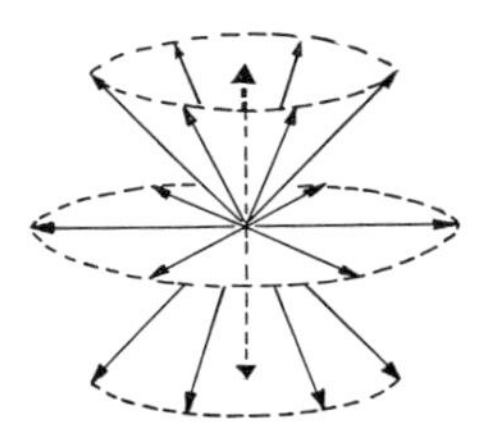

Fig. 42. Space quantization of angular momentum, when $l = 1$. l is restricted to lying on one of the cones or the disc.

It is worth noting from the uncertainty principle that a definite value of angular momentum in a particular direction requires complete uncertainty about the corresponding coordinate, which is in this case the angle of rotation about the axis. That is, the assertion that the total angular momentum vector lies at a definite angle to the axis requires that all positions attained by rotation about the axis are equally probable, as implied in Fig. 42. If all components of momentum are to be simultaneously specified, all corresponding angles of rotation about three mutually perpendicular axes are completely uncertain. The implied spherical symmetry corresponds to zero angular momentum, $l = 0$, and a spherical electron distribution.

Direct evidence for this interpretation of the quantum numbers l and m comes from the Zeeman effect and the Stern-Gerlach experiment, but these phenomena are not completely accounted for by l and m. There is a further contribution to the overall observable angular momentum which must first be described before we discuss the experiments.

Relativistic treatment

We have accepted without detailed derivation that the quantum mechanical postulates lead to a dynamical law, which when applied to the hydrogen atom, gives the results stated. We may carry this a stage further by stating the result which is obtained[2] (see also Ref. 14, Chap. XI) when the transformation is based on relativity mechanics rather than Newtonian.

In the first place it is found that the total angular momentum M is given by

$$M^2 = (k^2 - \tfrac{1}{4})\hbar^2 \text{ where } k = -l \text{ or } l+1 \neq 0 \qquad (2.34)$$

The quantum number k (not the same as the quantum number which governed the ellipticity of Bohr-Sommerfeld orbits in the old quantum theory) can evidently take the values $-1, -2, -3 \ldots -(n-1)$ and $1, 2, 3 \ldots n$. Thus, when:

$n = 1,\ k = 1$	corresponding to $l = 0$	and giving $\lvert k\rvert = 1$
$n = 2,\ k = -1, 1, 2$	corresponding to $l = 1, 0, 1$	giving $\lvert k\rvert = 1, 2$
$n = 3,\ k = -2, -1, 1, 2, 3$	corresponding to $l = 2, 1, 0, 1, 2$	giving $\lvert k\rvert = 1, 2, 3$

and so on.

The component of angular momentum in a specified direction z is M_z given by:

$$M_z = (m + \tfrac{1}{2})\hbar \qquad \begin{aligned} m &= 0, \pm 1, \pm 2 \ldots \pm l \\ &= 0, \pm 1, \pm 2 \ldots \pm (|k| - 1), -|k| \end{aligned} \tag{2.35}$$

In this term we see immediately the additional contribution to the angular momentum: in fact there is an extra $\frac{1}{2}\hbar$ to be added to the component of momentum given by (2.33).

The energy levels for the hydrogen-like atom turn out to depend principally upon n (as before) but also upon k, according to the equation:

$$T_{n,k} = -\frac{RZ^2}{n^2} - \frac{\alpha^2 RZ^4}{n^3}\left\{\frac{1}{|k|} - \frac{3}{4n}\right\} \tag{2.36}$$

α is the *fine structure constant* given by

$$\alpha = \frac{e^2}{\hbar c} = 7.2977 \times 10^{-3} \approx \frac{1}{137} \tag{2.37}$$

The energy levels according to (2.36) are lower than the non-relativistic levels (given by the first term) by an amount

$$\Delta T = \frac{\alpha^2 RZ^4}{n^3}\left(\frac{1}{|k|} - \frac{3}{4n}\right) \tag{2.38}$$

For the hydrogen atom $\alpha^2 RZ^4 = 5.84$ cm^{-1}, when $n = 1$, $|k| = 1$ (corresponding to $l = 0$ or an S state) and $\Delta T = 1.46$ cm^{-1}. There is still only one level for $n = 1$. However, when $n = 2$, $|k| = 1$ (corresponding to $l = 0$ and 1) or 2 (corresponding $l = 1$): $\Delta T = 0.456$ cm^{-1} in the first case and 0.091 cm^{-1} in the second. There are thus two levels for $n = 2$, the lower one being degenerate (S and P coincide), the upper one being a P level, and the difference between them is 0.365 cm^{-1}. This is shown in Fig. 43. The levels for $n = 3$ are three in number: $|k| = 1$ ($l = 0, 1$: S and P states) gives $\Delta T = 0.1622$ cm^{-1}; $|k| = 2$ ($l = 1, 2$: P and D states) gives $\Delta T = 0.0541$ cm^{-1}; $|k| = 3$ ($l = 2$: D state) gives 0.0180 cm^{-1}.

The relativistic treatment is amply justified spectroscopically apart from its prediction of additional angular momentum. As a consequence of the splitting of the hydrogen atom levels T_2 and T_3 we should expect a splitting

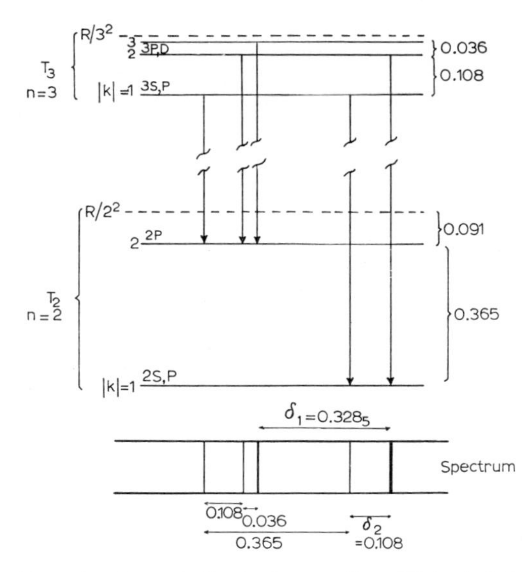

Fig. 43. Fine structure of the hydrogen line (Hα) at 6562.79 A.

of the transition $T_3 - T_2 = \omega_{3,2}$, the first member of the Balmer series. Indeed the H_α line was known to have fine structure before any explanation was forthcoming. The selection rule $\Delta k = 0, \pm 1$ should limit the transition $n = 3 \rightarrow 2$ to the five shown in Fig. 43.

Prior to 1937 a large number of determinations had been made of the separation between two lines which were not further resolved, one comprising the three lines at lower frequency and the other comprising two lines at high frequency. Calculation of the relative intensities of the components shows that the separation of the two composite lines should be about 0.317 cm^{-1}. The experimental results lie between about 0.30 and 0.34 cm^{-1}. Higher resolution enabled the lines to be observed whose predicted separations are marked in Fig. 40 as $\delta_1 = 0.3285$ cm^{-1} and $\delta_2 = 0.108$ cm^{-1}. For δ_1 the observed values [3,4] lie in the range 0.317 – 0.321 cm^{-1} and for δ_2 in the range 0.119—0.132 cm^{-1}.

Transitions between the two terms comprising T_2 have also been observed. The difference of 0.365 cm^{-1} corresponds to a frequency of 10,950 Mc/s and this transition has been detected in the microwave region. Measurements are complicated by a small shift in frequency (the *Lamb shift*) which arises from interaction between the atomic electron and its radiation field and is responsible for the discrepancies in δ_1 and δ_2. Nevertheless the basic explanation of the hydrogen fine structure remains.

References p. 127–128

Multiplet spectra

If the hydrogen spectrum is to be accounted for in this manner, it is reasonable to suppose that other spectra will be similarly influenced. In particular, the one-electron alkali metals should exhibit the doublet separation of P, D and F levels since for every l except $l = 0$ there are two values of k, $-l$ and $l + 1$. We have seen (p. 91) that this is the case. It is also found that the doublet separation is less in D and F than in P levels, and decreases as n increases. The doublet separations of the lowest P levels increase rapidly with Z, although not as rapidly as the term Z^4 might suggest on account of the screening effect of inner electrons. The separations, and the corresponding values of an effective charge

$$Z_{\text{eff}} = \left(\frac{2n^3 \Delta\omega}{\alpha^2 R}\right)^{\frac{1}{4}} \tag{2.39}$$

are as follows:

	Li	Na	K	Rb	Cs	
$\Delta\omega$	0.34	17.2	94.0	220	5540	cm^{-1}
Z_{eff}	0.98	3.55	6.8	9.9	25.4	
Z	3	11	19	37	55	

Although the relativistic treatment thus gives a qualitative explanation of the doubling of levels in the alkali spectra and was shown by Dirac in 1928 to provide the electron with an inherent angular momentum additional to that previously shown to be described by quantum number l, the first explanation of the doublet structure took a somewhat different course. Uhlenbeck and Goudsmit had already in 1925 introduced the idea of the spinning electron which proved so fruitful in explaining both the multiplet structure and the magnetic behaviour of the atomic spectra and in providing a basis for an account of valency and the periodic table.

Electron spin and the Vector model

The unit of orbital angular momentum, governed by l, is $\hbar = h/2\pi$. We now *assume* that the electron possesses an inherent angular momentum which is *one half* this unit in magnitude. It is further assumed that the result of vectorial addition of the orbital and spin components is restricted to those values of its quantum number which form a set of integers or of half-integers, differing by an integer. That is, if $\boldsymbol{l}$ is the vector of magnitude $\sqrt{l(l+1)}\hbar$, $l = 0, 1, 2, 3 \ldots$ and $\boldsymbol{s}$ is the vector of magnitude $\sqrt{s(s+1)}\hbar$, $s = \frac{1}{2}$, then

$$\boldsymbol{j} = \boldsymbol{l} + \boldsymbol{s} \tag{2.40}$$

is the resultant of magnitude $\sqrt{j(j+1)}\hbar$ where j is either $0, 1, 2, 3 \ldots$ or $1/2, 3/2, 5/2 \ldots$.

The result of some vectorial additions is shown in Fig. 44, and it can be seen that when $l = 0$, $j = \frac{1}{2}$ but when $l \neq 0$, $j = l \pm \frac{1}{2}$. The reason for the assumptions now becomes clear. They give a model of the observed situation in which the states for $l = 1, 2, 3 \ldots$ (the one-electron P, D and F states) are double if it is assumed that the energy depends upon the total angular momentum j as well as upon the orbital momentum, l. The S term, $l = 0$, remains single since the addition of $\boldsymbol{s}$ to $\boldsymbol{l}$ can then only yield one $\boldsymbol{j}$, quantum number $j = \frac{1}{2}$.

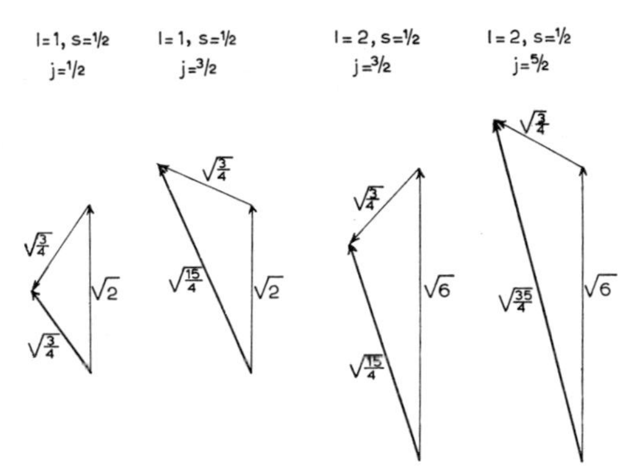

Fig. 44. Vectorial addition of l and s.

The spectral terms for the alkali metals, and indeed for hydrogen, now require extra distinguishing marks. The nP terms become two, $n^2P_{\frac{1}{2}}$ and $n^2P_{\frac{3}{2}}$; the nD terms become n^2D and $n^2D_{\frac{5}{2}}$; the nF terms become $n^2F_{\frac{5}{2}}$ and $n^2F_{\frac{7}{2}}$. The subscript gives the value of j and the superscript indicates the *multiplicity*, *i.e.* that in this case the levels go in pairs. $n^2D_{\frac{5}{2}}$, for example is pronounced "n-doublet-D-five-halves". For no very good reason it is customary to denote the S level as a doublet, thus $^2S_{\frac{1}{2}}$, although it has no partner. However, the point to be noted from all this is that the value $s = \frac{1}{2}$ is forced upon us by the doublet splitting of the one-electron levels: $s = 1$ would yield triplet splitting, as will appear in what follows.

In systems of more than one electron it is assumed, in order to obtain the number and type of terms, that each individual electron makes its own contributions $\boldsymbol{l}_i$ and $\boldsymbol{s}_i$ to the total angular momentum. The individual $\boldsymbol{l}_i$ and $\boldsymbol{s}_i$ no longer have precise physical meaning as angular momenta unless their interactions are small by virtue of, say, a large difference in n. The resultant vectors:

$$\boldsymbol{L} = \boldsymbol{l}_1 + \boldsymbol{l}_2 + \boldsymbol{l}_3 + \ldots \tag{2.41}$$

and

$$\boldsymbol{S} = \boldsymbol{s}_1 + \boldsymbol{s}_2 + \boldsymbol{s}_3 + \ldots \tag{2.42}$$

References p. 127–128

do, however, retain approximate significance as angular momenta. The vectorial addition is restricted to those vectors $\boldsymbol{L}$ which have magnitude $\sqrt{L(L+1)}\hbar$ where $L = 0, 1, 2, 3 \ldots$ and to those $\boldsymbol{S}$ which have magnitude $\sqrt{S(S+1)}\hbar$ where $S = 0, 1, 2, 3$ or $\frac{1}{2}, \frac{3}{2}, \frac{5}{2}, \ldots$ Then the total angular momentum $\boldsymbol{J}$ is determined from $\boldsymbol{L}$ and $\boldsymbol{S}$, in precisely the same manner as $\boldsymbol{j}$ from $\boldsymbol{l}$ and $\boldsymbol{s}$ in the case of one electron. That is:

$$\boldsymbol{J} = \boldsymbol{L} + \boldsymbol{S} \tag{2.43}$$

$\boldsymbol{J}$ having magnitude $\sqrt{J(J+1)}\hbar$ where J is either $0, 1, 2, 3 \ldots$ or $\frac{1}{2}, \frac{3}{2}, \frac{5}{2} \ldots$ The term designations are retained so that S, P, D and F states correspond to $L = 0, 1, 2, 3$. (Note the distinction between the term designation S for $L = 0$ and the quantum number S for total spin.)

We saw that for one electron $s = \frac{1}{2}$: j could then take two values except when $l = 0$. For two electrons $S = 0$ or 1. L may take any of the integral values between $l_1 + l_2$ and $|l_1 - l_2|$ as shown in Fig. 45 for $l_1 = 2$ and $l_2 = 1$.

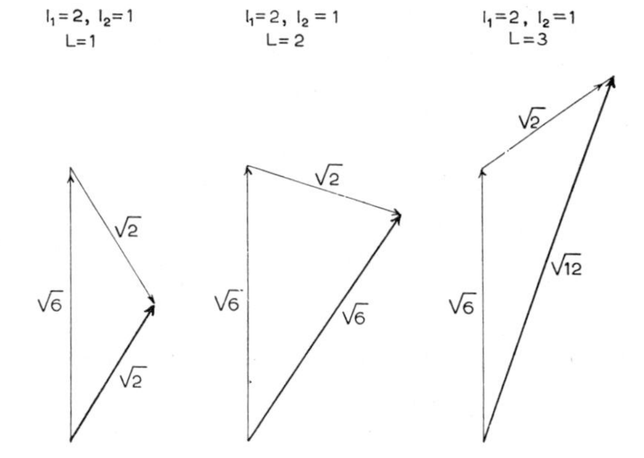

Fig. 45. Vector addition of $L = l_1 + l_2$ when $l_1 = 2$ and $l_2 = 1$, or of $J = L + S$ when $L = 2$ and $S = 1$.

Now, provided $L \neq 0$, vectorial addition of $\boldsymbol{L}$ and $\boldsymbol{S}$ gives $J = L$ for $S = 0$ and $J = L - 1$, L or $L + 1$ for $S = 1$. Fig. 45 equally well shows the coupling of $L = 2$ and $S = 1$. We thus arrive at the conclusion that for $S = 0$ there is a set of singlet states $^1P_1$, $^1D_2$, $^1F_3 \ldots$, and for $S = 1$ a set of triplet states $^3P_0$, $^3P_1$, $^3P_2$; $^3D_1$, $^3D_2$, $^3D_3$; $^3F_2$, $^3F_3$, $^3F_4 \ldots$; remembering that the subscript denotes J values.

The superscript denotes multiplicity and it can be seen that this is equal to $2S + 1$. In the one-electron case $S = s = \frac{1}{2}$ and the multiplicity is 2: the S state is labelled ^2S although the duplicity is not realised in that state since j cannot be $-\frac{1}{2}$. Similarly in the two-electron case the S state when $S = 1$ is labelled $^3S_1$, despite the fact that $J = 1$ is the only possible value for the total angular momentum quantum number. For three electrons, $S = \frac{1}{2}$ or $\frac{3}{2}$ and multiplicities 2 and 4 appear. For four electrons, $S = 0$, 1 or 2 and the possible multiplicities are 1, 3 and 5. It will be seen, however, that the higher multiplicities, arising from large S, can only be fully realised in the states for which $L > S$. The alternation of multiplicities across the periodic table is observed quite generally: atoms or ions with odd numbers of electrons have even multiplicities and atoms or ions with even numbers of electrons have odd multiplicities. Helium (p. 107) shows the expected singlet and triplet terms: so also do the alkaline earths in their arc spectra (Ca-I, Sr-I, . . .). On the other hand the first spark spectra of these elements (Ca-II, Sr-II, . . .) have the doublet structure appropriate to the arc spectra of the alkali metals since they arise from Ca^+, Sr^+, etc. which are ions with an odd number of electrons.

This treatment of electron spin in terms of the vector model amounts to an assumption that the wave function describing the motion of the electron can be written as the product of independent functions $\psi_{n,l}$, describing the orbital motion, and ψ_{spin}, called the *spin eigenfunction*. To the extent that the two are independent, the Schrödinger equation resolves into two equations, one in $\psi_{n,l}$ and one in ψ_{spin}. There thus arise a set of $\psi_{n,l}$ for each ψ_{spin} and hence the multiplicity $2S + 1$ of the $\psi_{n,l}$ functions. In reality interaction occurs and the spin must then be regarded as a perturbation of the equation with a consequent splitting of the $E_{n,l}$ levels. The greater the interaction the less do $\boldsymbol{L}$ and $\boldsymbol{S}$ have the significance of exact angular momenta but $\boldsymbol{J}$ remains as the angular momentum of magnitude $\sqrt{J(J+1)}\hbar$.

The fact that a large number of spectra can be accounted for in terms of the method of vector addition which we have described (the so-called *Russell-Saunders coupling*) means that normally the interaction between the various $\boldsymbol{l}_i$ is strong, leading to $\boldsymbol{L}$, and between the various $\boldsymbol{s}_i$ is also strong, leading to $\boldsymbol{S}$, but that the spin perturbation of $\psi_{n,l}$ is relatively weak, and interaction between the individual $\boldsymbol{l}_i$ and $\boldsymbol{s}_i$ is small.

Although Dirac's quantum mechanics makes no explicit reference to spin, it does show that the additional angular momentum which is referred to as electron spin is a phenomenon required by the theory of relativity. Where the equations in three dimensions require for each electron three quantum numbers, n, l and m, in four dimensions four quantum numbers are needed, n, l, m and s. The doublet splitting of one-electron P, D and F states is given by the relativistic theory (p. 93) in terms of the auxiliary quantum number, k, which appears to replace l without explicit mention of s. However, except when $l = 0$, there are two values of k for every l, which corresponds to two values of $j = l + s$, when $s = \pm\frac{1}{2}$. The level for $|k| = l$ lies below its double for $|k| = l + 1$, as the level for $j = l - \frac{1}{2}$ lies below its double for $j = l + \frac{1}{2}$.

Space quantization of $\boldsymbol{J}$

The solutions of the hydrogen-atom problem which do not take spin into account predict space quantization of the vector $\boldsymbol{l}$ (p. 84). Now the inclusion of spin requires that this space quantization be applied to $\boldsymbol{J}$ which is the proper angular momentum rather than to $\boldsymbol{l}$ or $\boldsymbol{L}$. Thus the total angular momentum vector $\boldsymbol{J}$

$$\text{magnitude of } \boldsymbol{J} = \sqrt{J(J+1)}\,\hbar \tag{2.44}$$

can take any such directions that the magnitude of its component $\boldsymbol{M}$ in one direction has certain values.

$$\text{magnitude of } \boldsymbol{M} = M\hbar \text{ where } M = -J, -J+1 \ldots J-1, J \tag{2.45}$$

The number of values which $\boldsymbol{M}$ can take is $2J + 1$, but it should be noted that, since J may have half-integral values by virtue of the inclusion of spin, $2J + 1$ may be even.

This conclusion is strikingly confirmed by the experiment by Stern and Gerlach in which a beam of one-electron atoms is passed through an inhomogeneous magnetic field acting perpendicular to the beam. The effect of the field is to split the beam into as many components as there are possible values of M, *i.e.* $2J + 1$. For $M = 0$ there should be an unperturbed beam and deflections become greater the greater M. The ground state for silver atoms, $^2S_{\frac{1}{2}}$, has $L = 0$ and $J = \frac{1}{2}$. In the Stern-Gerlach experiment a beam of silver atoms splits into two, corresponding to $M = \pm \frac{1}{2}$, and there is no unperturbed beam. Had L determined the total angular momentum there would have been only the unperturbed beam for $L = 0$. For fluorine ($^2P_{\frac{3}{2}}$) $L = 1$, and hence space quantization of L would predict an undeflected beam for $M = 0$ and two others symmetrically placed for $M = -1$ and $M = +1$. In fact the beam splits into four with no undeflected beam,

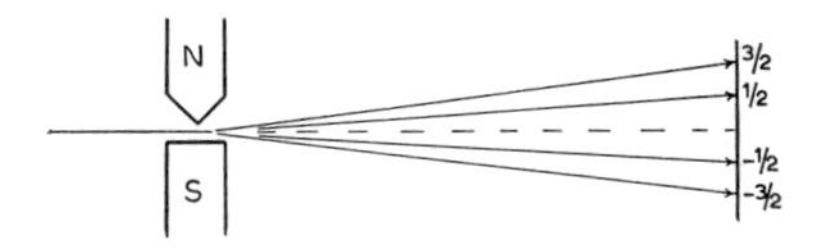

Fig. 46. Stern-Gerlach experiment illustrated for atoms with $j = 3/2$.

corresponding to the four values of $M = \frac{3}{2}, \frac{1}{2}, -\frac{1}{2}$ and $-\frac{3}{2}$ arising from space quantization of $J = \frac{3}{2}$. This fact has been used as a test for fluorine atoms in an experiment[5] on the thermal dissociation of fluorine molecules.

Zeeman effect

When a source of atomic spectral lines is subjected to a magnetic field it is found that the lines split into a number of components at slightly different frequencies. This implies that the energy levels are themselves split by the magnetic field. If this effect, known as the *Zeeman effect,* were to be accounted for in terms of classical theory concerning a revolving negative charge with angular momentum $\boldsymbol{J}$, the magnetic moment $\boldsymbol{\mu}$ would be given by:

$$\boldsymbol{\mu} = -\frac{e}{2mc}\boldsymbol{J} \tag{2.46}$$

where e and m are the charge and mass of the electron and the negative sign indicates that the direction of magnetic moment is opposite to that of angular momentum. Accordingly we might expect that, in magnitude:

$$\mu = \frac{e\hbar}{2mc}\sqrt{J(J+1)} = \beta\sqrt{J(J+1)} \tag{2.47}$$

The unit

$$\beta = \frac{e\hbar}{2mc} = 9.2732 \times 10^{-21} \text{ erg gauss}^{-1} \tag{2.48}$$

is called the *Bohr magneton.*

A magnetic field H defines the direction in which the component of angular momentum is quantized and in view of (2.47) it might be expected that the component $\boldsymbol{\mu}_z$ of magnetic moment would be similarly quantized so that, in magnitude:

$$\mu_z = M\beta \text{ where } M = -J, -J+1 \ldots J-1, J \tag{2.49}$$

The energy perturbation arising from interaction of magnetic moment and field is the product of μ_z and the field strength. The expected perturbation of the energy levels is therefore:

$$U_M = \mu_z H = M\beta H \tag{2.50}$$

To take into account deviations from (2.50) it is customary to write

$$U_M = gM\beta H \tag{2.51}$$

where g is a factor introduced by Landé. We must now consider two examples of the Zeeman effect to show how the deviations from (2.50) contributed to the theory of electron spin.

Where the transition under observation is between singlet states ($S = 0$, $J = L$) it is found that the splitting of both levels is the same in magnitude but that the number of components depends upon L or J. A $^1D_2$ level splits into five components in agreement with the five values of M (-2, -1, 0, $+1$, $+2$) and a $^1P_1$ level splits into three, (Fig. 47). Transitions being per-

References p. 127–128

mitted only where $\Delta M = 0, \pm 1$, there are three sets of equal transitions. Thus the single transition $^1P_1$—$^1D_2$ splits into three components under the influence of the magnetic field, the central line remaining at the same frequency as in the absence of the field. This behaviour is observed, for example,

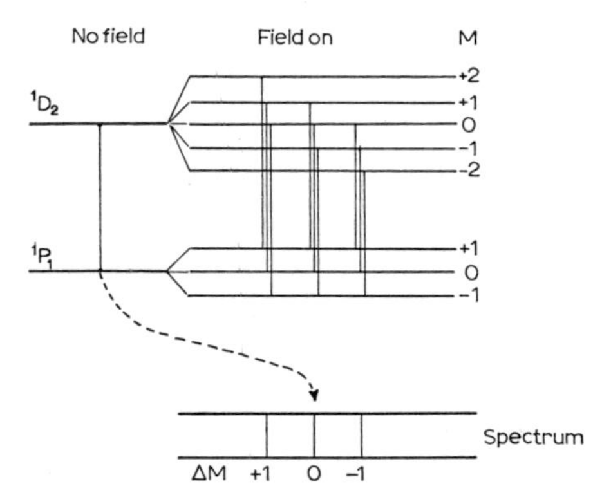

Fig. 47. Normal Zeeman effect in $^1D_2$—$^1P_1$ transition.

in the cadmium red line at 6438.47 Å and being in accordance with equation (2.50) is called the *normal Zeeman effect.*

The sodium yellow doublet arises from transitions $^2P_{\frac{1}{2}}$—$^2S_{\frac{1}{2}}$ and $^2P_{\frac{3}{2}}$—$^2S_{\frac{1}{2}}$. The Zeeman effect on the first of them is shown in Fig. 48. The fact that the

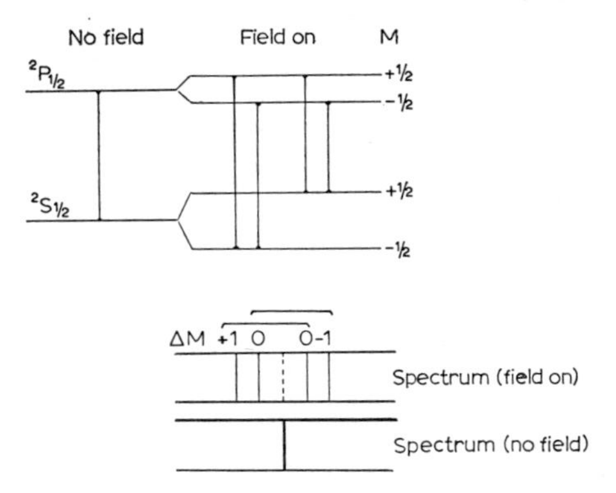

Fig. 48. Anomalous Zeeman effect in $^2P_{\frac{1}{2}}$—$^2S_{\frac{1}{2}}$ transition.

line splits into four components shows that the energy perturbations of the upper and lower levels in the same field are different, although $J = \frac{1}{2}$ in both cases. This is not in accordance with (2.50) and is therefore an example of an *anomalous Zeeman effect.* It is not immediately clear that it is the S level that has the larger magnetic perturbation but comparison with the Zeeman effect on the other line of the doublet shows that Fig. 48 is correct. Hence the equal separation of the two bracketted pairs of components in the figure gives the splitting of the lower $^2S_{\frac{1}{2}}$ level. In this case $L = 0$ and $J = \frac{1}{2}$ arises solely from the spin of one electron, $S = s = \frac{1}{2}$. It is found that the splitting in this case is exactly the same for the same field strength as for a $^1P_1$ level in which $L = 1$ and $S = 0$. Hence, the magnetic moment of an electron due to its spin is β although the angular momentum is only $\frac{1}{2}\hbar$. Or, the magnetic effect of the spin, $\boldsymbol{S}$, is twice as great as that of the orbital angular momentum, $\boldsymbol{L}$.

It was a matter of considerable importance when it was recognised that this property of electrons is implied by the Zeeman effect. The interpretation is consistent with the relativistic theory. The same conclusion is also to be drawn from the magnitude of the deflection in the Stern-Gerlach experiment with $^2S_{\frac{1}{2}}$ silver atoms where again $L = 0$ and $S = \frac{1}{2}$.

The interpretation of Landé's g-factor now becomes apparent. In the normal Zeeman effect in singlet-singlet transitions $S = 0$, $J = 1$ and $g = 1$. For S states, $L = 0$, $J = S$ and $g = 2$. When $L \neq 0$ and $S \neq 0$ the magnitude of the effect depends upon how J is made up. Theory gives:

$$g = 1 + \frac{J(J+1) + S(S+1) - L(L+1)}{2J(J+1)} \tag{2.52}$$

For $^2P_{\frac{1}{2}}$, for example, $g = 2/3$ as may be deduced from the fact that separation of the two closest pairs of lines in Fig. 48 is one third the separation of the bracketted pairs.

Spin Resonance

Two important spectroscopic methods have been developed in the last fifteen years which depend upon the interaction of spin and magnetic field. We introduce the theory here and discuss applications in Chapter IV.

Paramagnetic resonance

From equation (2.51) we have different energy levels in a magnetic field depending upon the orientation of the electronic magnets with respect to the field. If the field is a weak one and insufficient to break down the coupling between L and S a state with a given value of J splits into $2J + 1$ sublevels: if the field is strong enough to destroy the coupling, the number of levels, $2S + 1$, is determined by S. Whatever the case the spacing between the sublevels (*i.e.* the Zeeman splitting of the atomic lines) with the magnetic fields usually available is such that transition between the sublevels themselves ($\Delta M = \pm 1$) falls in the region of microwaves. The magnitude of the

Bohr magneton is such that a field of 10,000 gauss makes $\beta H = 9.27 \times 10^{-17}$ erg. If $g = 2$, then the quantum of magnitude $g\beta H$ has frequency 28,000 Mc/s; and microwave power is absorbed at that frequency. This is the phenomenon of *paramagnetic resonance* or *electron spin resonance* (ESR).

Whereas the Zeeman effect on atomic spectral lines refers to free atoms paramagnetic resonance is normally observed in solids at low temperatures. The expression (2.52) for g does not then hold. Rather do the values of the sublevels and the number of such levels which are produced by an external magnetic field, lead to information about the electric field within the crystal.

Nuclear spin resonance

It is appropriate to note at this point some of the parallel phenomena associated with *nuclear spin*. Under very high resolution atomic spectral lines are found to have a *hyperfine structure*, part of which can be accounted for by the presence of isotopes, *i.e.* different nuclear masses. A complete explanation however, requires the concept of nuclear spin, just as electron spin was required to account for the coarse multiplet structure. The implication of the smaller spacing caused by coupling with nuclear spin is that the nuclear magnetic moment is smaller than the electron moment (presumably because of the larger mass).

Nuclear angular momentum may be defined by vector $\boldsymbol{I}$ such that:

$$\text{magnitude of } \boldsymbol{I} = \hbar\sqrt{I(I+1)} \tag{2.53}$$

where I is the spin quantum number (often referred to as the nuclear spin) for the nucleus in question. It has one value (as does the spin quantum number for one electron) and values for some nuclei are given in Table 8. By analogy with electronic angular momentum we may note that the component $\boldsymbol{I}_z$ of $\boldsymbol{I}$ is quantized so that

$$\text{magnitude of } \boldsymbol{I}_z = M\hbar \tag{2.54}$$

where $M = -I, -I+1 \ldots 0 \ldots I-1, I$ or $-I, -I+1, \ldots -\frac{1}{2}, +\frac{1}{2} \ldots I-1, I$. There are thus $2I+1$ possible orientations of the axis of nuclear spin and the maximum observable value of the components is $I\hbar$.

The magnetic moment μ of a nucleus is not, in general, a predictable quantity: nor is its spin quantum number I. Both are experimental quantities. By analogy with the electron (equation 2.48) we may define a unit of nuclear magnetic moment, the *nuclear magneton*, in terms of the mass, m_p of the proton, thus:

$$\mu_n = \frac{e\hbar}{2m_p c} = 5.0493 \times 10^{-24} \text{ erg gauss}^{-1}$$

Now for any nucleus we write

$$\mu = g\mu_n\sqrt{I(I+1)} \tag{2.55}$$

and the set of $(2I + 1)$ permitted components:

$$\mu_z = Mg\mu_n \text{ where } M = -I, -I+1 \ldots I-1, I. \tag{2.56}$$

The maximum observable component, μ_0 is clearly

$$\mu_0 = gI\mu_n \tag{2.57}$$

and this is generally referred to as the *magnetic moment* of the nucleus (and often given the unqualified symbol μ). If measured in terms of nuclear magnetons the magnetic moment is just gI.

The nuclear g-factor used here is a dimensionless number, analogous to the spectroscopic g-factor for electrons, but it takes care of a somewhat different imperfection in theory. It does not so much represent the unpredictable part of magnetic coupling with the environment, but the lack of theory for magnetic moment of a free nucleus. It also includes the fact that we have used for all nuclei a unit, the nuclear magneton, defined in terms of the proton. It is frequent practice to absorb g and μ_n into a quantity called the *gyromagnetic ratio* defined by $\mu_0 = \gamma I\hbar$, so that

$$\gamma = \frac{g\mu_n}{\hbar} = \frac{ge}{2m_p c} \tag{2.58}$$

This factor, γ, has the same status as g in regard to the present inability of nuclear theory to account for the magnitude of the magnetic moment.

According to this discussion a nucleus is sufficiently characterized if the spin quantum number I and one of the quantities μ_0, g or γ is given. In Table 8 there is a selection of common particles and their spin quantum numbers. Magnetic moments are quoted in nuclear magnetons and in erg gauss^{-1}. These values have been obtained initially from hyperfine structure of atomic spectral lines. Nuclear spins can be obtained from the hyperfine structure of electron paramagnetic resonance when a paramagnetic ion is observed in sufficient magnetic dilution in an isomorphous crystal lattice of a diamagnetic salt. They are also obtained from the interaction with other nuclear motions to which we refer in the next chapter (p. 211). The nuclear moments are obtained from the magnitudes, rather than the numbers, of these various interactions.

In a magnetic field of strength H the energy of the nuclear magnet relative to that in the absence of the field is (compare 2.51)

$$U_M = \mu_z H = gM\mu_n H = M\hbar\gamma H = \frac{M\mu_0 H}{I} \tag{2.59}$$

The effect of the magnetic field H is thus to produce $2I + 1$ equidistant levels between which transitions may occur, limited to $\Delta M = \pm 1$. The absorption of a quantum to effect this transition is *nuclear magnetic resonance* (NMR), occurring at a frequency,

$$\nu = \frac{g\mu_n H}{h} = \frac{\gamma H}{2\pi} = \frac{\mu_0 H}{Ih} \tag{2.60}$$

For a field of 10,000 gauss $\mu_n H = 5.05 \times 10^{-20}$ erg and consequently *proton* magnetic resonance might be expected to occur at 7.65 Mc/s if $I = \frac{1}{2}$ and $g = 1$. In fact proton magnetic resonance occurs at 42.6 Mc/s under a field of 10,000 gauss and hence $g = 5.56$. Consequently $\gamma = 2.68 \times 10^4$ radians gauss^{-1} sec^{-1}.

TABLE 8

SPIN QUANTUM NUMBER AND MAGNETIC MOMENT OF SOME COMMON PARTICLES

Particle	Spin I	Magnetic moment μ_0		Resonance frequency (Mc sec^{-1})
		(nm)	(10^{24} erg gauss^{-1})	for 10 k gauss field
e (electron)	$\frac{1}{2}$	—	−9270	28,003
^{1}n (neutron)	$\frac{1}{2}$	−1.9130	−9.66	29.1
^{1}H (proton)	$\frac{1}{2}$	2.79270	14.1	42.6
^{2}H (deuteron)	1	0.85738	4.33	6.5
^{7}Li	$\frac{3}{2}$	3.2560	16.4	16.5
^{9}Be	$\frac{3}{2}$	−1.1774	−5.95	6.0
^{11}B	$\frac{3}{2}$	2.6880	13.6	13.7
^{12}C	0	0	0	0
^{13}C	$\frac{1}{2}$	0.70216	3.55	10.7
^{14}N	1	0.40357	2.04	3.1
^{15}N	$\frac{1}{2}$	−0.28304	−1.43	4.3
^{16}O	0	0	0	0
^{17}O	$\frac{5}{2}$	−1.8930	−9.56	5.7
^{19}F	$\frac{1}{2}$	2.6273	12.8	40.1
^{23}Na	$\frac{3}{2}$	2.2161	11.2	11.3
^{29}Si	$\frac{1}{2}$	−0.55477	−2.80	8.5
^{31}P	$\frac{1}{2}$	1.1305	5.71	17.2
^{32}S	0	0	0	0
^{35}Cl	$\frac{3}{2}$	0.82089	4.14	4.2
^{37}Cl	$\frac{3}{2}$	0.68329	3.45	3.5

The figures in the last two columns are given merely to two or three significant figures since they are intended to indicate only relative positions.

Table 8 shows that the resonance frequencies for all nuclei fall in the region 3—43 Mc/s for a 10,000 gauss field. Nuclear magnetic resonance thus calls for short-wave radio techniques whereas electron spin resonance falls in the microwave region.

The value of nuclear magnetic resonance from the chemical point of view lies in the fact that the nuclear magnet serves as a sensitive probe for the local magnetic fields within a crystal or molecule. Electron spin resonance performs a similar function and both these effects will be illustrated further in Chapter 4. It should be noted from Table 8 that three commonly occurring nuclei, ^{12}C, ^{16}O and ^{32}S have zero spin and therefore no magnetic moment. The table also distinguishes between those particles, such as the electron, for which the magnetic moment lies in the opposite direction (negative) to the angular momentum and those in which the two moments lie in the same direction (positive). The sign of μ_0, however, has no direct effect on the resonant frequency.

Helium

The simplest system containing two electrons is the helium atom and its spectrum is already much richer than the hydrogen atom spectrum. It provides the simplest illustration of the interaction of two indistinguishable particles, a real appreciation of which is essential for an understanding of chemical behaviour in electronic terms.

Fig. 49 shows an incomplete energy level diagram and some of the transitions which occur. There is one series in the far ultra-violet (584.4 Å is the

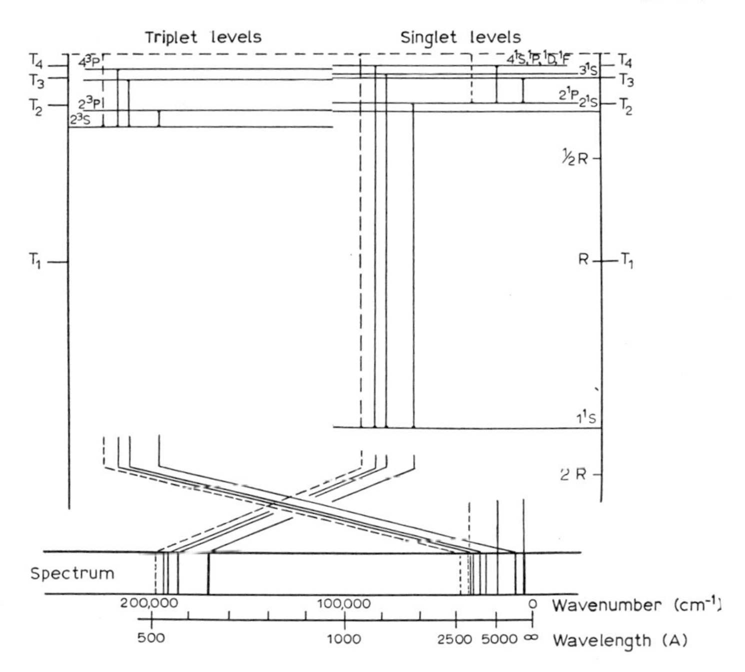

Fig. 49. Helium atom spectrum and term scheme. The diagram is simplified somewhat so as not to obscure the analysis into two non-combining systems.

References p. 127–128

lowest wavelength of the series) which is also found in absorption. It is identified as a $nP - 1S$ series and gives ground state as $1S$. In the near ultra-violet, visible and infra-red there are, as in the alkali metals, principal diffuse, sharp and fundamental series. There are, however, two of each and they have different series limits. In order to represent this complexity it is found necessary to have two sets of terms which are illustrated in the figure on either side of the centre line. Within either set there is the expected series of transitions conforming to $\Delta L = \pm 1$ but no spectral lines corresponding to transitions from one set to the other have been observed. There is, in one set, a $1S$ level and the lowest P, D and F levels are $2P$, $3D$ and $4F$. The same statement holds in the other set except that there is no $1S$ state. From the position of the hydrogen levels $T_n = -R/n^2$ shown in the figure it can be seen that the $1S$ level lies below $T_1 = -R$ but above $-4R$, which is the lowest level of the He^+ ion. The quantum defect in the $1S$ state is 0.25 and $Z_{eff} = 1.34$.

Singlet and triplet levels

From the discussion of the vector model it can be seen that the following six terms are possibilities for two electrons having values of $n = 1$ or 2:

n_1	l_1	s_1	n_2	l_2	s_2	L	S	J	state
1	0	$\frac{1}{2}$	1	0	$-\frac{1}{2}$	0	0	0	1^1S_0
					$+\frac{1}{2}$		1	1	(1^3S_1)
1	0	$\frac{1}{2}$	2	0	$-\frac{1}{2}$	0	0	0	2^1S_0
					$+\frac{1}{2}$		1	1	2^3S_1
1	0	$\frac{1}{2}$	2	1	$-\frac{1}{2}$	1	0	0	2^1P_0
					$+\frac{1}{2}$		1	1	2^3P_3

and these are indeed correct designations for the terms. The levels of the left hand set of spectral lines in Fig. 49 are all singlets and those of the right hand set are all triplets. We now wish to know why triplet and singlet levels do not combine, why they differ in energy for the same principal quantum number, and why no 1^3S_1 term is found.

The hydrogen-like atomic wave functions and eigenvalues are obtained after putting the potential energy term $V = -Ze^2/r$ into the Schrödinger equation (p. 73). For the helium atom ($Z = 2$) the potential energy takes the form:

$$V = -\frac{2e^2}{r_1} - \frac{2e^2}{r_2} + \frac{e^2}{r_{12}} \tag{2.61}$$

where the two electrons are at distances r_1 and r_2 from the nucleus and r_{12} from each other. If the mutual repulsion term e^2/r_{12} is neglected in the first instance the potential energy is

$$V = -\frac{2e^2}{r_1} - \frac{2e^2}{r_2} \tag{2.62}$$

This, because it contains no r_{12} term, enables the Schrödinger equation, whose solution in general should be $\psi(r_1 r_2)$, to be separated into two equations, one in $\phi_{n_1}(r_1)$ and one in $\phi_{n_2}(r_2)$ so that

$$\psi(r_1 r_2) = \phi_{n_1}(r_1) \cdot \phi_{n_2}(r_2) \tag{2.63}$$

$\phi_{n_1}(r_1)$ and $\phi_{n_2}(r_2)$ are hydrogen-like wave functions having identical form when $n_1 = n_2$ but differing when $n_1 \neq n_2$. Therefore when $n_1 = n_2 = n$, interchange of the two electrons makes no difference to the wave function ψ, thus:

$$\psi(r_1 r_2) = \phi_n(r_1) \cdot \phi_n(r_2) = \phi_n(r_2) \cdot \phi_n(r_1) = \psi(r_2 r_1) \tag{2.64}$$

On the other hand, when $n_1 \neq n_2$,

$$\phi_{n_1}(r_1) \cdot \phi_{n_2}(r_2) \neq \phi_{n_1}(r_2) \cdot \phi_{n_2}(r_1) \tag{2.65}$$

which implies that the probability of finding simultaneously electron 1 at radius r_1 and electron 2 at radius r_2 is not the same as that of finding simultaneously electron 2 at r_1, and electron 1 at r_2. A difference in the probability of the configuration of the two electrons when they are interchanged means that the two configurations are distinguishable. This cannot be so since the energy remains unaltered, *viz.* $E_{n_1} + E_{n_2}$. *A wave-function of a system of two or more electrons* (or other identical particles) *which implies that the individual electrons are distinguishable is unacceptable as a description of the configuration.*

It is possible however to construct acceptable solutions by linear combination, thus:

$$\psi_s(r_1 r_2) = \phi_{n_1}(r_1)\phi_{n_2}(r_2) + \phi_{n_1}(r_2)\phi_{n_2}(r_1) \tag{2.66}$$

and

$$\psi_a(r_1 r_2) = \phi_{n_1}(r_1)\, \phi_{n_2}(r_2) - \phi_{n_1}(r_2)\, \phi_{n_2}(r_1) \tag{2.67}$$

If the individual products are solutions of the Schrödinger equation so also are the combinations and it can be seen that interchange of electrons now gives:

$$\psi_s(r_1 r_2) = \psi_s(r_2 r_1); \qquad \psi_s^2(r_1 r_2) = \psi_s^2(r_2 r_1) \tag{2.68}$$

and

$$\psi_a(r_1 r_2) = -\psi_a(r_2 r_1); \quad \psi_a^2(r_1 r_2) = \psi_a^2(r_2 r_1) \tag{2.69}$$

That is to say, the probability density is unchanged by interchanging electrons but the wave function itself either remains unchanged (ψ_s, symmetric) or changes sign (ψ_a, antisymmetric).

The same considerations apply if more than one quantum number is required to describe the state of each electron. Hence, if the spectral terms for

References p. 127–128

helium are due to the excitation of an electron (without saying which) to states with n and l other than $n = 1$, $l = 0$ while the other electron remains in the state $n = 1$, $l = 0$, we find that for each state of the hydrogen-like atom with given n and l there are two states, one symmetric and one anti-symmetric, in the excited helium atom. The ground state, however, is acceptably described by the symmetric solution $\phi_1(r_1) \cdot \phi_1(r_2)$ and there is no degeneracy. The observation of only one $1S$ term, but of two terms for every higher value of n and l, is in exact agreement with this description.

Symmetry of observed states

To identify the singlet and triplet levels it is necessary to take account of the spin. If the total eigenfunction (p. 76) can be written as the product $\psi_s \cdot \psi_{spin}$ or $\psi_a \cdot \psi_{spin}$, ψ_{spin} is the spin function for two electrons. Denoting a spin function for one electron by α if $s = +\frac{1}{2}$ and by β if $s = -\frac{1}{2}$, we may write as possibilities for ψ_{spin} for two electrons:

$$\alpha(1)\alpha(2) \qquad \beta(1)\beta(2) \qquad \alpha(1)\beta(2) \qquad \alpha(2)\beta(1)$$

Of these the first two are acceptable but the last two violate the principle of indistinguishability. They must be replaced by linear combinations and thus we have:

$$\begin{array}{cl} \alpha(1)\alpha(2) & \left.\begin{array}{r}\text{symmetric}\\ \text{symmetric}\\ \text{symmetric}\end{array}\right\} \psi_{spin,\,s} \\ \beta(1)\beta(2) & \\ \alpha(1)\beta(2) + \alpha(2)\beta(1) & \\ \alpha(1)\beta(2) - \alpha(2)\beta(1) & \text{antisymmetric} \quad \psi_{spin,\,a}. \end{array}$$

Three of these are symmetric ($\psi_{spin,\,s}$) and one is antisymmetric ($\psi_{spin,\,a}$) with respect to interchange of electrons. This description of a triplet state in a two-electron system corresponds to both electrons having $s = +\frac{1}{2}$, so that $S = 1$, and $2S + 1 = 3$. A total wave function $\psi_s \cdot \psi_{spin,\,s}$ would be symmetric and a triplet state. Four possibilities can now be envisaged for the total eigenfunction, *viz.*

$\psi_s \cdot \psi_{spin,\,s}$	symmetric, triplet
$\psi_a \cdot \psi_{spin,\,s}$	antisymmetric, triplet
$\psi_s \cdot \psi_{spin,\,a}$	antisymmetric, singlet
$\psi_a \cdot \psi_{spin,\,a}$	symmetric, singlet.

In fact, the helium term scheme admits only two possibilities, one set of singlet states and one of triplets. Furthermore the singlet set is seen from the nature of the ground state 1^1S_0 to be associated with the symmetric orbital function ψ_s. Thus the singlet terms correspond to $\psi_s \cdot \psi_{spin,\,a}$. The triplet set is antisymmetric in its orbital functions, *i.e.* $\psi_a \cdot \psi_{spin,\,s}$.

We thus arrive at the conclusion that, in the helium spectrum, only those total eigenfunctions which are antisymmetric to exchange of electrons are observed. Those which are symmetric do not occur. It can be shown that there is a total prohibition against any form of transition between states with antisymmetric total eigenfunctions and those with symmetric total eigenfunctions. There is also a prohibition against transitions between states with orbital function ψ_s and ψ_a so long as the approximation holds that the total eigenfunction can be written as the product of an orbital function and a spin function. We have seen already (p. 84) that there is justification for this in the applicability of Russell-Saunders coupling, and it is further borne out by the absence of spectral combination between the singlet and triplet levels in helium. That some inter-combination must nevertheless occur is shown by the existence of both states.

If, by means of absorption of radiation, an atom or molecule is excited from a (usually) singlet ground state and then by radiationless transfer of energy falls to the lowest triplet state, it may, by reason of the partial prohibition, remain for some time in this state. The lowest triplet level thus represents a *metastable state*.

Energy and electron correlation

Although the existence of two-term systems is a consequence of spin degeneracy, the energy differences do not arise from this cause. The values of the S, P, D and F levels are chiefly governed by the same considerations as in the alkali metals, namely the spatial distribution and interaction of electrons. In this case, since there are only two electrons, a more detailed account can be given of the energetics of the interaction. By ignoring the mutual repulsion between electrons we have seen that every state of the helium atom in which $n_1 = 1$ and $n_2 \neq 1$ is degenerate. If the actual forms of the hydrogen-like atomic orbitals ϕ_n are taken into account and ψ_s^2 and ψ_a^2 calculated the following facts emerge[5].

(i) states $1s^1$, $1s^1$ (*i.e.* both electrons have $n = 1$), symmetric 1^1S: the configuration of maximum probability is that in which both electrons are at the same distance from the nucleus.

(ii) states $1s^1$, $2s^1$ (*i.e.*, $n_1 = 1$; $n_2 = 2$ and $l_2 = 0$): in the symmetric $\psi_s(2^1S)$ state there is high probability that both electrons are at the same distance from the nucleus; in the antisymmetric $\psi_a(2^3S)$ state the probability of both electrons being at the same distance from the nucleus is zero. Both configurations have spherical symmetry but in ψ_s the electrons are close together and in ψ_a they stay apart.

(iii) states $1s^1$, $2p^1$ (*i.e.*, $n_1 = 0$; $n_2 = 1$ and $l_2 = 1$): the configurations of high probability for the symmetric $\psi_s(2^1P)$ state are those in which both electrons are on the same side of the nucleus and for the $\psi_a(2^3P)$ state those in which both electrons are on opposite sides of the nucleus.

References p. 127–128

The symmetric singlet states are those in which the electrons are close together and we should therefore expect the repulsion energy of the electrons to be greater than in the corresponding antisymmetric triplet states in which the electrons tend to stay apart. Thus the term 2^1S lies higher than 2^3S, and 2^1P higher than 2^3P. If there were no repulsion between electrons but only attraction between the nucleus and each electron the energies of the $1S$, $2S$ and $2P$ states would be $-8R$, $-5R$ and $-5R$. These would be the energies for complete removal of both electrons from the atom in the given states. The actual energies can be derived from the term schemes by adding the term value (removal of one electron) to $-4R$ (for removal of the second electron). The repulsion energy is then obtained by difference. The results are as follows:

	1^1S	2^1S	2^3S	2^1P	2^3P
total energy	-5.78	-4.29	-4.35	-4.24	$-4.27 \times R$
repulsion energy	2.22	0.71	0.65	0.76	$0.73 \times R$

Repulsion energy can also be calculated approximately for the electron distribution given by ψ_s and ψ_a. For 2^1S (electrons together) the calculated value is 0.93 R and for 2^3S (electron apart) the calculated value is 0.75 R.

It follows from the observation that ψ_s is associated with an antisymmetric spin function and from the calculation of the electron distribution described by ψ_s that in a configuration in which electrons tend to come together the spins of the electrons are opposed, *i.e.*, $s_1 = +\frac{1}{2}$, and $s_2 = -\frac{1}{2}$. Conversely the association of ψ_a with symmetric triplet spin functions shows that configurations in which electrons tend to stay apart have the spin of the electrons parallel, *i.e.* $s_1 = \frac{1}{2}$ and $s_2 = \frac{1}{2}$. (But notice that one of the $\psi_{\text{spin, s}}$ is a linear combination of the products of opposed spins, $\alpha(1)\beta(2)$ and thus the picture of parallel spin is more convenient than real). It is indeed a common spectroscopic observation, known as Hund's rule, that equivalent states of lowest electronic energy (minimum electron repulsion energy because electrons are as far apart as possible) are associated with the maximum multiplicity (all spins parallel) allowed by the Pauli principle. It is not permissible to ascribe the energy lowering to the spin orientation any more than it is proper to ascribe the stability of a covalent bond to spin pairing.

Pauli's Principle

Although Pauli's principle seems to have been proposed primarily in order to explain the periodic table it can be seen, in its most general form, as a simple extension of conclusions drawn from the helium spectrum. The principle may be stated as follows:

The total eigenfunction (including space and spin) for electrons is always antisymmetric.

We have already seen that the helium atom spectrum requires for its interpretation only two sets of terms, each representing total eigenfunctions which are antisymmetric. Some observations of molecular spectra and the phenomenon of ortho- and para-states of hydrogen show that the same antisymmetry principle applies to protons, neutrons and odd nuclei. The opposite principle of symmetric total eigenfunctions seems to apply to even nuclei (*e.g.* deuterons) and to protons. The principle is not derivable, but must be accepted as a law of nature.

The general statement can, however, be reduced to the particular one appropriate to the description of atom building. We have seen that if an electron 1 in an atom has quantum numbers n, m, l and s defining the atomic orbital $\phi(1)$ and electron 2 in the same atom has quantum numbers n', m', l' and s' defining the atomic orbital $\phi'(2)$, the only acceptable total eigenfunctions are the linear combinations:

$$\psi_a = \phi(1)\phi'(2) - \phi(2)\phi'(1) \tag{2.70}$$

$$\psi_s = \phi(1)\phi'(2) + \phi(2)\phi'(1) \tag{2.71}$$

Pauli's principle now rejects (2.71) as not observed in the behaviour of electrons. Now suppose the two electrons to have the same set of the four quantum numbers. The functions ϕ' and ϕ become identical and

$$\psi_a = \phi(1)\phi(2) - \phi(2)\phi(1) = 0 \tag{2.72}$$

Hence: *states of an atom in which any two electrons have the same set of four quantum numbers do not exist.*

Alternatively: *No more than two electrons in an atom may have the same set of three quantum numbers, n, l and m, and they must have opposed spins.*

The periodic table

The chemical significance of Pauli's principle follows from the last statement of it. The available hydrogen-like atomic orbitals, with their quantum numbers, n, l, m and s, are enumerated up to $n = 3$ in the Table 9. The states with $n = 1$, being of lowest energy will be expected to take, and are observed to take, the first two electrons as we start from hydrogen and proceed from element to element by unit increments of atomic number. But when two electrons have taken the two sets of quantum number $(1, 0, 0, +\frac{1}{2})$ and $(1, 0, 0, -\frac{1}{2})$ no more may enter the same atom with $n = 1$. The third electron takes the next lowest energy level. If we assume in the first instance that electron interaction is only sufficient to introduce a slight gradation in S, P, D and F levels, which otherwise are hydrogen-like, the third and fourth

References p. 127–128

TABLE 9

OPERATION OF PAULI'S PRINCIPLE WHEN $n = 1, 2,$ AND 3

n	l	m	s	Multiplicity of n, l states	Multiplicity of n states
1	0	0	$\pm\frac{1}{2}$	2 (1s)	2
2	0	0	$\pm\frac{1}{2}$	2 (2s)	8
	1	−1	$\pm\frac{1}{2}$	6 (2p)	
	1	0	$\pm\frac{1}{2}$		
	1	+1	$\pm\frac{1}{2}$		
3	0	0	$+\frac{1}{2}$	2 (3s)	18
	1	−1	$\pm\frac{1}{2}$	6 (3p)	
	1	0	$\pm\frac{1}{2}$		
	1	+1	$\pm\frac{1}{2}$		
	2	−2	$\pm\frac{1}{2}$	10 (3d)	
	2	−1	$\pm\frac{1}{2}$		
	2	0	$\pm\frac{1}{2}$		
	2	+1	$\pm\frac{1}{2}$		
	2	+2	$\pm\frac{1}{2}$		

etc.

electrons (in lithium and beryllium) take the sets of quantum numbers $(2, 0, 0, +\frac{1}{2})$ and $(2, 0, 0, -\frac{1}{2})$ to give S-states and the fifth (in boron) takes $n = 2$, $l = 1$.

It is customary to indicate values of n and l by a number for n and the letters s (not to be confused with spin quantum number) for $l = 0$, p for $l = 1$, d for $l = 2$ and f for $l = 3$. From the table it can be seen that for any given n the number of s-electrons cannot exceed 2; of p-electrons, 6; of d-electrons, 10; and of f-electrons, 14. In the absence of any greater interaction than we have already assumed the orbitals would therefore be occupied in the order:

1s	2s	2p	3s	3p	3d	4s	4p	4d	4f	5s	5p	5d	5f	5g
2	2	6	2	6	10	2	6	10	14	2	6	10	14	18

But it can be seen from the alkali metal spectra and from the discussion of the helium atom, that electron interaction is by no means negligible even for those atoms in which all but one of the electrons form an atomic core. The order in which the orbitals are in fact occupied is, with minor variation:

1s	2s	2p	3s	3p	4s	3d	4p	5s	4d	5p	6s	4f	5d	6p	7s	5f
2	2	6	2	6	2	10	6	2	10	6	2	14	10	6	2	14
2	8		8		18			18			32					

TABLE 10

OPERATION OF PAULI'S PRINCIPLE IN RELATIVISTIC (n, k, m) ATOMIC STATES

n	k	m	Multiplicity of n, k states	Multiplicity of n states
1	+1	−1	2 (s)	2
		0		
2	+1	−1	2 (s)	
		0		
	−1	−1	2 (p)	8
		0		
	+2	−2		
		−1		
		0	4 (p)	
		+1		
3	+1	−1		
		0	2 (s)	
	−1	−1		
		0	2 (p)	
	+2	−2		
		−1	4 (p)	
		0		
		+1		
	−2	−2		
		−1	4 (d)	18
		0		
		+1		
	+3	−3		
		−2		
		−1	6 (d)	
		0		
		+1		
		+2		

etc.

On p. 87 we saw that the degeneracy of the nth state, according to the restriction upon l and m, was n^2. In order to give a multiplicity of n states which agrees with the periodic table it now appears that the concept of electron spin permits a doubling of the number of states to $2n^2$. Clearly the observed periodicity of chemical behaviour, as described in any book of inorganic chemistry, is adequately accounted for. The account, however, depends upon the theoretical restriction on n, l, and m in the quantum mechanical treatment of the hydrogen atom; upon the additional spin which doubles the number of available orbitals; and upon the Pauli principle, which was evidently first put forward for that purpose.

It is interesting to note the way in which the same numbers of available orbitals arise from the quantal restrictions in the relativistic treatment. Given (p. 94) that:

References p. 127–128

$$k = \pm 1, \pm 2 \ldots \pm (n - 1), + n \tag{2.73}$$

and

$$m = 0, \pm 1, \pm 2 \ldots \pm (|k| - 1), - |k| \tag{2.74}$$

the available orbitals can be enumerated. This is done for $n = 1$, 2 and 3 in Table 10. Equation 2.36 shows that energy increases with n and also that, for a given n, it increases with $|k|$. It is evident from Table 10 that the relativistic treatment leads to the correct numbers without explicit mention of electron spin.

Deducing configuration from spectral terms

Knowledge of the electron configuration of atoms is firmly based on spectroscopic evidence. It might be thought that with a large number of electrons outside the atomic core a bewilderingly large number of possible states would be predicted by application of the vector model. This is indeed so for excited states of the atom, although many spectra can be explained in terms of the excitation of one electron. In the ground state, however, the possibilities for coupling are restricted by the Pauli principle. Electrons having the same n, l designation are called *equivalent electrons* and it is with these that we are normally concerned in the ground state.

The important example of this restriction is the assumption which we have made about the "closedness" of certain configurations; particularly of the configuration which corresponds to chemical inertness in the inert gases. In neon, for example, the configuration is $1s^2\,2s^2\,2p^6$. Each pair of s electrons has $L = 0$ and, with antiparallel spins, $S = 0$. Since there are six p-electrons the Pauli principle does not permit independent choice of m (direction of $\boldsymbol{l}$) or of s. They must each have one of the six different (m, s) sets shown in Table 9. The resultant $S = 0$ and, since the possible orientations of $\boldsymbol{l}$ are all equally represented, $L = 0$. In this way any completed subgroup forms a $^1S_0$ state and can be ignored in assessing numbers and types of terms, if not their energies.

The simplest closed shell is that of helium in the ground state. On p. 110 two states ($^1S_0$ and $^3S_1$) were postulated for the pair of equivalent electrons, $1s^2$. We now see that the non-existence of the ^3S state is an example of the Pauli principle. For the non-equivalent pair of electrons, $1s^1$, $2s^1$ the Pauli principle is not violated in the ^3S state. The ground state term for lithium is $^2S_{\frac{1}{2}}$ ($L = 0$, $S = \frac{1}{2}$, $J = \frac{1}{2}$) which indicates the configuration $1s^2$, $2s^1$. For beryllium $^1S_0$ indicates the closed group $1s^2 2s^2$. One might, on this account, suppose beryllium to be as inert as helium. That it is not so is reflected in the first ionization potentials (9.32 eV to remove one electron from beryllium in its $^1S_0$ ground state; 24.58 eV for helium) and is no doubt due to the screening by the $1s^2$ electrons. Excitation to the $2s^1 2p^1$ configuration which is the preparatory step for formation of covalent bonds requires 5.3 eV, whereas excitation of helium to $1s^1 2s^1$ involving a change in n, requires some 20.5 eV.

Boron has five electrons of which four only can be accommodated in the $1s$ and $2s$ subgroups. The ground term is in fact $^2P_{\frac{1}{2}}$ ($L = 1$, $S = \frac{1}{2}$, $J = \frac{1}{2}$) and demonstrates the configuration $1s^2 2s^2 2p^1$. Carbon has the ground term $^3P_0$ ($L = 1$, $S = 1$, $J = 0$) indicating for the two equivalent electrons not in the closed sub-groups $l_1 = 1, m_1 = +1$, $s_1 = \frac{1}{2}$ and $l_2 = 1$, $m_2 = 0$, $s_2 = \frac{1}{2}$. It is sometimes useful to distinguish between the p orbitals with different m as p_x, p_y and p_z. The configuration in carbon is therefore $2p_x{}^1 2p_y{}^1 2p_z{}^0$ since in the alternative configuration $2p_x{}^2 2p_y{}^0 2p_z{}^0$ the spins would necessarily be antiparallel by the Pauli principle and the appropriate term would be ^1D. This term is observed but is not the ground term. The configuration $2s^1 2p_x{}^1 2p_y{}^1 2p_z{}^1$ is also conceivable and excitation to this state is the imagined first step in the formation of four covalent bonds. The lowest term for such a configuration is $^5S_2$ which has not been observed spectroscopically.

In nitrogen the ground term $^4S_{\frac{3}{2}}$ ($L = 0$, $S = \frac{3}{2}$, $J = \frac{3}{2}$) indicates that the configuration is $2s^2 2p_x{}^1 2p_y{}^1 2p_z{}^1$, the three $\boldsymbol{l}$-vectors resulting in $L = 0$ since each orientation is present. The three parallel spins result in maximum S and maximum multiplicity. The oxygen atom is normally in the state $^3P_2$ ($L = 1$, $S = 1$, $J = 2$) corresponding to $2s^2 2p_x{}^2 2p_y{}^1 2p_z{}^1$. Two electrons must necessarily have the same m and are therefore spin-paired. S therefore cannot exceed 1. With two electrons having $m = 1$, one with $m = 0$ and one with $m = -1$, it follows that $L = 1$. It is noticeable that here the $J = 2$ level ($^3P_2$) is the lowest member of the triplet whereas in carbon the $J = 0$ level ($^3P_0$) lies lowest. The same reversal of the multiplet splitting occurs in fluorine where the ground term is $^2P_{\frac{3}{2}}$ ($L = 1$, $S = \frac{1}{2}$, $J = \frac{3}{2}$) rather than $^2P_{\frac{1}{2}}$. Either term would accord with the configuration $2s^2 2p_x{}^2 2p_y{}^2 2p_z{}^1$. The period is completed with neon ($^1S_0$) in which all electrons are in closed subgroups (or shells), $1s^2 2s^2 2p^6$. The first ionization potential of neon (21.56 eV) indicates that the removal of one electron from this closed configuration is almost as difficult as from the helium $1s^2$ group. The excitation potential to $1s^2 2s^2 2p^5 3s^1$, since it involves a change in n, is also large.

The next period, through to argon ($Z = 18$), has the same sequence of ground state terms. Potassium and calcium also conform, with electrons going into the $4s$ subgroup. The next element, scandium (21) has a ground term $^2D_{\frac{3}{2}}$ ($L = 2$, $S = \frac{1}{2}$, $J = \frac{3}{2}$) which cannot be due to one electron with $l = 1$ but indicates that the third electron has $l = 2$, thus $1s^2 2s^2 2p^6 3s^2 3p^6 4s^2 3d^1$. This is the start of the first transition group, comprising ten elements from scandium to zinc. At chromium a small variation is shown by the ground term since the configuration is expected to be $4s^2 3d^4$, and hence a term $^5D_0$ ($m = 2, 1, 0$ and -1, and four parallel spins), whereas in fact the

References p. 127–128

ground term is $^7S_3$ ($L = 0$, $S = 3$, $J = 3$). The multiplicity indicates the uncoupling of the $4s^2$ pair to permit six electrons to contribute to the total spin. Thus the configuration is $4s^1 4d^5$. The energy differences between various l values are becoming so small that the excitation $s \rightarrow d$ is energetically overcome by the stability conferred by the partial "closedness" of the $4d^5$ group. In the next element, manganese (25) the term $^6S_{\frac{5}{2}}$ ($J = S = \frac{5}{2}$) agrees with the restoration of the $4s^2$ subgroup and the retention of five electrons in five different d orbitals. Similar behaviour in the second transition group, but not the third, is clearly demonstrated by identification of the ground spectral terms.

It is important for chemists to realise that their knowledge of electronic configuration comes directly from atomic spectra. It is true that, given the general principles of atom building, the electron distribution in an atom may be deduced from the element's chemical behaviour or, what comes to the same thing, from its position in the periodic table. Nevertheless, the electronic configurations which are to be found tabulated in most modern inorganic chemistry texts are deduced from ground state terms of the spectra of free atoms.

Electronic States in Molecules

Unambiguous and complete characterization of electronic states in molecules has not progressed much beyond diatomic molecules, either in theory or in spectroscopic practice. In a few cases quantum mechanical calculations guided by principles of symmetry, have enabled certain electronic transitions in polyatomic molecules to be reliably interpreted but for the most part empirical correlations serve to give a partial characterization. This important aspect will be further discussed in Chapter 4. On the other hand detailed interpretations have been given of the electronic spectra of diatomic molecules and they form the starting point for the theory of molecular orbitals and the quantum mechanical interpretation of the covalent bond. The following account can only show some of the important features.

Molecular orbitals

It has proved possible to develop the ideas used in the interpretation of atomic spectra and to apply them to molecules. The basis for discussion is the molecule-ion H_2^+ since it contains one electron. This system of two separated protons plus one electron may be formed in theory by bringing together a proton and a hydrogen atom from a large distance apart or it may be regarded as developed from the helium ion He^+ (two united protons plus two electrons: the neutrons do not affect the present considerations).

Hydrogen-like states (1*s*, 2*s*, 2*p*, 3*s*, 3*p*, 3*d*, . . .) are possibilities for the helium ion. The application of an electric field (sufficient to uncouple the electron spin from the orbital angular momentum) simulates the development of the axially symmetric H_2^+ ion from the united atom. Under these conditions the component of orbital angular momentum, $\boldsymbol{\lambda}$, about the nuclear axis is the molecular analogue of the atomic orbital angular momentum $\boldsymbol{l}$. $\boldsymbol{\lambda}$ has magnitude $\lambda\hbar$, where $\lambda = 0, 1, 2, 3, \ldots l$. By analogy with the atomic case, the designations $\sigma, \pi, \delta, \phi, \gamma \ldots$ are used for $\lambda = 0, 1, 2, 3, 4 \ldots$ Since $\boldsymbol{\lambda}$ may take either of two directions along the axis the states for which $\lambda \neq 0$ are doubly degenerate on that account. Provided the inter-nuclear distance of the molecule is small, quantum numbers n and l still have approximately the same significance as in the united atom. The energy levels are largely governed by n, but the degeneracy in the spherically symmetrical He^+ system is split by the electric field, or the formation of the molecule, and somewhat different levels arise for different l and λ. Accordingly the states available to one electron in the H_2^+ ion are, in order of increasing energy:

$1s\sigma$	$2s\sigma$	$2p_x\sigma$	$2p_y\pi =$	$2p_z\pi$	$3s\sigma$	$3p_x\sigma$	$3p_y\pi =$	$3p_z\pi$	$3d_{xy}\pi$	$3d_{xz}\pi$	$4p_x\sigma$
0	0	1	1	1	0	1	1	1	2	2	1
0	0	0	−1	+1	0	0	−1	+1	−1	+1	0

The symbols represent *molecular orbitals* and the energies associated with each are spectral terms which would be used in the interpretation of the electronic spectrum of H_2^+. However that spectrum has never been observed.

The same molecular orbitals can be obtained by modification of the atomic orbitals of the separated atoms as they are brought together. Such a process should lead to the same number of molecular orbitals but they are differently labelled to indicate a different origin. It can be also assumed that a "separated-atom molecular orbital" goes over smoothly into a corresponding "united-atom molecular orbital" on passing from large inter-nuclear distance to small. A pictorial representation of the generation of molecular orbitals in this way, from *s* and *p* atomic orbitals is given in Fig. 50. The molecular orbitals depicted in the centre column are obtained by addition or subtraction of the atomic orbitals of the separate atoms on the left. They are labelled accordingly and the subscripts *g* and *u*, appropriate only to molecular orbitals derived from like atoms, indicate that the orbital remains unchanged (gerade) or changes sign (ungerade) on inversion through the centre of symmetry. In the right hand column are the obviously related atomic orbitals of the united atom.

It can be seen, in the simple illustration, that in some orbitals there is a

References p. 127–128

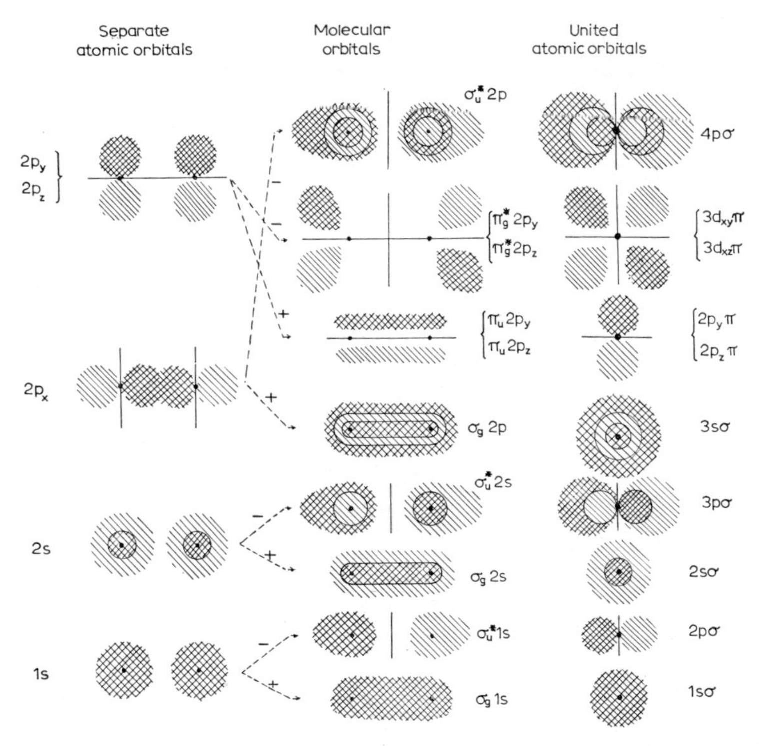

Fig. 50. Molecular orbitals from the atomic orbitals of equal separated atoms and from the united atom. The wave function has positive values in the cross-hatched areas and negative values in the diagonally shaded areas. Nodal surfaces are indicated by lines.

high electron density between the nuclei and in others there is not. Those with the accumulation of electrons between the nuclei may be expected to represent a stable molecule in which the attraction between nuclei and the electron cloud between them outweighs the mutual repulsion of the nuclei. Such orbitals are called *bonding orbitals*. Those in which there is a much smaller probability of finding the electron between the nuclei are likely to be unstable and the orbitals are called *anti-bonding orbitals*, and marked with an asterisk.

By analogy with the atomic case we may now regard those one-electron molecular orbitals as available to accommodate electrons in accordance with the Pauli principle as molecules are built up by the addition of electrons. As in the atomic case, relative energy may be expected to be modified by the presence of other electrons. In addition the energy order varies with internuclear distance. In hydrogen two electrons would naturally go into the $\sigma_g 1s^2$ configuration and yield a stable molecule. Further electrons must go into higher orbitals and thus in the molecule He_2 the configuration is $\sigma_g 1s^2 \cdot \sigma_n^* 1s^2$, comprising an anti-bonding and a bonding orbital. He_2 is unstable in the ground electronic state but excited states are stable and have been observed. The configuration of the He_2 ground state $\sigma_g 1s^2 \cdot \sigma_n^* 1s^2$ is often abbreviated to KK in giving the molecular orbitals of higher species, since the lower orbitals approximate to simple 1s atomic orbitals when higher molecular orbitals are filled. The order of the molecular orbitals is now slightly changed (to the one shown in Fig. 50) and this is illustrated by the configuration of the molecules N_2, O_2 and F_2. Three different designations are given for the orbitals, the third being ascribed to Mulliken.

"United atom"	$1s\sigma$	$2p\sigma$	$2s\sigma$	$3p\sigma$	$3s\sigma$	$2p_y\pi$	$=2p_z\pi$	$3d_{xy}\pi$	$=3d_{xz}\pi$	$4p\sigma$
"separate atom"	$\sigma_g 1s$	$\sigma_u^* 1s$	$\sigma_g 2s$	$\sigma_u^* 2s$	$\sigma_g 2p$	$\pi_u 2p_y$	$=\pi_u 2p_z$	$\pi_g^* 2p_y$	$=\pi_g^* 2p_z$	$\sigma_u^* 2p$
"Mulliken"	K	K	$z\sigma$	$y\sigma$	$x\sigma$	$v\pi$		$v\pi$		$u\sigma$
N_2	2	2	2	2	2	2	2			
O_2	2	2	2	2	2	2	2	1	1	
F_2	2	2	2	2	2	2	2	2	2	

In many such cases, Russell-Saunders coupling satisfactorily explains the electronic state. $\boldsymbol{\Lambda} = \boldsymbol{\lambda}_1 + \boldsymbol{\lambda}_2 + \ldots$ and $\boldsymbol{S} = \boldsymbol{s}_1 + \boldsymbol{s}_2 + \ldots$ The vectors $\boldsymbol{\Lambda}$ and $\boldsymbol{S}$, however, are not coupled (strong field) and they lie along the internuclear axis. Their total effect is thus a vector of magnitude $\Lambda \pm S$. Two equivalent electrons in the same σ orbital have $\lambda_1 = \lambda_2 = 0$ and hence $\Lambda = 0$. By Pauli's principle the spins must be paired and $S = 0$. Thus two equivalent σ electrons form a closed subgroup. So also do four equivalent π-electrons since, by Pauli's principle, two $\boldsymbol{\lambda}$-vectors must lie in one direction and two in the other. Hence $\Lambda = 0$ and $S = 0$. Such closed groups have terms $^1\Sigma_0$ (by analogy with the atomic term for closed shells, $^1S_0$). If there are two equivalent π electrons $\lambda_1 = 1$ and $\lambda_2 = 1$ and thus $\Lambda = 0$ or 2. In the case $\Lambda = 2$, the λ_1 and λ_2 are parallel and therefore the spins must be antiparallel. $S = 0$, and a $^1\Delta$ state (by analogy with ^1D for the atomic case $L = 2$) results. When $\Lambda = 0$ no such restriction exists and $S = 0$ or 1 resulting in $^1\Sigma$ and $^3\Sigma$ states. The state of maximum multiplicity has lowest energy and thus oxygen has the ground term $^3\Sigma$. This is observed spectroscopically and is also in agreement with the paramagnetism of oxygen which corresponds to two unpaired electrons.

In hydrogen the configuration $1s\sigma^2$ of the ground state has the term $^1\Sigma_g$. $^3\Sigma_g$ is impossible for this configuration for precisely the same reason as the 1^3S term does not occur in helium: all other quantum numbers being equal the spins must be antiparallel. In the excited configuration both $^1\Sigma$ and $^3\Sigma$ terms occur. In so far as the electrons in nitrogen occupy 5 bonding and 2 anti-bonding orbitals, the bond may be called a triple-bond, $N \equiv N$. Similarly fluorine is singly-bonded, F—F. Oxygen cannot be quite so simply

References p. 127–128

described but approximates to a double bond.

The ideas contained in Fig. 50 can also be applied qualitatively to molecules containing unlike nuclei. Some modification is required. The subscripts *g* and *u* no longer apply. The electron distribution will also be unsymmetrical with consequent effect on the bonding energies. The third (Mulliken) method of designating molecular orbitals has the advantage that it may be applied to such heteronuclear molecules. The notation $z\sigma$ is meant to imply the lowest molecular orbital of σ-type and is not restricted to those which are, as in the homonuclear molecule, compounded wholly of $2s$ atomic orbitals.

The theoretical prediction of the term manifold for hydrogen, along lines similar to those described above, made it possible for the very complex many-line molecular spectrum of hydrogen to be analyzed satisfactorily. Analysis of a large number of other diatomic species, many of which exist only in a discharge tube or a flame, has been carried out: the methods and resultant molecular parameters are fully discussed by Jevons[21] and by Herzberg[22].

The hydrogen molecule-ion, H_2^+

The addition or subtraction of $1s$ orbitals shown in Fig. 50 is represented by the expressions

$$\sigma_g 1s : \psi_g = \phi_A + \phi_B \tag{2.75}$$

$$\sigma_u 1s : \psi_u = \phi_A - \phi_B \tag{2.76}$$

where the new gerade and ungerade functions (even or odd with respect to inversion) are obtained from the hydrogen $1s$ atomic orbitals ϕ centred on nuclei A and B. The two atomic orbitals are equally represented in the new molecular orbital.* In $H_2^+ \psi_g(1) = \phi_A(1) + \phi_B(1)$ defines an equal probability of finding the one electron close to either nucleus. $\psi_u(1) = \phi_A(1) - \phi_B(1)$ also defines a probability density which is symmetrical with respect to the two nuclei. The two states ψ_g and ψ_u are of equal energy when the internuclear distance is very large. Their energies differ as the nuclei are brought together because $\psi_g(1)$ gives a high probability and $\psi_u(1)$ a low probability of the electron being between the nuclei. Both states are doublet levels, $^2\Sigma_g$ being stable and $^2\Sigma_u$ unstable.

Hydrogen molecule

We may approach the *two-electron* problem of the neutral hydrogen molecule by making use of the molecular orbitals defined in 2.75 and 2.76. In

* On this account the electronic configuration of the H_2^+ is sometimes referred to as a *resonance* between the two states $H_AH_B^+$ and $H_A^+H_B$ which would be described approximately by molecular orbitals identical with the atomic orbitals, namely ϕ_A and ϕ_B respectively.

doing so we may make use of the principles which were found to apply on proceeding from the one-electron hydrogen atom to the two-electron helium atom. Whereas the atomic orbitals $1s$ and $2s$ of hydrogen allowed us to envisage configurations $1s^2$, $1s^1 2s^1$, and $2s^2$ for helium, the molecular orbitals $\sigma_g 1s$ and $\sigma_u{}^* 1s$ for $H_2{}^+$ permit consideration of configurations $\sigma_g 1s^2$, $\sigma_g 1s^1 \cdot \sigma_u{}^* 1s^1$ and $\sigma_u{}^* 1s^2$ for H_2.

The configuration $\sigma_g 1s^2$ implies the symmetric wave function

$$\Psi_s = \psi_g(1)\psi_g(2) \tag{2.77}$$

where the 1 and 2 refer to the coordinates of the two electrons. Being a symmetric function, it can by Pauli's principle only combine with an antisymmetric spin function, $S = 0$. Hence the ground term $^1\Sigma_g$. Expansion of Ψ_s in terms of ϕ_A and ϕ_B gives:

$$\Psi_s(\sigma_g 1s^2) = \phi_A(1)\phi_B(2) + \phi_A(2)\phi_B(1) + \phi_A(1)\phi_A(2) + \phi_B(1)\phi_B(2) \tag{2.78}$$

In the configuration $\sigma_g 1s^1 \cdot \sigma_u 1s^1$ the wave function $\psi_g(1) \cdot \psi_u(2)$ is not acceptable since it implies that the electrons can be distinguished. The linear combinations

$$\begin{aligned}\Psi_s(\sigma_g 1s \cdot \sigma_u 1s) &= \psi_g(1) \cdot \psi_u(2) + \psi_g(2) \cdot \psi_u(1)\\ &= 2\phi_A(1)\phi_A(2) - 2\phi_B(1)\phi_B(2)\end{aligned} \tag{2.79}$$

$$\begin{aligned}\Psi_a(\sigma_g 1s \cdot \sigma_u 2s) &= \psi_g(1) \cdot \psi_u(2) - \psi_g(2) \cdot \psi_u(1)\\ &= 2\phi_A(2)\phi_B(1) - 2\phi_A(1)\phi_B(2)\end{aligned} \tag{2.80}$$

are acceptable. The function Ψ_s (2.79) is symmetric with respect to interchange of electrons and therefore must be associated with an antisymmetric spin function $S = 0$. It is however antisymmetric (ungerade) with respect to inversion or to interchange of nuclei: hence the term $^1\Sigma_u$. The function Ψ_a (2.80) is antisymmetric with respect to interchange of electrons and therefore combines with a symmetric spin function. It remains ungerade with respect to interchange of nuclei and hence defines a $^3\Sigma_u$ state.

In the configuration $\sigma_u{}^* 1s^2$ the function $\psi_u(1) \cdot \psi_u(2)$ is satisfactory and has the term $^1\Sigma_g$. Hence:

$$\Psi_s(\sigma_u{}^* 1s^2) = \phi_A(1)\phi_A(2) + \phi_B(1)\phi_B(2) - \phi_A(1)\phi_B(2) - \phi_A(2)\phi_B(1) \tag{2.81}$$

The other approach to the problem of the hydrogen molecule is that due to Heitler and London. It starts from the atomic orbitals and writes the product $\phi_A(1)\phi_B(2)$ of finding one electron near to nucleus A and the other near to nucleus B. This violates the principle of indistinguishability and must be replaced by

$$\Psi_s(1s_A \cdot 1s_B) = \phi_A(1)\phi_B(2) + \phi_A(2)\phi_B(1) \quad {}^1\Sigma_g \tag{2.82}$$

and

$$\Psi_a(1s_A \cdot 1s_B) = \phi_A(1)\phi_B(2) - \phi_A(2)\phi_B(1) \quad {}^1\Sigma_u \tag{2.83}$$

References p. 127–128

The first is symmetric with respect to interchange of electrons, and therefore singlet: it is symmetric with respect to inversion, hence gerade. The second is antisymmetric and ungerade.

Comparison of the two methods of forming approximate wave functions shows that the Ψ_a functions are identical (apart from a normalizing constant) and that Ψ_s $(1s_A \cdot 1s_B)$ (2.82) and $\Psi_s(\sigma_g 1s^2)$ (2.78) differ in that the latter contains the "ionic" terms $\phi_A(1)\phi_A(2)$ and $\phi_B(1)\phi_B(2)$, which increase the probability of finding both electrons near the same nucleus. The $\Psi_s(\sigma_g 1s \cdot \sigma_u 1s)$ (2.79) is wholly composed of these terms. All the functions contain the essential idea that the probability density is not altered by interchange of electrons. The symmetric wave functions Ψ_s give a high probability and the antisymmetric functions Ψ_a a low probability of finding both electrons between the nuclei. It remains to see how the energy of the described configurations varies with inter-nuclear distance.

Explicit forms of the atomic orbitals, ϕ, can be used to calculate the probability densities for the various functions and thus to obtain the energies of Coulombic interaction between electrons and nuclei (attractive), and between electrons (repulsive) and between nuclei (repulsive) at various inter-nuclear distances. The sum of these, $U(r) = E_\varepsilon + V_n$ (see p. 125), is plotted against r and if a minimum occurs in the curve which is less than the energy of the infinitely separated atoms the electronic state is stable with respect to dissociation. Calculations show that $\Psi_s(\sigma_g 1s^2)$ or $\Psi_s(1s_A \cdot 1s_B)$ for hydrogen are the lowest states in the two approximations and give the curve labelled $^1\Sigma_g$ in Fig. 51. On the other hand the first excited state Ψ_a in both approximations, is repulsive or unstable at all nuclear separations, as shown by the curve $^3\Sigma_u$. It is also possible to calculate $U(r)$ for the unacceptable function $\phi_A(1) \cdot \phi_B(2)$ which gives the broken-line curve in Fig. 51. Although it predicts a stable molecule it is clear that the acceptable functions lead to an increase or a corresponding decrease in the electron density between the nuclei and hence to an increase or decrease in the Coulombic attraction between electrons and nuclei.

Because an additional energy term arises when the principle of indistinguishability is taken into account by way of functions which differ only in exchange of electrons the additional energy term is referred to as *exchange energy*. Like *resonance energy* which appears (p. 219) when a slightly different device is used to satisfy the principle of indistinguishability, exchange energy is no more than the measure of improvement in our calculations when we take steps to set up acceptable wave functions.

Neither of the approximations gives an accurate value of the dissocation energy or the equilibrium inter-nuclear separation of hydrogen. This is, in part, due to an overemphasis by the molecular orbital treatment and an un-

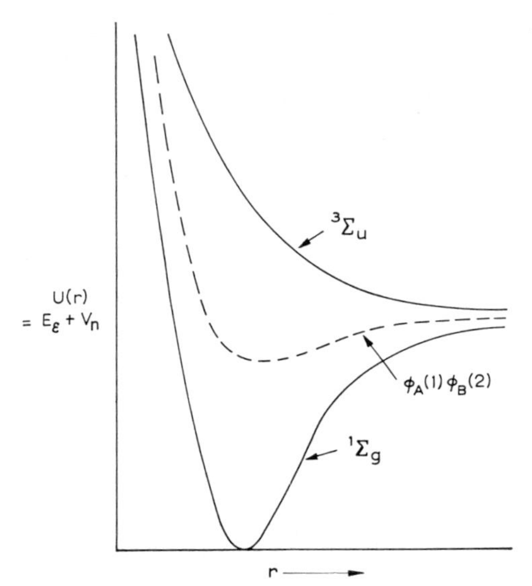

Fig. 51. Potential curves for the lowest $^1\Sigma_g$ and $^3\Sigma_u$ states of hydrogen. The energy corresponding to the simple product function $\varphi_A(1) \cdot \varphi_B(2)$ is shown as the broken curve.

deremphasis by the Heitler-London treatment of the contribution of "ionic" terms. Better approximations take this into account and lead to good agreement with the observed molecule. The application of the principle of indistinguishability clearly provides a general explanation of covalent bonds and of bonding and antibonding orbitals. However, further extension of the discussion of valence would go beyond the scope of this book[6,7,23,24].

The Heitler-London treatment does not naturally give an account of excited states of molecules in the way that molecular orbital theory does. It does not, for example, give the other two electronic states, $\psi_s(\sigma_g 1s\sigma_u{}^* 1s)^1\Sigma_u$ and $\psi_s(\sigma^* 1s^2)^1\Sigma_g$, both of which have been detected in the analysis of the hydrogen molecule spectrum. The first behaves spectroscopically in accordance with the predominance of the "ionic" terms. The second is of interest in that a pair of electrons in the antibonding orbital ($\sigma^* 1s^2$) give a weakly stable state.

Multiplicity of molecular states

In all cases orbital wave functions which have been formed so as to satisfy the principle of indistinguishability (exchange degeneracy) occur in pairs, one symmetric and one antisymmetric. The symmetric functions bring the electrons together and antisymmetric functions keep them apart. As a consequence the electron repulsion energy in the symmetric states is larger than in the antisymmetric but the effect on the total energy is small compared with the effect on the attractive force. The symmetric functions in molecules not only bring the electrons together but bring them together in the region *between the nuclei*. The symmetric functions are therefore those which lead to stronger chemical bonding. By Pauli's principle they are brought together in pairs (antisymmetric spin function, $S = 0$, singlet states) with antiparallel spins. This is the basis for the idea that covalent bonds are formed by the sharing of pairs of electrons with opposed spin and that the valency of an element is equal to the number of unpaired electrons in the atom. However the spin pairing is not to be thought of as conferring stability upon the molecule. It is rather an observed characteristic, as expressed in the Pauli principle, of those electronic configurations which give stable states.

Most chemical species contain an even number of electrons. When they are all paired off the ground term is singlet, usually $^1\Sigma$. Triplet states then arise for uncoupling of an electron pair. As in the near prohibition of combination between singlet and triplet levels in helium, so in molecules triplet states are generally only reached by radiationless transitions from excited singlet states. When the lowest triplet state is formed it does not immediately lose energy because of the near prohibition against transition to the ground singlet state. The lowest triplet state is thus a *metastable state*, which has some important consequences in certain chemical phenomena (p. 107). There are a few cases, of which oxygen is the most prominent example, in which the ground state is triplet. Then the lowest singlet state is metastable.

A few molecules known to the chemist are odd. In the technical jargon an *odd molecule* is one with an odd number of electrons. Nitric oxide is a well-known example and its ground state is the doublet $^2\Pi_{\frac{3}{2}}$, $^2\Pi_{\frac{1}{2}}$ corresponding to a single unpaired electron in the configuration:

$$KK \cdot \sigma 2s^2 \cdot \sigma^* 2s^2 \cdot \sigma 2p^2 \cdot \pi 2p^4 \cdot \pi^* 2p^1$$

Some isolable polyatomic molecules are odd and most of the free radicals have a single unpaired electron, and hence a doublet ground state. There are also a large number of "spectroscopic" molecules and ions (those which are only observed spectroscopically in discharge tubes, flames, or comets), some of which are odd molecules. The observation and characterization of these have played a large part in establishing the energy order of molecular orbitals.

REFERENCES

[1] H. E. WHITE, *Phys. Rev.*, 37 (1931) 1416.
[2] P. A. M. DIRAC, *Proc. Roy. Soc. (London)*, A 117 (1928) 610.
[3] W. E. LAMB, *Repts. Progr. in Phys.*, 14 (1951) 19.
[4] G. W. SERIES, *The Spectrum of Atomic Hydrogen*, Oxford Univ. Press, 1957.
[5] T. A. MILNER and P. W. GILLES, *J. Am. Chem. Soc.*, 81 (1959) 6115.
[6] P. G. DICKENS and J. W. LINNETT, *Quart. Revs. (London)*, 11 (1957) 291.
[7] C. A. COULSON, *Quart. Revs. (London)*, 1 (1947) 144.

Of the numerous books dealing with fundamental quantum mechanics there are two which introduce the subject without making very sophisticated mathematical demands:

[8] J. FRENKEL, *Wave Mechanics (Elementary Theory)*, 2nd ed., Oxford Univ. Press, 1936.
[9] L. PAULING and E. B. WILSON, *Introduction to Quantum Mechanics*, McGraw-Hill, 1935.

Some of the larger physical chemistry texts give many of the solutions of the standard problems, *e.g.*

[10] E. A. MOELWYN-HUGHES, *Physical Chemistry*, 2nd ed., Pergamon Press, 1957.
[11] J. R. PARTINGTON, *An Advanced Treatise on Physical Chemistry*, Vol. I, Longmans, Green and Co., 1949.

Important advanced monographs in correspondingly advanced mathematical language are:

[12] W. HEISENBERG, *The Physical Principles of the Quantum Theory*, Dover Publs., 1930.
[13] H. WEYL, *The Theory of Groups and Quantum Mechanics*, 2nd English ed., Dover Publs., 1950.
[14] P. A. M. DIRAC, *The Principles of Quantum Mechanics*, 3rd ed., Oxford Univ. Press, 1947.
[15] J. FRENKEL, *Wave Mechanics (Advanced General Theory)*, Oxford Univ. Press, 1934.

The author has found particularly clarifying the formal approach of Chapters VII and VIII in:

[16] R. C. TOLMAN, *The Principles of Statistical Mechanics*, Oxford Univ. Press, 1938.

For an account of the development of quantum mechanics and a non-mathematical discussion of some of the concepts, see:

[17] M. BORN, *Experiment and Theory in Physics*, Cambridge Univ. Press, 1943.
[18] M. BORN, *The Restless Universe*, Dover Publs., 1951.

Necessary texts for an understanding of atomic spectra, and naturally containing much related discussion of the theoretical interpretation, are:

[19] H. E. WHITE, *Introduction to Atomic Spectra*, McGraw-Hill, 1934.
[20] G. HERZBERG, *Atomic Spectra and Atomic Structure*, 2nd ed., Dover Publs., 1944.

The electronic states of diatomic molecules are fully covered by:

[21] W. JEVONS, *Band Spectra of Diatomic Molecules*, Phys. Soc., 1932.
[22] G. HERZBERG, *Spectra of Diatomic Molecules*, Van Nostrand, 1950.

The latter carries a masterly summary of quantum mechanical principles and an account, firmly related to the spectroscopic evidence, of the theory of chemical bonding. The discussion of valence in this chapter should be extended by reference to:

[23] C. A. Coulson, *Valence*, Oxford Univ. Press, 1952.
[24] J. A. A. Ketelaar, *Chemical Constitution*, Elsevier, 1958.

Full tabulation of atomic spectral lines is given by:

[25] H. Kayser, *Tabelle der Hauptlinien der Linienspektra aller Elementen*, Springer Verlag, 1926.
[26] H. Kayser, *Tabelle der Schwingungszahlen*, see *Phys. Rev.*, 48 (1935) 98.
[27] W. Grotrian, *Graphische Darstellung der Spektren von Atomen und Ionen mit ein, zwei und drei Valenzelektronen*, Springer Verlag, 1928.

Chapter 3

CHARACTERIZATION OF STATES: NUCLEAR MOTION

We have considered in the previous chapter the general process of characterization of states which is, in essence, the assignment of observed spectroscopic transitions to theoretically predicted stationary states. The process was illustrated by reference to the electronic states of atoms and diatomic molecules. We turn now to the spectroscopic characterization of states of nuclear motion.

The justification for the separate consideration of electronic and nuclear states of molecules was discussed on p. 76. So far as electronic states were concerned it was assumed that nuclei remained essentially stationary during the transitions by which electronic states were characterized. It is assumed now that changes in nuclear configuration are accompanied by a rapid adjustment of electron distribution. This redistribution is reflected in such quantities as force constant and dissociation energy — a fact which should be borne in mind even though it may not always be made explicit.

The nuclear motions with which the present chapter is concerned are vibration, rotation and nuclear spin. The theoretical predictions include adjustable structure parameters, such as bond lengths, bond angles, and factors which take account of symmetry. Consequently, once it has been established that the theory is generally valid, the spectroscopic characterization of nuclear states leads to detailed information about molecular structure, mainly in the ground state but also in excited states.

Vibrational States of Diatomic Molecules

The potential energy curve for a diatomic molecule gives the variation with internuclear distance r of the potential energy of the molecule $U(r) = E_\varepsilon + V_n$ in a given electronic state denoted by ε. It is the potential energy function required for the Schrödinger equation for vibrational motion of the molecule. The assumption that the equation for the total description of the molecular motion can be separated into nuclear and electronic parts (equation 2.20) implies here that the electronic configuration adjusts itself instantaneously as the nuclei move. Since the electrons in the molecule form the substance of the chemical bond, changes in the electronic configuration with changes in internuclear distance result in restoring forces (in the classical mechanical sense). Such forces, attractive if the bond is extended beyond its equilibrium distance and repulsive if the bond is compressed, are implied by the form of $U(r)$ for a stable molecule. In a different electronic state the electron configuration differs, as does the way it changes with internuclear distance. There is, in consequence, a different $U(r)$ for each electronic state: the curves for three such states of the bromine molecule are shown in Fig. 52.

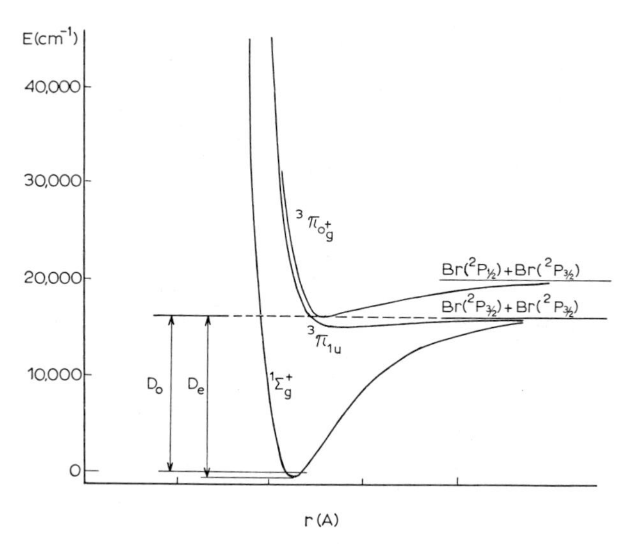

Fig. 52. Potential energy curves for the ground state and first two excited electronic states of bromine.

The shape of the potential energy curve can be determined from spectroscopic measurement, as we shall see. After laborious calculations involving improvements on the simple quantum mechanical description outlined on p. 123, the potential energy curve for hydrogen and a few other simple molecules has been fairly accurately reproduced. These calculations are, however, essentially numerical and are not suitable for including in the Schrödinger equation for vibration so as to obtain general vibrational eigenfunctions. For that purpose a number of approximate algebraic functions for $U(r)$ have been suggested from time to time. From them vibrational eigenfunctions and energy levels can be predicted in terms of parameters related to the chemists' idea of the strength of the chemical bond. Contrariwise the evaluation of these parameters (force constants and anharmonicity constants) from the spectroscopic data makes it possible to attempt to correlate them with other chemical or structural information. Spectroscopic transitions are readily observed between vibrational energy levels in the same electronic state (usually the ground state) and between different vibrational levels in different electronic states. In the first case there can only result information about one electronic state, but in the second case vibrational states can be characterized for both upper and lower electronic states.

Harmonic diatomic oscillator

The simplest form of $U(r)$ which has any similarity to the curve for a reasonably stable molecule (*e.g.* the ${}^1\Sigma_g$ state of bromine, Fig. 52) is the simple Hooke's law expression:

$$U(r) = \tfrac{1}{2}k(r - r_e)^2 = \tfrac{1}{2}kx^2 \tag{3.1}$$

where r is the internuclear distance, r_e its equilibrium value, and $x = r - r_e$ is the displacement from the equilibrium distance. Fig. 53 shows two such

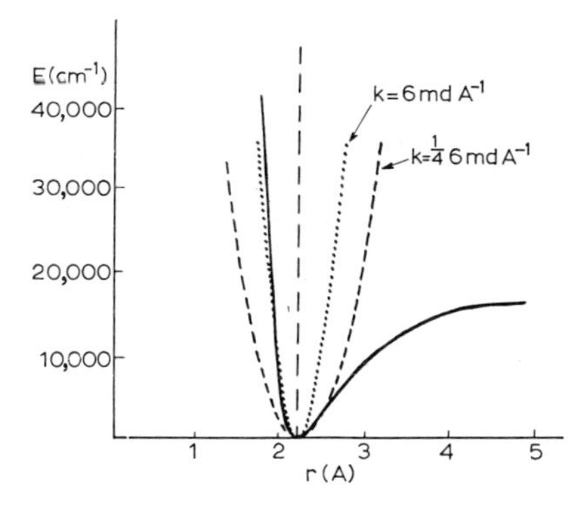

Fig. 53. Hooke's law approximation to the ground state potential curve for bromine. Harmonic oscillator potential energy functions are drawn for two different values of the force constant.

Hooke's law parabolas as approximations to the ground-state potential curve for bromine. The values of the force constant k for the two curves are 6 and 1.5×10^5 dyne cm^{-1}, or 6 and 1.5 millidyne Å^{-1}. It is evident that the approximation is only valid, if at all, for small values of x, and that the parabola is steeper for larger values of k.

In classical mechanics two masses, m_1 and m_2, connected by a Hooke's-law spring of constant k, vibrate under a potential energy (3.1) with a frequency

$$\nu_c = \frac{1}{2\pi}\sqrt{\frac{k(m_1 + m_2)}{m_1 m_2}} = \frac{1}{2\pi}\sqrt{\frac{k}{m}} \tag{3.2}$$

References p. 216–217

where m is the reduced mass. With the larger of the above two force constants a bromine molecule, having $m = \frac{1}{2}(81 \times 1.66 \times 10^{-24})$ g would be expected to vibrate with frequency $\nu_c(Br_2) = 1.5 \times 10^{13}$ sec^{-1}.

In the quantum mechanical solution the vibrational energy levels of the molecule are:

$$E_v = \hbar \sqrt{\frac{k}{m}}\,(v + \tfrac{1}{2}) \qquad (3.3)$$

or as *term values* (in cm^{-1})

$$G_v = \frac{1}{2\pi c} \sqrt{\frac{k}{m}}\,(v + \tfrac{1}{2}) \qquad (3.4)$$

where v is the vibrational quantum number $= 0, 1, 2, 3 \ldots$. The eigenfunctions associated with the first four values of v are:

$$\psi_0 = \left(\frac{\alpha}{\pi}\right)^{\frac{1}{4}} e^{-\frac{1}{2}\alpha x^2}$$

$$\psi_1 = \frac{1}{\sqrt{2}} \left(\frac{\alpha^3}{\pi}\right)^{\frac{1}{4}} 2x e^{-\frac{1}{2}\alpha x^2}$$

$$\psi_2 = \frac{1}{\sqrt{8}} \left(\frac{\alpha}{\pi}\right)^{\frac{1}{4}} (4\alpha x^2 - 2) e^{-\frac{1}{2}\alpha x^2}$$

$$\psi_3 = \frac{1}{\sqrt{48}} \left(\frac{\alpha^3}{\pi}\right)^{\frac{1}{4}} (8\alpha x^3 - 12x) e^{-\frac{1}{2}\alpha x^2}, \quad \text{where } \alpha = \sqrt{\frac{mk}{\hbar}} = 2\pi\nu_c \frac{m}{\hbar} \qquad (3.5)$$

The variation of ψ_n^2 with x for a hypothetical molecules in which $k = 4$md Å^{-1} and $m = 4.5 \times 10^{-23}$ g $= 27$ a.m.u. is drawn in Fig. 54.

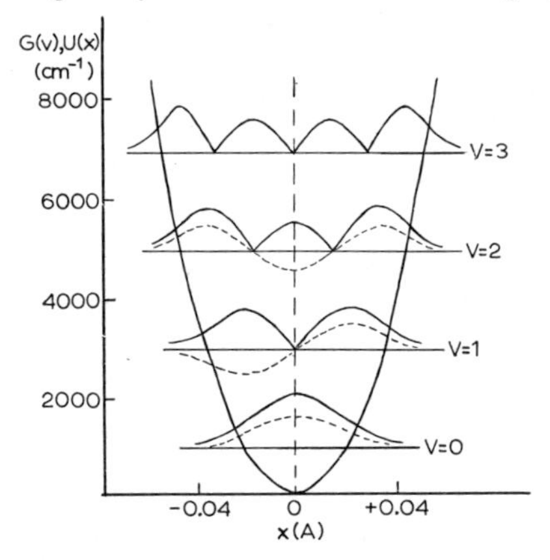

Fig. 54. Hooke's law potential energy curve and resultant vibrational eigenvalues and probability densities for a molecule with $k = 16$ md A^{-1} and $m = 7$ a.m.u.

The potential curve is, so far as quantum mechanics is concerned, merely the primary datum from which the eigenfunctions and energy levels are deduced. In classical mechanics it marks the limit of the extension and compression of the molecule possessing a given amount of energy. Quantum mechanics only permits the molecule to have energy values indicated by the horizontal lines $v = 0$, $v = 1$, $v = 2, \ldots$, and hence in classical mechanics a molecule oscillating with total energy indicated by those lines would have its turning points at values of x given by the intersections of the lines and the parabola.

Although it is still convenient to think of the molecule vibrating in this way it is not a valid description in quantum mechanics. All that can be said with precision is that the probability of the molecule having a particular internuclear distance is given by the square of the eigenfunction. Eigenfunctions and probability densities are drawn in Fig. 54 as functions of x with an arbitrary ordinate based on each level in turn. In the zero level the probability density is a maximum at $x = 0$ but in the first excited level there is zero probability of the molecule having $r = r_e$ and equal probability of it being extended or compressed to the same magnitude of x. This is the nearest that a valid quantum mechanical description can get to depicting the molecule as vibrating.

The state indicated by the minimum in $U(r)$, by which the equilibrium distance r_e is at present defined, is one in which both position and momentum of the nuclei would be exactly known. This situation is contrary to the uncertainty principle and thus the state $v = 0$ is the lowest accessible state. Its energy relative to the minimum in the potential curve is called *residual energy*, or often *zero point energy*.

Vibrational spectrum

Transitions are normally observed spectroscopically (p. 82) only when they involve a change in dipole moment (dipole transitions in absorption or emission) or a change in polarizability (Raman effect). A homonuclear diatomic molecule which has no permanent dipole will not acquire one by vibration and therefore transitions between vibrational levels in the same electronic state are not observed either in absorption or emission for such molecules as H_2, O_2, N_2, Cl_2, Br_2. Vibration of these molecules may, however, alter the polarizability and therefore appear in the Raman spectrum. Closer examination of the transition probabilities shows that both in dipole transitions and in the Raman effect, transitions in the same electronic state are limited by the selection rule:

$$\Delta v = \pm 1 \tag{3.6}$$

As an illustration of this selection rule it is instructive to examine the probabilities of dipole transitions from the level $v = 0$ to $v = 1$, 2, or 3. The dipole moment of a heteronuclear molecule must be zero at very small and at very large internuclear separations, and is therefore likely to vary with internuclear distance in the manner shown in Fig. 55. The equilibrium distance

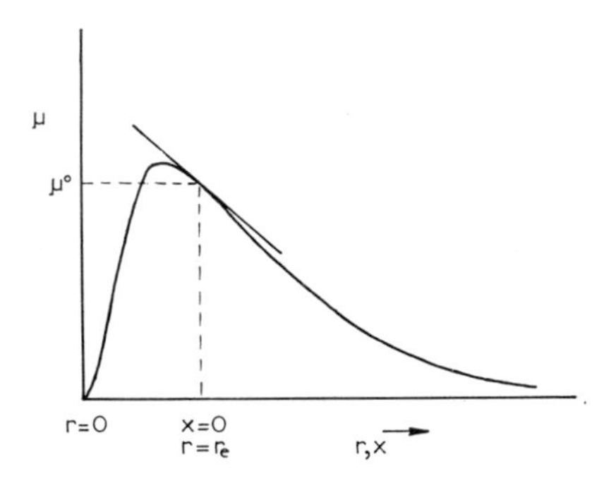

Fig. 55. Possible variation of dipole moment with internuclear distance.

may be greater than or less than the value of r for maximum dipole but provided the two distances are not equal it is possible as a first approximation, to put

$$\mu = \mu^0 + \frac{d\mu}{dx} \cdot x \tag{3.7}$$

where μ^0 is the permanent dipole moment. Now the intensity of a dipole transition between states v' and v'' depends upon the integral:

$$R_x^{v'v''} = \int_{-\infty}^{\infty} \psi^*_{v'}(x)\mu\psi_{v''}(x)dx \tag{3.8}$$

If the expressions (3.5) for the vibrational eigenfunctions ψ_0, ψ_1, ψ_2, and ψ_3 are introduced along with the linear function (3.7) it can readily be shown that $R_x^{02} = R_x^{03} = R_x^{13} = 0$ in agreement with (3.6) but that, for example

$$R_x^{01} = \frac{1}{\sqrt{2\alpha}} \frac{d\mu}{dx} \tag{3.9}$$

The quantum mechanical theory of the interaction of radiation with a system of quantized energy states gives the transition probability for absorptions as

$$P_{01} = \frac{2\pi}{3\hbar^2} |R_x^{01}|^2 = \frac{\pi}{3h\nu_c m} \left(\frac{d\mu}{dx}\right)^2 \tag{3.10}$$

where P_{01} is directly related to the observed intensity of the transition (p. 323). (3.10) shows immediately the requirement that $d\mu/dx \neq 0$ or that the dipole moment must change with vibration (in the classical sense). This further implies that a diatomic molecule must have a permanent dipole since $d\mu/dx$ would be unlikely to differ from zero if the molecule itself were non-polar, *i.e.* if $\mu^0 = 0$.

It should be noted that the prohibition on changes of vibrational quantum number greater than one depends upon (i) the assumption that the potential energy is a function of x^2 and not higher powers of x (*harmonic oscillator*) and (ii) the assumption that μ is a linear function of x. When higher powers of x are introduced into $U(r)$ (*anharmonic oscillator*, p. 137) or into μ (*electrical anharmonicity*) then transitions $\Delta v = \pm 2, \pm 3 \ldots$ become possible.

For the allowed transitions, $\Delta v = \pm 1$, it follows that:

$$\omega = G_{v+1} - G_v = \frac{1}{2\pi c}\sqrt{\frac{k}{m}} = \frac{\nu_c}{c} \tag{3.11}$$

Thus in the harmonic oscillator, in which the energy levels are equally spaced, all transitions correspond to the same frequency. The frequency of the classical oscillation is ν_c and, according to classical theory, the dipole moment oscillating with this frequency should interact with light of the same frequency. It is now shown that the quantum of light required to effect a transition from any vibrational level to an adjacent one has a corresponding frequency (or wave-number value, ω) equal to that of the classical oscillation. Accordingly the radiation frequency associated with the transition $v = 0 \rightarrow 1$ is often loosely referred to as "the vibrational frequency" of the molecule and the absorption of the appropriate quantum is spoken of as "exciting the vibration of the molecule". It should be remembered that in the quantum mechanical description the molecule can hardly be regarded as either vibrating or not vibrating in either of the states $v = 0$ or $v = 1$ and consequently cannot be asigned a vibrational frequency except by reference to the classical mechanical result. The term values (3.4) may however be simply expressed as:

$$G_v = \omega(v + \tfrac{1}{2}) \qquad v = 0, 1, 2, 3 \ldots \tag{3.12}$$

Typical of the experimental observations on heteronuclear diatomic molecules is the absorption spectrum of hydrogen chloride gas in the infra-red (Fig. 56). There is a band or collection of lines disposed almost symmetrically about a wave-number value of 2886 cm^{-1}. Another collection of lines occurs in the far infra-red (50—100 μ) with about the same spacing between the lines as occurs in the band near 3μ. It is difficult to show the similar behaviour of molecules such as CO and NO on one diagram since the lines comprising the band in the near infra-red are closer together (and if observed under sufficiently low resolution appear as a continuous broad absorption or band). The lines at low frequency are crowded much closer to zero than in HCl and cannot be distinctly drawn on the same scale. On the other hand

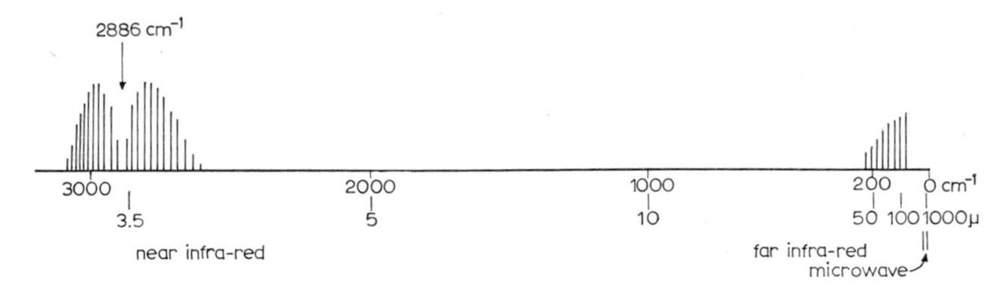

Fig. 56. Absorption spectrum of HCl (diagrammatic).

they do not form a band since the very high resolution of microwave spectrometers enables the individual lines to be separated with ease. According to (3.12) and (3.6) any allowed vibrational transition from whatever level occurs at the same wave-number ω: that is, neglecting rotation, theory suggests the occurrence of a single line although at different places for different molecules. As we shall see, rotational transitions are expected to give rise to many equally spaced lines. It is therefore reasonable, and all subsequent discussion bears this out, to identify the low frequency lines with rotational motion. The single high-frequency band arises from the single vibrational transition with rotational structure superimposed. The value corresponding to the vibrational transition alone is the *band origin* (see p. 78) located by the gap at 2886 cm^{-1} (Fig. 56). The same behaviour is observed in other heteronuclear molecules, and the following values of ω are obtained. Corresponding force constants are also given:

HF	HCl	HBr	HI	NO	CO	
3958	2886	2559	2230	1877	2145	cm^{-1}
8.85	4.82	3.85	2.93	15.5	18.6	millidyne Å^{-1}

Raman spectrum

The Raman spectrum of a diatomic molecule may also arise from transitions between vibrational states, subject to the selection rule (3.6). In this case it is the polarizability of the molecule which determines the scattering. We have seen (Fig. 35, p. 64) that an incident quantum of radiation, ω^*, may suffer an "elastic" collision and be scattered unchanged (ω^*, Rayleigh scattering) or may suffer an "inelastic" collision and be scattered with more or less energy ($\omega^* \pm \omega_v$, Raman scattering). Whereas a permanent dipole of a molecule arises from an *existing* separation between the centres of positive and negative charge, the polarizability, α, measures the dipole moment μ which is *induced* in a molecule by the action of an electric field, F; thus:

$$\mu = \alpha F \tag{3.13}$$

By analogy with (3.7) the polarizability may be expected to change with internuclear distance according to:

$$\alpha = \alpha^0 + \left(\frac{d\alpha}{dx}\right) x \qquad (3.14)$$

The intensity of the Raman transitions involving states v' and v'' depends upon the integral

$$[\alpha]^{v'v''} = F \int \psi_{v'}{}^*(x)\alpha\psi_{v''}(x)dx \qquad (3.15)$$

The integral simplifies to one involving only α^0 when $v' = v''$: it then gives the Rayleigh scattering at unchanged frequency. For $v' \neq v''$ the part of the integral containing α^0 vanishes and leaves an integral which depends upon the change of polarizability with x. This gives the Raman scattering at frequencies displaced from the incident frequency by $(E_{v'} - E_{v''})/hc$. Such scattering occurs (a) if v' and v'' differ by one unit (harmonic oscillator selection rule 3.6) and (b) if $d\alpha/dx$ differs from zero. Although homonuclear diatomic molecules have $\mu^0 = 0$ and hence $d\mu/dx = 0$, the polarizability $\alpha^0 \neq 0$ and, in general, the change with vibration $d\alpha/dx \neq 0$. Consequently the vibrations of homonuclear molecules, as well as of heteronuclear, can be observed as Raman displacements.

The same pattern of behaviour can be observed in the Raman spectrum of HCl as in the infra-red: the high-frequency displacement centred on 2886 cm^{-1} is assigned to the vibrational transition, in agreement with the infra-red line. Values of ω for other heteronuclear molecules agree with those given on p. 136 and for some homonuclear molecules the following values of ω and k are obtained (except those quoted for bromine and iodine which come from electronic spectra).

	H_2	N_2	O_2	F_2	Cl_2	Br_2	I	
ω	4160	2331	1555	892	556	322	213	cm^{-1}
k	5.15	22.4	11.4	4.46	3.19	2.45	1.70	millidyne $Å^{-1}$

The appearance of these values in the interpretation of electronic spectra is discussed on p. 141 and the rotational structure of the infra-red and Raman bands on p. 185.

Anharmonic oscillator

It is clear from Fig. 53 that the Hooke's law harmonic oscillator is not adequate for large displacements from the equilibrium distance. That it is not adequate for an accurate description of higher vibrational levels is seen from the fact that higher terms values are not the exact multiples of ω required by (3.12). Weak lines arising from transitions $v = 0 \rightarrow 2, 0 \rightarrow 3, \ldots$

can be detected in absorption by sufficiently high pressures of gases such as hydrogen chloride. The measured wave-numbers for these overtone lines in HCl are as follows:

v	0—1	0—2	0—3	0—4	0—5	
ω	2885.90	5668.05	8346.98	10923.11	13396.55	cm^{-1}
$\Delta v \cdot \omega_0$	0	5772	8658	11544	14430	cm^{-1}
difference	0	104	311	621	1033	cm^{-1}

These, and similar results, led to the use of the empirical expression:

$$G_v = \omega_e(v + \tfrac{1}{2}) - \omega_e x_e(v + \tfrac{1}{2})^2 + \omega_e y_e(v + \tfrac{1}{2})^3 + \ldots \qquad (3.16)$$

which, since it can also be derived from a potential energy function of the form:

$$U(r) = c_2(r - r_e)^2 + c_3(r - r_e)^3 + c_4(r - r_e)^4 + \ldots \qquad (3.17)$$

describes the vibrational levels of an *anharmonic oscillator*. The levels G_v are no longer equally spaced but converge as v increases. ω_e is an "equilibrium" value in the sense that (3.17) is most nearly harmonic when $(r - r_e)$ is small. $\omega_e x_e$ is always less than ω_e and usually positive. $\omega_e y_e$ is often negligible.

The transitions $\Delta v = \pm 1$ are no longer at precisely the same wave number: thus

$$\omega_0 = \omega_e - 2\omega_e x_e + \tfrac{13}{4}\,\omega_e y_e \qquad (v = 0 \rightarrow 1)$$

$$\omega_1 = \omega_e - 4\omega_e x_e + \tfrac{49}{4}\,\omega_e y_e \qquad (v = 1 \rightarrow 2)$$

and generally

$$\omega_v = \omega_e - 2(v + 1)\omega_e x_e + \left(3v^2 + 6v + \tfrac{13}{4}\right)\omega_e y_e \quad (v = v \rightarrow v + 1) \qquad (3.18)$$

Successive lines lie $2\omega_e x_e - 3(2v + 1)\omega_e y_e$ apart.

In pure vibrational spectra in absorption transitions from $v = 1, 2, \ldots$ are not strong even when $\Delta v = 1$ on account of the low population of those levels. They are more likely to be observed at high temperatures and in cases where ω_e is small. If $\omega_e x_e$ is large enough they may be readily resolved as so-called "hot bands" but in other cases they may merely appear as unresolved shoulders or contribute to the broadening of a band envelope. On the other hand transitions in which $\Delta v = \pm 1$ occur from many different values of v in electronic spectra (p. 141).

The selection rule $\Delta v = \pm 1$ no longer holds strictly for an anharmonic oscillator and weak transitions may be expected for $\Delta v = 2, 3, 4, 5 \ldots$. For absorption from the zero level

$$\omega(0-1) = \omega_e - 2\omega_e x_e + \tfrac{13}{4}\,\omega_e y_e$$
$$\omega(0-2) = 2\omega_e - 6\omega_e x_e + \tfrac{31}{2}\,\omega_e y_e$$
$$\omega(0-3) = 3\omega_e - 12\omega_e x_e + \tfrac{171}{4}\,\omega_e y_e$$
$$\omega(0-4) = 4\omega_e - 20\omega_e x_e + 91\,\omega_e y_e$$
$$\omega(0-5) = 5\omega_e - 30\omega_e x_e + \tfrac{665}{4}\,\omega_e y_e \qquad (3.19)$$

and the observed values for HCl can be represented by

$$\omega_e = 2990.18, \qquad \omega_e x_e = 52.22, \qquad \omega_e y_e = 0.079\ \text{cm}^{-1}$$

A somewhat better fit results from the values:

$$\omega_e = 2989.74, \qquad \omega_e x_e = 52.05, \qquad \omega_e y_e = 0.056\ \text{cm}^{-1};$$

and the comparison gives an idea of the accuracy of the empirical expression (3.16) containing cubic terms.

A similar analysis of the carbon monoxide absorption spectrum in the near infra-red leads to values:

$$\omega_e = 2170.21, \qquad \omega_e x_e = 13.461, \qquad \omega_e y_e = 0.0308\ \text{cm}^{-1}$$

Vibration in electronic transitions

Vibrational structure in electronic transitions can now be given a quantitative interpretation. An upper electronic state (denoted by a single prime) has electronic energy and vibrational energy expressed as the sum of the terms T'_e (electronic) and G' (vibrational):

$$T'_e + G'(v) = T'_e + \omega'_e(v' + \tfrac{1}{2}) - \omega'_e x'_e(v' + \tfrac{1}{2})^2 + \ldots \qquad (3.20)$$

Vibrational levels are sometimes referred to as *vibronic levels* when different electronic states are involved.

We are, for the present, ignoring rotational energy. It is usually sufficiently accurate to limit G to two terms: at any rate, introduction of higher terms in $(v + \frac{1}{2})$ does not introduce any new gross features in the spectrum. T'_e is the energy of the minimum in the potential curve and is generally measured from the minimum of the ground electronic state, *i.e.* $T_{\text{ground}} = 0$. A similar expression holds for the lower electronic state (denoted by double primes) and usually the vibrational parameters are different in the two electronic states because changes in the electronic distribution affect the strength of the bond between the atoms.

Provided the electronic change leads to a change in dipole moment a transition $T'_e - T''_e = \omega_{00}$ is permitted and furthermore there is now *no restriction upon Δv*. This immediately leads to complexity in the spectrum. In general, transitions between vibronic levels v' in the upper state and v'' in the lower state occur at wave number values:

References p. 216–217

$$\begin{aligned}\omega &= T'_{\rm e} - T''_{\rm e} + G'(v') - G''(v'') \\ &= \omega_{00} + (\omega'_{\rm e}v' - \omega''_{\rm e}v'') - (\omega'_{\rm e}x'_{\rm e}v' - \omega''_{\rm e}x''_{\rm e}v'') \\ &\qquad - (\omega'_{\rm e}x'_{\rm e}v'^2 - \omega''_{\rm e}x''_{\rm e}v''^2)\end{aligned} \tag{3.21}$$

Each transition thus represented is in fact accompanied by rotational change and therefore appears not as a line but as a band. The location of the origin of the band is not always easy and we shall see (p. 211) that the origin does not correspond to the point of maximum intensity in a photograph of the unresolved band. However, for a vibrational analysis it is sufficient, at least for a first attempt, to assume that the points of maximum intensity give the wave-number values which have to be fitted with (3.21).

The whole complex of bands is known as a band system and it has its origin at:

$$\omega_{00} = \omega_{\rm e1} + \tfrac{1}{2}(\omega'_{\rm e} - \omega''_{\rm e}) - \tfrac{1}{4}(\omega'_{\rm e}x'_{\rm e} - \omega''_{\rm e}x''_{\rm e}) \tag{3.22}$$

$\omega_{\rm e1} = T'_{\rm e} - T''_{\rm e}$ is not directly observed because the lowest level in each state is not $T_{\rm e}$ but the vibrational level $v = 0$, which includes the residual energy. If, however, $\omega'_{\rm e}$, $\omega''_{\rm e}$, $\omega'_{\rm e}x'_{\rm e}$ and $\omega''_{\rm e}x''_{\rm e}$ are obtained from the vibrational analysis the $\omega_{\rm e1}$ may be calculated from ω_{00}.

It can be seen from Fig. 57(a) that if both upper and lower states were harmonic oscillators with the same frequency, *i.e.* $\omega'_{\rm e} = \omega''_{\rm e} = \omega_{\rm e}$ and $\omega'_{\rm e}x'_{\rm e} = \omega''_{\rm e}x''_{\rm e} = 0$, a simple series of bands would appear:

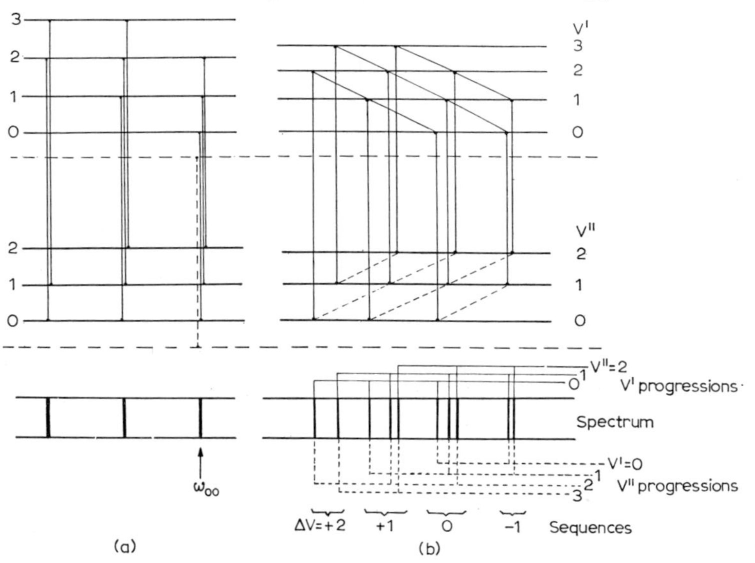

Fig. 57. Vibrational structure of electronic; spectrum (a) same, and (b) different vibrational levels in upper and lower electronic states. Progressions and sequences are indicated.

$$\omega = \omega_{00} + \omega_e(v' - v'') \tag{3.23}$$

That this is not the case is evident from the complexity of band systems. Nevertheless regularities can be noted in the vibrational structure of a band system and analysis is greatly assisted by recognition of progressions and sequences.

A *progression* is a series of bands having one level (upper or lower) in common, while a *sequence* is a group of bands arising from transitions in which Δv is constant. Fig. 57(b) shows how the transitions between the first three or four vibronic levels can be arranged as v''-progressions (each progression having a different common upper v'—level) or as v'-progressions (each progression having a different common lower v''-level) or as sequences, $\Delta v = -1, 0, +1$ or $+2$. It can be seen from the general formula (3.20) that differences between successive members of a v''-progression, *i.e.* between bands arising from the transitions v'', v' and $v'' + 1$, v', are given by the formula:

$$\Delta G(v'') = \omega''_e - 2\omega''_e x''_e(v'' + 1) \tag{3.24}$$

and similarly between successive members of a v'-progression

$$\Delta G(v') = \omega'_e - 2\omega'_e x'_e(v' + 1) \tag{3.25}$$

Second differences are:

$$\Delta G^2(v'') = \Delta G(v'') - \Delta G(v'' + 1) = 2\omega''_e x''_e \tag{3.26}$$

and

$$\Delta G^2(v') = \Delta G(v') - \Delta G(v' + 1) = 2\omega'_e x'_e. \tag{3.27}$$

In absorption at low temperatures (when all the molecules are initially in the state $v'' = 0$) the v'-progression on $v'' = 0$ should be observed and this is clearly of considerable assistance in the assignment of quantum numbers to transitions. Such a progression should conform to:

$$\omega_{0v'} = \omega_{00} + \omega'_e v' - \omega'_e x'_e(v' + v'^2) \tag{3.28}$$

Analysis of a band system

An example of the method of analysis is in absorption and emission by carbon monoxide. A band system is observed in the vacuum ultraviolet which is called the fourth positive group of carbon monoxide (the name has no special significance) and in Table 11 is drawn up a partial list* of the bands of this system. The values in square brackets have not been observed and should be ignored for the moment. The values in round brackets are observed but have not been measured with the same accuracy (1—2 cm^{-1}) as the others. The wave number values are arranged in accordance with their assignments to various v'—v'' transitions and such a table is known as a

* Taken from a full table given by Herzberg, p. 156 of Ref. 30.

TABLE 11

PART OF DESLANDRES TABLE FOR THE FOURTH POSITIVE GROUP OF CO, WITH ΔG AND ΔG^2 VALUES

v'	$\Delta G(v')$	v'' 0	$\Delta G(v'')$ 0	1	$\Delta G(v'')$ 1	2	$\Delta G(v'')$ 2	3
0		(64,703)	(2101)	62.601.8	2117.1	60,484.7	2091.5	58,393.2
	0	(1528)		1485.8		[1486.3]		1488.4
1		66,231.3	2143.7	64,087.6	[2116.6]	[61,971.0]	[2089.4]	59,881.6
	1	1443.5		1445.5		[1445.1]		1443.6
2		67,674.8	2141.7	65,533.1	2117.0	63,416.1	2090.9	61,325.2
	2	1413.0		1411.2		1412.0		[1411.6]
3		69,087.8	2143.5	66,944.3	2116.2	64,828.1	[2091.3]	[62,736.8]
	3	1381.7		1379.1		(1371)		[1379.7]
4		70,469.5	2146.1	68,323.4	(2124)	(66,199)	(2082)	64,116.5
	4	1337.7		1342.7		(1351)		1341.7
5		71,807.2	2141.1	69,666.1	(2116)	(67,550)	(2092)	65,458.2

v'	$\Delta G^2(v'')$ (0″)	$\Delta G^2(v'')$ (1″)
0	(−16)	25.6
1	[27.1]	[27.2]
2	24.7	26.1
3	27.3	[24.9]
4	(22)	(42)
5	(25)	(24)

Mean $\Delta G^2(v'') = 26.3$

$\Delta G^2(v')$	v'' 0	1	2	3
(0′)	(85)	40.3	[41.2]	44.8
(1′)	30.5	34.3	[33.1]	[32.0]
(2′)	31.3	32.1	(41)	[31.9]
(3′)	44.0	36.4	(20)	[38.0]

Mean $\Delta G^2(v') = 36.1$

Deslandres table. It will be seen that any horizontal row comprises a v''-progression, any vertical column a v'-progression and any diagonal a sequence.

The assignments are based on a number of considerations. In the first place the bands in the first column appear in absorption (Fig. 58) and since the lowest frequency band is quite intense there is little doubt that it is the lowest and that there is not another at still lower frequency too weak to be observed. Hence the first column is assigned with reasonable certainty.

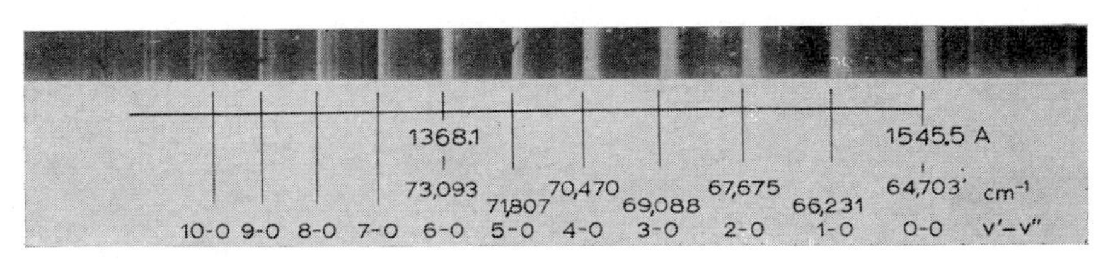

Fig. 58. Absorption spectrum of CO in vacuum ultraviolet.

According to the general formula (3.21) ω should increase initially as v' increases with constant v'' (v'-progressions), and decrease as v'' increases with constant v' (v''-progressions). The first differences $\Delta G(v'')$ (3.24) should decrease with increasing v'' but should not depend upon v'. Likewise the $\Delta G(v')$ (3.25) should decrease with increasing v' but should not depend upon v''.

Given the data of the first column the second column is obtained by picking out of the spectrum another progression whose members lie at the smallest constant wave-number interval from the v'-progression based on $v'' = 0$. The third column (v'-progression based on $v'' = 2$) is then found in the same way, and in similar manner the rest of the table is built up. Provided the first few rows and columns can be written down correctly the assignment of higher transitions is not difficult. The position of the figures in the $v'' = 1$ column is fixed by the observation of 62,601.8 cm^{-1} as the first band in the progression. If the first member ($v' = 0$) had not been found (as is the case in this band system in progressions based on $v'' \geqq 6$, *i.e.* in the seventh column and beyond) it might be thought possible to assign 64,087.6 cm^{-1} to $v' = 0$ under $v'' = 1$ and to move all the other members of that progression up one place. However, the result would not be satisfactory since the differences $\Delta G(v'')$ would not then be constant but would change with v', thus: 615, 698.2, 730.5, 764.4, 803.4 cm^{-1}.

An assignment based on such considerations is justified by its overall consistency and by the further tests that intensities of both v' and v'' progressions should vary smoothly with v' and v'' and that second differences ΔG^2 should be constant throughout the table. Table 11 includes ΔG^2 values for CO, and mean values are calculated from the more accurate differences as

$$\Delta G^2(v'') = 26.3 \text{ cm}^{-1}, \qquad \Delta G^2(v') = 36.1 \text{ cm}^{-1}.$$

The figures in the table in round brackets represent less accurate wave-number measurements. The first and second differences which depend upon them are similarly indicated, and it can be seen that one less accurate band position can affect four values of ΔG and four values of ΔG^2. The reader is invited to test the improvement in difference values when $\omega_{4,2}$ is changed from 66,199 to 66,204 cm^{-1}. It is evident that reasonably accurate parameters will only be obtained if the ω-values used are band origins (relating to rotationless states) and not band heads (see p. 140). The data given here for CO are in fact band origins.

A missing band in the Deslandres table is not necessarily an embarrassment provided the assignment of quantum numbers is possible without it. Thus in Table 11 the wave-numbers values in square brackets (for bands 1—2 and 3—3) are not observed. ω_{12} can however be deduced from the facts that:

$$\Delta G(1'') + \Delta G(2'') = \omega_{11} - \omega_{13}$$

and

$$\Delta G^2(0'') = \Delta G(0'') - \Delta G(1'') = \Delta G(1'') - \Delta G(2'') = \Delta G^2(1'')$$

The values of $\Delta G(1'') + \Delta G(2'')$ so obtained are seen to be in satisfactory agreement with other values of these differences. The calculated ω_{12} then yield satisfactory $\Delta G(0')$ and $\Delta G(1')$ values. $\Delta G(2')$ and $\Delta G(3')$ for $v'' = 3$, ω_{33} and $\Delta G(2'')$ for $v' = 3$ can be similarly calculated. Six further values of $\Delta G^2(v')$ and three of $\Delta G^2(v'')$, are obtained from these interpolations and have been used in obtaining the mean value given above.

The fourth positive group of bands in the CO emission spectrum has been identified as arising from the $A-X$ electronic transition. That is to say, the lower vibronic levels $G''(v'')$ belong to the ground state $X(^1\Sigma)$ and the upper levels $G'(v')$ belong to the first excited electronic state $A(^1\Pi)$. The parameters ω''_e and $\omega''_e x''_e$ should be directly comparable with those obtained from pure vibrational spectra in the infra-red. From (3.24) and (3.26):

$$\omega''_e x''_e = \tfrac{1}{2}\Delta G^2(v'') \qquad (3.29)$$

and

$$\omega''_e = \Delta G(v'') + (v'' + 1)\Delta G^2(v'') \qquad (3.30)$$

Referring again to Table 11, and taking in turn the averages of the accurate values of $\Delta G(0'')$, $\Delta G(1'')$ and $\Delta G(2'')$, we obtain:

$$\left.\begin{aligned} \Delta G(0'') + \Delta G^2(v'') &= 2143.2 + 26.3 = 2169.5 \\ \Delta G(1'') + 2\Delta G^2(v'') &= 2116.7 + 52.6 = 2169.3 \\ \Delta G(2'') + 3\Delta G^2(v'') &= 2090.8 + 78.9 = 2169.7 \end{aligned}\right\} = \omega''_e$$

Hence, from our analysis of the electronic band system we obtain for CO:

$$\omega''_e = 2169.5 \text{ and } \omega''_e x''_e = 13.15 \text{ cm}^{-1}$$

which compare favourably with values obtained from the pure vibrational spectrum on p. 139.

Sequences

Further assistance in the analysis of vibronic spectra comes from sequences when these are recognisable. When $\Delta v = v' - v'' =$ constant, the bands in a sequence are given by the formula:

$$\begin{aligned} \omega = \omega_{00} &+ \omega'_e \Delta v - \omega'_e x'_e(\Delta v + \Delta v^2) \\ &+ [(\omega'_e - \omega''_e) - (\omega'_e x'_e - \omega''_e x''_e) - 2\omega'_e x'_e \Delta v]v'' \\ &- (\omega'_e x'_e - \omega''_e x''_e)v''^2 \end{aligned} \qquad (3.31)$$

The wave-number values in a given sequence clearly vary with v'' but if the differences in the square bracket are small the variation with v'' will be small and all the bands of a sequence will lie close together. In the limit, when

$\omega'_e = \omega''_e = \omega_e$ and $\omega'_e x'_e = \omega''_e x''_e = 0$, all bands in a sequence merge into one band, $\omega = \omega_{00} + \omega_e \Delta v$, as depicted in Fig. 57(a). In practice sequences are observed as prominent groups of bands in a number of emission spectra. Some molecules which give rise to recognisable sequences are listed in Table 12*.

TABLE 12

VIBRATIONS PARAMETERS FROM EMISSION SPECTRA OF SOME DIATOMIC MOLECULES

		$\omega'_e - \omega''_e$	$\omega'_e x'_e - \omega''_e x''_e$	$\omega'_e x'_e$
CN (violet bands)	$B(^2\Sigma^+) \leftrightarrow X(^2\Sigma^+)$	95.42	7.37	20.25
CN (red bands)	$A(^2\Pi) \leftrightarrow X(^2\Sigma^+)$	− 254.27	− 0.26	12.88
C_2 (Swan bands)	$A(^3\Pi_g) \leftrightarrow X(^3\Pi_u)$	146.87	4.77	16.44
NO (γ-bands)	$A(^2\Sigma^+) \leftrightarrow X(^2\Pi)$	467.4	0.51	14.48
N_2 (2nd positive)	$C(^3\Pi_u) \leftrightarrow B(^2\Sigma_g^+)$	301.0	2.61	17.08
PN	$A(^1\Pi) \leftrightarrow X(^1\Sigma^+)$	− 234.15	0.239	7.222
N_2^+	$B(^2\Sigma_u^+) \leftrightarrow X(^2\Sigma_g^+)$	212.65	7.05	23.19
		all values in cm^{-1}		

The CN bands and the so-called Swan bands of C_2 occur in the spectrum of a carbon arc in air. The Swan bands also appear in the spectrum of the blue cone of the Bunsen flame or of a candle flame. Fig. 59 shows clearly

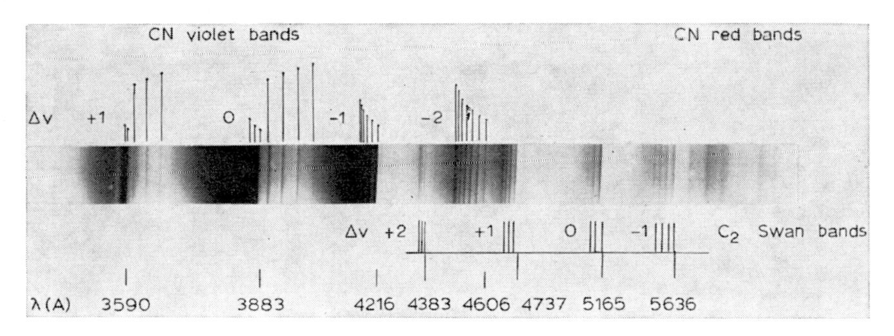

Fig. 59. Swan bands of C_2; and CN violet and red bands. Wavelengths (spectrum by Professor R. T. Birge) are given of the origins of the sequences and Δv is indicated for each sequence.

the appearance of the $\Delta v = +2, +1, 0, -1, -2$ and -3 sequences which make up the Swan bands. In most of the transitions listed the difference $\omega'_e - \omega''_e$ is considerably larger than either the difference $\omega'_e x'_e - \omega''_e x''_e$ or the absolute value of $2\omega'_e x'_e$ so that for low Δv and low v'' the direction and magnitude of change of ω with v'' is governed by $\omega'_e - \omega''_e$. Where this is positive ω increases with v''. Since the transitions from low values of v'' are

* Taken from Herzberg, appendix to Ref. 30

References p. 216–217

usually the most intense in these molecules which have well-defined sequences, the sequence for positive $\omega'_e - \omega''_e$ appears to fade away in the direction of increased frequency, or shorter wavelength. This *degrading-to-the-violet* appears again in rotational structure. It can be seen in Fig. 59 and in the other band-systems where $\omega'_e - \omega''_e$ is positive. In the spectrum of PN however, the sequences extend towards lower frequency or longer wavelength, in agreement with the negative value of $\omega'_e - \omega''_e$. In all cases, of course, the sequency $\Delta v = 1$ occurs at higher frequency than the sequences $\Delta v = 0$.

In a few special cases, the values of the vibrational parameters are such that a sequence, after starting out in one direction with increase in v'', converges sharply and then begins to move in the opposite direction. This reversal should occur when $d\omega/dv'' = 0$ or when

$$v'' = \frac{(\omega'_e - \omega''_e) - (\omega'_e x'_e - \omega''_e x''_e) - 2\omega'_e x'_e \Delta v}{2(\omega'_e x'_e - \omega''_e x''_e)} \tag{3.32}$$

In practice reversal is only observed if the relative values of $(\omega'_e - \omega''_e)$ and $(\omega'_e x'_e - \omega''_e x''_e)$ are such as to permit its occurrence within about the first ten members of the sequence. The turning point depends upon Δv and thus appears at different places in different sequences from the same molecule. The vibrational parameters for the two electronic states of CN which give rise to the violet bands are of the right magnitude for reversal of the sequences. Thus for the upper $^2\Sigma^+$ state $\omega'_e = 2164.13$, $\omega'_e x'_e = 20.25$ cm^{-1}; for the ground $^2\Sigma^+$ state $\omega''_e = 2068.705$, $\omega''_e x''_e = 13.144$ cm^{-1}; and the band system origin is $\omega_{00} = 25797.85$ cm^{-1}. Fig. 60 has been calculated from these data, and shows the variation of predicted wave-number values (open

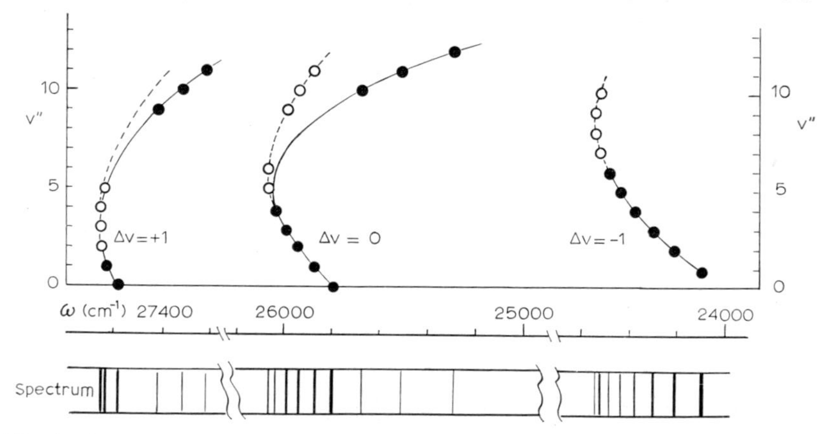

Fig. 60. Reversal of sequences in the CN violet bands. Analysis of the $\Delta v = +1$, 0 and -1 sequences shown in Fig. 59.

circles) with v'' for the three sequences $\Delta v = -1$, 0 and $+1$. In the observed spectrum[1] the reversal is sharper than is predicted: the full circles indicate the assignments for the observed bands, including the weak *tail bands* which appear on the low frequency side of the 0 and $+1$ sequences. The discrepancy between the observed and predicted values indicates the need for including a $\omega'_e y'_e$ term, which we have not taken into account.

Potential energy curves

The expression of $U(r)$ as a polynomial in x

$$U(r) = c_2 x^2 + c_3 x^3 + c_4 x^4 + \ldots \qquad (3.33)$$

provides a convenient means of tabulating spectroscopic data on diatomic molecules since it gives $G(v)$ to any required power of $(v + \frac{1}{2})$. Herzberg (Ref. 29, Table 39, pp. 501—581) has compiled a list of values of ω_e, $\omega_e x_e$ and $\omega_e y_e$, (occurring in equation 3.13) along with other structural parameters, for all the observed electronic states of all known diatomic species. On the other hand a cubic or quartic curve such as (3.33) is clearly not adequate to give the general shape of the potential energy curve in Fig. 52.

It ought to be stressed that it is not immediately obvious from the vibrational energy levels what is the precise shape of the curve. A method has been developed by Rydberg and Klein[2,3] (also described by Gaydon, Ref. 40, p. 34) by which the curve can be drawn, without any assumptions about the algebraic form, from measured vibrational and rotational levels. The method is, in essence, the reverse of the normal theoretical procedure. That is to say, it uses the quantum levels and deduces from them the classical mechanical potential energy curves. The classical turning points, r_{max} and r_{min}, are obtained for each level by a numerical procedure which takes account of the variation of both vibrational and rotational energy levels with quantum number. r_{max} and r_{min} are then marked off on each horizontal line representing a vibrational level and the potential curve is drawn through these points as in Fig. 61. Objections to the procedure are that the classical mechanical approach makes the method unreliable at the lower levels (below about $v = 5$) and, since the method also depends upon graphical estimation of slopes, the accuracy of the potential curve is limited. The first objection is not important since it is mainly at higher levels that the curve is required and the second objection is met to some extent by the analytical formulation which Rees[4] has given. However, the point is that a potential energy curve can be drawn from the data. There exists, therefore, an empirical basis for testing the accuracy of the various functions which have been proposed to represent the potential energy curves.

Any mathematical representation of a potential curve of a stable state should fulfil the following conditions:

References p. 216–217

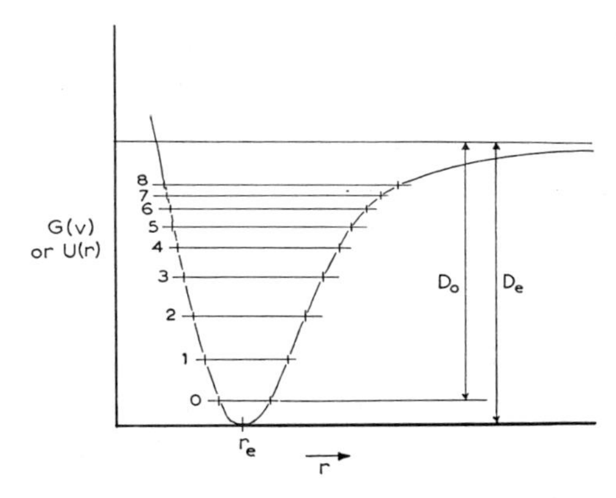

Fig. 61. Potential curve drawn through the classical turning point derived by the Rydberg-Klein method from the vibrational levels (shown) and rotational levels (not shown).

(i) at $x = 0$ or $r = r_e$; $dU/dx = dU/dr = 0$ and $d^2U/dx^2 > 0$ *i.e.* minimum potential energy at equilibrium distance.

(ii) at $x = \infty$ or $r = \infty$; $U(\infty)$ differs from the energy at the minimum by D_e, the dissociation energy of the molecule.

(iii) at $x = -r_e$ or $r = 0$; $U \rightarrow \infty$

To this may be added the convention that $U = 0$ at $x = 0$ or $r = r_e$ and hence $U(\infty) = D_e$. Of the many functions which have been proposed the following give good approximations to the actual potential energy curves of diatomic molecules:

$$U(x) = D_e \{1 - e^{-\beta x}\}^2 \qquad \text{Morse}^5 \qquad (3.34)$$

$$U(x) = D_e \{1 - (1 + \alpha x)e^{-\alpha x}\} \qquad \text{Rydberg}^6 \qquad (3.35)$$

$$U(r) = D_e \{1 - e^{-n(r-r_e)^2/2r}\} \qquad \text{Lippincott}^7 \qquad (3.36)$$

$$U(r) = D_e + \frac{a}{r^m} - be^{-nr} \qquad \text{Linnett}^8 \qquad (3.37)$$

The first three functions satisfy conditions (i) and (ii) and are defined so that $U = 0$ at the minimum and $U = D_e$ at infinite internuclear separation. They do not give $U = \infty$ at $r = 0$, but the value of $U(0)$ is usually sufficiently large for the lower part of the repulsive curve to be well represented. They each contain one arbitrary constant, in addition to D_e and r_e, and in Lippincott's function the constant n is to be evaluated from the ionization potentials, I_A and I_B, of the two atoms A and B, and the ionization potentials I_A^0 and I_B^0 of the corresponding atoms in the same row and first column of the periodic table, thus:

$$n = 6.03 \times 10^8 \left[\frac{I_A I_B}{I_A^0 I_B^0}\right]^{\frac{1}{2}} \text{cm}^{-1} \qquad (3.38)$$

Thus for AB = ClF, $I_A = 13.01$, $I_{A^0} = 5.14$ (for Na), $I_B = 17.42\, I_{B^0} = 5.39$ (for Li) (all in eV) and $n = 6.30 \times 10^8 \times 2.85 = 17.2 \times 10^8$ cm^{-1}. For hydrogen atoms I_H/I^0_H is given the value of 0.88 rather than 1.

Linnett's function is of a different type. It contains a term representing the repulsion between the atomic cores which is the same for all electronic states of a given molecule. The exponential term represents the electronic attraction and its constants (b and n) are different in different electronic states of the same molecule. The constants are not all independent for while the function gives $U = D_e$ at $r = \infty$ and $U = \infty$ at $r = 0$ it does not give $U = 0$ and $dU/dr = 0$ at $r = r_e$. Imposing that condition replaces the constants a and b by functions of r_e, so that equation (3.34) now takes the form (containing two adjustable parameters apart from D_e and r_e):

$$U(r) = D_e \left[1 + \frac{r_e}{\frac{m}{n} - r_e} \left(\frac{r_e}{r}\right)^m - \frac{m}{\frac{m}{n} - r_e} e^{-n(r-r_e)} \right] \tag{3.39}$$

The observed quantities in spectroscopy are energies of transitions between levels and from them can be deduced the relative positions of the levels themselves. We need therefore to relate the parameters of the potential functions to the empirical constants ω_e, $\omega_e x_e$, $\omega_e y_e$ of the vibrational levels $G(v)$. In this respect the Morse function is particularly convenient since the Schrödinger equation containing that form of $U(r)$ can be rigorously solved. The result is a set of vibrational terms which depend upon $(v + \frac{1}{2})$ to the first and second power and no higher:

$$G(v) = \beta \sqrt{\frac{D_e h}{2\pi^2 cm}} (v + \tfrac{1}{2}) - \frac{h\beta^2}{8\pi^2 cm} (v + \tfrac{1}{2})^2 \tag{3.40}$$

where D_e is measured, as is usual, in cm^{-1} and m is the reduced mass. Comparing this equation with the empirical expression (3.16), we see that:

$$\omega_e = \beta \sqrt{\frac{D_e h}{2\pi^2 cm}}, \qquad \omega_e x_e = \frac{h\beta^2}{8\pi^2 cm}, \qquad \omega_e y_e = 0 \tag{3.41}$$

The need to include the term $\omega_e y_e (v + \frac{1}{2})^3$ in many vibronic levels indicates that the Morse function is not entirely satisfactory. To illustrate the measure of agreement obtainable we may calculate, using (3.41), the value $\beta = 0.555 \times 10^8$ cm^{-1} = 0.555 Å^{-1}, for HCl from the observed $\omega_e x_e$ (p. 139). From the observed ω_e and the dissociation energy D_e(HCl) = 37,212 cm^{-1} obtained from thermochemical measurements, $\beta = 0.596$ Å^{-1}. From Lippincott's function

$$\omega_e = \frac{1}{2\pi} \sqrt{\frac{D_e h n}{r_e c m}} \tag{3.42}$$

and

References p. 216–217

$$\omega_e x_e = \frac{3h}{64\pi^2 cm}\left(\frac{n}{r_e} + \frac{1}{r_e^2}\right) \tag{3.43}$$

Given $r_e = 1.275 \times 10^{-8}$ cm for HCl (from rotational spectra, p. 185) and calculating $n = 9.53 \times 10^8$ cm^{-1} from equation (3.38), the anharmonicity constant for HCl is $\omega_e x_e = 50.7$ cm^{-1}, using (3.43), compared with the observed value 52.1 cm^{-1}. From n, r_e and the dissociation energy D_e(HCl) = 37,212 cm^{-1}, equation (3.42) gives $\omega_e = 3094$ cm^{-1}, compared with the observed value 2990 cm^{-1}.

It is interesting to compare the connections between ω_e and $\omega_e x_e$ which the Morse and Lippincott functions predict.

Morse:
$$\omega_e x_e = \frac{1}{4}\frac{\omega_e^2}{D_e} \tag{3.44}$$

Lippincott:
$$\begin{aligned}\omega_e x_e &= \frac{3\omega_e^2}{16 D_e} + \frac{3\omega_e^4 \pi^2 cm}{4hD_e^2 n^2} \\ &= \frac{3\omega_e^2}{16 D_e} + \frac{3h}{64\pi^2 mc r_e^2}\end{aligned} \tag{3.45}$$

The Lippincott function evidently gives a better agreement between values of $\omega_e x_e$ and D_e than does the Morse function, at the price of some loss of simplicity. The extra quantity required (r_e or n) is always either known of calculable.

The polynomial (3.33) yields:

$$\omega_e = \frac{1}{2\pi}\sqrt{\frac{2c_2}{m}} \quad \text{and} \quad \omega_e x_e = \left[\frac{15}{8}\left(\frac{c_3}{c_2}\right)^2 - \frac{3}{2}\left(\frac{c_4}{c_2}\right)\right]\frac{h}{8\pi^2 mc} \tag{3.46}$$

but naturally contains no reference to D_e.

Force constants in diatomic molecules

Anharmonicity in molecular vibrations requires a modification of the idea of force constant as it was used on p. 131. In the Hooke's law expression the force constant can readily be shown to be the proportionality constant relating the restoring force to the displacement from the equilibrium position. This is retained as the definition of the force constant, now written k_e, with the proviso that the displacement from the equilibrium position is infinitesimal. Thus:

$$\frac{dU}{dx} = k_e x \quad \text{or} \quad \frac{d^2U}{dx^2} = k_e \tag{3.47}$$

For the harmonic oscillator (3.1), $k_e = k$, independent of x. For the cubic function (3.33), $d^2U/dx^2 = 2c_2 + 6c_3 x$, and thus the restoring force is not proportional to x. At the equilibrium position however,

$$k_e = 2c_2 \tag{3.48}$$

For the Morse function:

$$k_e = 2hcD_e\beta^2 \tag{3.49}$$

For the Rydberg function:

$$k_e = 2hcD_e\alpha^2 \tag{3.50}$$

For the Lippincott function:

$$k_e = \frac{hcD_e n}{r_e} \tag{3.51}$$

Each of these equations assumes D_e in cm^{-1}

In each case

$$\omega_e = \frac{1}{2\pi c}\sqrt{\frac{k_e}{m}} \tag{3.52}$$

and thus ω_e bears the same relation to k_e as ω does to k in the harmonic approximation. It is common practice to calculate approximate force constants k_0 from the fundamental wave-number value ω_0, as we did on p. 137. To obtain k_e it is necessary to measure at least one overtone, or other vibrational transition, in order to derive both ω_e and $\omega_e x_e$. Then

$$k_0 = k_e\left[1 - \frac{2\omega_e x_e}{\omega_e}\right]^2 \tag{3.53}$$

The difference between k_e and k_0 is in the range 1 %.

From its definition it can be seen that a force constant gives a measure of the strength of the bond, but only in the region of the equilibrium position. Thus for N≡N, $k_e = 22.96$; for O=O, $k_e = 11.77$; and for F—F, $k_0 = 4.45$ *md* Å^{-1}: for P_2, S_2 and Cl_2 the corresponding values are 5.56, 4.96 and 3.29 *md* Å^{-1}. From the chemist's point of view the strength of a bond is the energy required to break it and involves the force field of the molecule at internuclear distances considerably greater than the equilibrium distance. The dissociation energy is the chemical strength of the bond and we now turn to spectroscopic methods of measuring it.

Dissociation Energy

The vibrational levels $G''(v'')$ in the ground state of a stable molecule evidently converge to a limit. This can be seen in the convergence of progressions in electronic spectra. Inspection of Fig. 61 shows that the convergence limit must correspond to the minimum number v_m of vibrational quanta required completely to overcome the restoring force and to make the

References p. 216–217

nuclei fly apart, or to make the molecule dissociate. The idea is similarly expressed in those potential curves which tend to a limiting value $U(\infty)$ at large internuclear distances. Measuring from the minimum of the potential curve we have

$$D_e = U(\infty) - U(r_e) \tag{3.54}$$

but relative to the ground vibrational state:

$$D_0 = G(v_m) - G(0) \tag{3.55}$$

Which of these two dissociation energies is measured depends upon the method. They are, of course, related by:

$$D_e = D_0 + \tfrac{1}{2}\omega_e - \tfrac{1}{4}\omega_e x_e + \tfrac{1}{8}\omega_e y_e \ldots \tag{3.56}$$

The ideal method of obtaining D_0 would be to measure the overtones — the $\omega(0, v)$ — in the vibrational absorption spectrum of the molecule up to the convergence limit. It is not possible to do this in fact because the intensity of such transitions falls off very rapidly as v increases. For a different reason, to wit the low population of levels other than $v = 0$, the intensities of the $v \rightarrow v + 1$ transitions in purely vibrational spectra are also low for $v > 0$. We require $G(v)$ values up to as high v as possible and we have seen that electronic spectra yield information about higher vibronic levels. However they necessarily also involve $G'(v')$ levels of the upper electronic state and we wish first to confine attention to the ground state.

If all levels $G(v)$ were known up to the limiting level $G(v_m)$, D_0 would be immediately found. The difference between $G(0)$ and $G(v_m)$ is also, of course, the sum of all the differences between consecutive levels. This is

$$\Delta G(v) = G(v + 1) - G(v) \tag{3.57}$$

then:

$$D_0 = \sum_{0}^{v_m - 1} \Delta G(v) = \Delta G(0) + \Delta G(1) + \qquad + \Delta G(v - 1) \tag{3.58}$$

When all the $G(v)$ are not known but only the first few this formula provides a basis for extrapolation, as was first suggested by Birge and Sponer[9] in 1926.

If the $G(v)$ can be satisfactorily represented by the formula (3.20), then it follows that:

$$\Delta G(v) = \omega(v) = \omega_e - 2\omega_e x_e (v + 1) \tag{3.25}$$

Since the limit is reached when $\Delta G(v_m) = 0$

$$v_m = \frac{\omega_e}{2\omega_e x_e} - 1 \tag{3.59}$$

and

$$D_0 = \sum_{0}^{v_m - 1} [\omega_e - 2\omega_e x_e (v + 1)] = \frac{{\omega_e}^2}{4\omega_e x_e} - \tfrac{1}{2}\omega_e \tag{3.60}$$

Hence:

$$D_e = \frac{\omega_e^2}{4\omega_e x_e} \tag{3.61}$$

a result which we have already derived on p. 150 from the expressions for ω_e and $\omega_e x_e$ which are obtained from a Morse potential function.

The algebraic procedure can be carried out graphically. For example, from the values of the fundamental and the first, second, third and fourth harmonics for HCl given on p. 138 we can deduce:

$$\begin{aligned}
\omega(0) &= \omega(0-1) &&= 2885.90 \\
\omega(1) &= \omega(0-2) - \omega(0-1) &&= 2782.15 \\
\omega(2) &= \omega(0-3) - \omega(0-2) &&= 2678.93 \\
\omega(3) &= \omega(0-4) - \omega(0-3) &&= 2576.13 \\
\omega(4) &= \omega(0-5) - \omega(0-4) &&= 2473.44
\end{aligned}$$

These values of $\omega(v)$ are plotted against v in Fig. 62(a) and the linear extrapolation gives $v_m = 27.6$. This in turn gives

$$\begin{aligned}
D_e(\text{HCl}) = G(v_m) &= 2990.18 \times 28.1 - 52.22 \times (28.1)^2 \\
&= 42{,}770 \text{ cm}^{-1} = 66.72 \text{ kcal/mole.}
\end{aligned}$$

and also

$$D_0(\text{HCl}) = 41{,}273 \text{ cm}^{-1}$$

The graphical linear extrapolation has no advantage over the algebraic except that it enables the nature of the assumption involved to be clearly seen. The extrapolation in Fig. 62(a) is a long one but the objection to its use is not that its length makes it inaccurate: a difference of one unit in the estimation of v_m in this case leads to a 0.1 % difference in D_e. A more fundamental objection is that the linear extrapolation assumes that the expression for $G(v)$ found for the first five vibrational levels satisfactorily describes all levels up to the dissociation limit. That this assumption is probably not valid in HCl is suggested by the fact that the combination of spectroscopic values of $D_0(H_2)$, $D_0(Cl_2)$ and $\Delta H_f(\text{HCl})$ gives $D_0(\text{HCl}) = 37{,}212$ cm^{-1}. This lower value implies that the upper vibrational levels converge more rapidly than do the lowest five levels.

To test the validity of expressions for $G(v)$ to higher values of v and in order to obtain better values of D_0 it is necessary to make use of band spectra in which many more vibrational levels are usually identified. In the emission spectrum of molecular hydrogen in the ultraviolet some fourteen transitions have been identified as arising from a common upper vibrational level and the ground state levels $v'' = 0$ to $v'' = 13$ (a v''-progression). Successive values of $\Delta G(v'')$ are plotted against v'' in Fig. 62(b). Observation of the first nine points would have indicated a linear extrapolation to $v_m = 18$ and

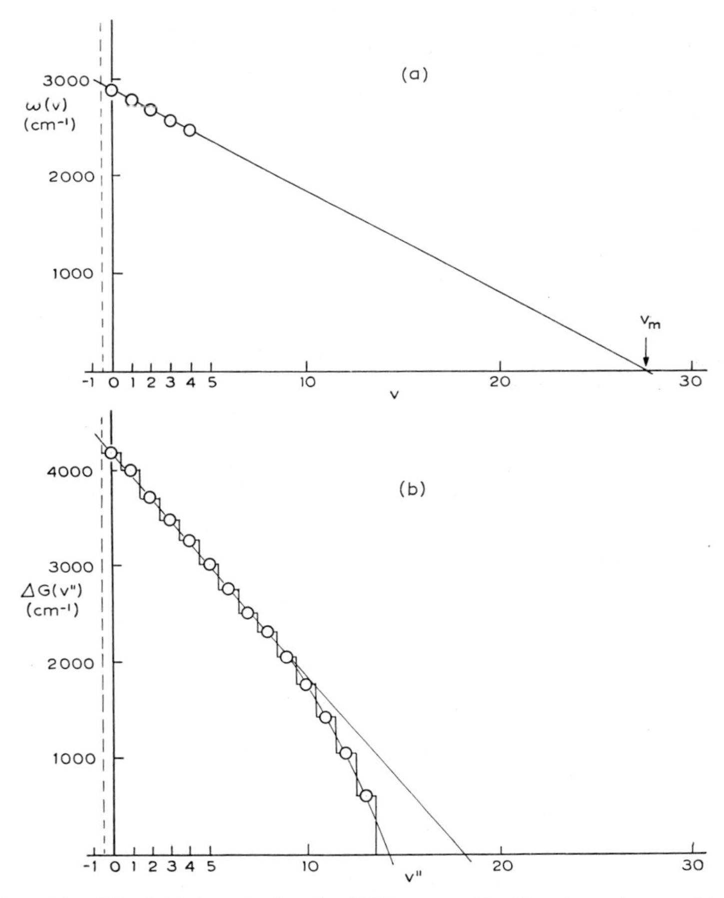

Fig. 62. (a) $\omega(v)$ plotted against v for HCl, pure vibrational spectrum. (b) $\Delta G(v'')$ plotted against v'' for the ground state of hydrogen.

$D_0(H_2) = 39{,}400$ cm^{-1}. However, the next few points show a more rapid convergence to $v_m = 14$ and hence a smaller value of D_0. Since the curvature invalidates the formula (3.59) the dissociation energy must now be obtained either graphically or with the help of a potential function which gives a better representation of the upper vibrational levels than does the Morse function. From Lippincott's function:

$$D_e = \frac{3\omega_e{}^2 \cdot 4\pi^2 cmr_e{}^2}{16\omega_e x_e \cdot 4\pi^2 cmr_e{}^2 - 3h} \tag{3.62}$$

and the same data for HCl, with the additional value of $r_e = 1.275$ Å, yield D_e (HCl) $= 34{,}870$ cm^{-1} and D_0(HCl) $= 33{,}375$ cm^{-1}. These values are too low by almost as much as the linear extrapolation is too high.

The graphical method of assessing the dissociation energy is to make use of the equation (3.58). Noting that the points lie at unit values of v it can be seen that D_0 is obtained by measuring the area under a smooth curve drawn through the points, between the limits $v = -\frac{1}{2}$ and $v = v$. The area under the curve in Fig. 62(b) gives $D_0(H_2) = 36{,}100\ cm^{-1}$. This value compares favourably with $D_0(H_2) = 36{,}120\ cm^{-1} = 4.476\ eV = 103.2$ kcal/mole obtained from direct observation of the dissociation limit. Graphical extrapolations are by no means all as well defined as Fig. 62(b). In many cases the data are sufficient to indicate a departure from linearity but still too far removed from the v-axis to enable a curve to be extrapolated with any certainty.

In some favourable cases electronic spectra are observed up to a convergence limit. The transitions which are, in principle, possible between two stable electronic states are further illustrated in Fig. 63. We postpone (p. 318) consideration of the factors which determine how much of the spectrum

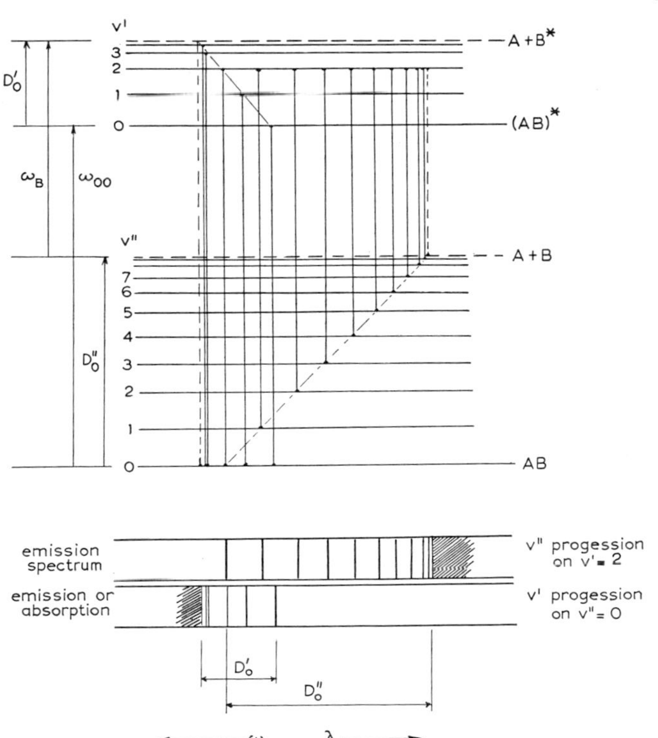

Fig. 63. Two possible progressions in transitions between stable electronic states showing dissociation limits.

is observed, and note first the information which is obtained if a complete progression can be measured. A complete v''-progression clearly spans the vibrational levels of the ground state. The convergence limit is at the low frequency (long wavelength) end and the wave-number interval between the first member of progression and the convergence limit is thus D_0'', the dissociation energy of the lower electronic state. Any emission of light of frequency lower than the convergence limit corresponds to the molecule retaining more than enough energy for dissociation. The excess of energy over D_0'' goes into kinetic energy of the dissociating fragments and is not quantized. Thus, there is a continuum of allowed energies above the dissociation limit and this is reflected in the continuum of transitions at frequencies lower than the convergence limit. A complete v'-progression, on the other hand, spans the vibrational levels of the upper state: the convergence limit occurs at the high frequency (short wavelength) end of the spectrum and the interval from the first member to the convergence limit gives D'_0, the dissociation energy of the upper state. The continuum at higher frequencies than the convergence limit now corresponds to absorption of more energy than is required to dissociate the upper state, the excess energy again going to provide the kinetic energy of the fragments. It is evident that if the convergence limits of the two sets of energy levels are different the dissociation products of the upper state must be themselves in excited states relative to the products of the lower state. In Fig. 63 we have shown the ground state as dissociating into normal ground state atoms A and B and the upper state dissociating into a normal A atom and an excited B atom, B*.

The distribution of intensity in progressions depends upon the Franck-Condon principle (p. 318) and the shapes and positions of the potential energy curves relative to one another. It is indeed unusual for the intensities to be such that a complete progression can be observed. However, so long as the convergence limit is detected and is sharp enough to be measured with precision, means can often be found of using it to assess D_0. Inspection of Fig. 63 shows that

$$\begin{aligned}\omega \text{ (convergence limit in absorption)} &= D''_0 + \omega_B \\ &= D'_0 + \omega_{00}\end{aligned} \qquad (3.63)$$

where ω_{00} is the origin of the band system and ω_B is the excitation energy of atom B to the state B*. The accepted values of the dissociation energies of oxygen, chlorine, bromine and iodine depend upon the observation of convergence limits in absorption.

In the case of oxygen the limit is at 56,850 cm^{-1} and the lowest lying levels of atomic oxygen are ^3P (ground state), ^1D (15,868 cm^{-1}) and ^1S (33,793 cm^{-1}). The ground molecular state dissociates into normal atoms but the products of dissociation of the upper molecular state must be excited. The products must be in atomic states which will combine to yield the term $^1\Sigma_u$ which has been identified as applying to the upper molecular state in this particular absorption. This restricts the possibilities to one normal atom (^3P) and either a ^1D or a ^1S oxygen atom. D_0'' (O_2) is therefore either 56,850 − 15,868 = 40,743 cm^{-1} or 56,850 − 33,793 = 23,057 cm^{-1}. Vibrational levels in the ground state have however been identified up to about 4,000 cm^{-1} above the lower of these two values, which cannot therefore correspond to D_0''. Hence D_0'' (O_2) = 40,743 cm^{-1} = 5.084 eV = 117.2 kcal/mole. Since ω_{00} = 49,363 cm^{-1} for this oxygen band system it follows that D'_0 (O_2) = 8,620 cm^{-1}.

A convergence limit in emission has been observed for hydrogen giving directly D_0'' (H_2) = 36,120 cm^{-1} = 103.2 kcal/mole. The most exact value for hydrogen comes however from a sharp long-wavelength edge to a continuous absorption at 850 Å. This continuous absorption arises merely from transition from $v'' = 0$ to the continuum above the dissociation limit and the minimum frequency (maximum wavelength) of this band of continuous absorption corresponds to D_0'' (H_2) = 36,116 cm^{-1} = 103.24 kcal/mole.

Other spectral methods of determining dissociation energies include those which depend upon the phenomenon of predissociation (important in photochemistry but not discussed here) and those in which photo-dissociation is deduced from the chemical activity of the products. In this case the maximum wavelength with which such photodissociation can be induced gives a probable upper limit to D_0''. It can happen that alternative values can be determined with great precision while the choice between the two values cannot be made with certainty. Such has been the case with nitrogen, N_2, where the choice lay between 7.373 and 9.756 eV. Electron impact studies now make it reasonably certain that the higher value is correct, *viz.* $D_0(N_2)$ = 225.8 kcal/mole.

A detailed discussion of the various spectroscopic determinations of dissociation energies is given by Herzberg[29] and by Gaydon[40]. Cottrell[39] reviews values obtained by all methods: the second edition of his book however should be consulted since it makes use of revised values of the heat of vaporization of carbon (170.9, not 138 kcal/mole) and of the heat of dissociation of nitrogen (quoted above).

Vibrational State of Polyatomic Molecules

In principle the description of the vibrational states of a polyatomic molecule follows the same lines as for a diatomic. A function is required which gives the potential energy of the mechanical model when its various atoms

References p. 216–217

are shifted from their equilibrium positions. The potential energy function completes the Hamiltonian, and enables the appropriate Schrödinger equation to be set up. In practice the potential energy function for the simplest molecule is too complex for this direct method of attack. The most general function takes account of the mutual interactions of every displacement of every atom. Even if the terms are restricted to quadratic terms and take no account of anharmonicity the number of terms is enormous, thus:

$$\begin{aligned} 2V = f_{11}x_1^2 &+ f_{22}y_1^2 + f_{33}z_1^2 + f_{44}x_2^2 + f_{55}y_2^2 + f_{66}z_2^2 + \ldots \\ &+ 2f_{12}x_1y_1 + 2f_{13}x_1z_1 + 2f_{14}x_1x_2 + 2f_{15}x_1y_2 + 2f_{16}x_1z_1 + \ldots \\ &+ 2f_{23}y_1z_1 + 2f_{24}y_1x_2 + 2f_{25}y_1y_2 + 2f_{26}y_1z_2 + \ldots \\ &+ 2f_{34}z_1x_2 + 2f_{35}z_1y_2 + 2f_{36}z_1z_2 + \ldots \\ &+ 2f_{45}x_2y_2 + 2f_{46}x_2z_2 + \ldots \\ &+ 2f_{56}y_2z_2 + \ldots \\ &+ \ldots \end{aligned} \qquad (3.64)$$

The f_{11}, f_{22}, . . . are constants and the x_1, y_1, z_1 are the three components of the displacement of one atom. Even for a diatomic molecule there are 21 terms. Many of the constants are zero because the potential energy depends only upon internal coordinates (relative displacement) and not upon the rotary and translatory motion of the molecule as a whole. This has the effect of reducing (3.64) for a diatomic molecule to:

$$2V = f_{11}x_1^2 + f_{44}x_2^2 + 2f_{14}x_1x_2 \qquad (3.65)$$

It can be shown that, in general, the number of terms is reduced to $\frac{1}{2}(3N - 6)(3N - 5)$.

Consideration of symmetry further reduces the complexity of (3.64). If $f_{11} = f_{44} = -f_{14} = k$, equation (3.65) becomes

$$2V = k(x_1 - x_2)^2 \qquad (3.66)$$

which is the same as the equation (3.1) for the harmonic oscillator.

An N-atomic molecule, considered as a collection of point masses obeying Newtonian mechanics, has $3N$ degrees of freedom, of which three are taken up with the translation of the whole molecule and another three with its rotation. (If the molecule is linear only two degrees of freedom are taken up with rotation.) There are thus $3N - 6$ (or $3N - 5$ if the molecule is linear) degrees of freedom whose exercise constitutes a distortion of the molecule. Since there are evidently restoring forces which resist such distortion, the mechanical model will vibrate. In the application of the classical equations of motion to the $3N$ particles, there are $3N - 6$ (or $3N - 5$) solutions representing periodic motion of the particles and six (or five) which have no periodicity.

Normal coordinates

For setting up the Schrödinger equation the general potential field represented by equation (3.64) or even the simplest expression (3.66), has the disadvantage that it contains cross-terms: that is, terms such as x_1x_2 or x_1y_1 occur as well as the square terms x_1^2 or x_2^2. It is desirable that new coordinates should be chosen in order that both the kinetic energy T and the potential V contain only square terms. If it is possible to choose coordinates, $q_1, q_2, q_3 \ldots q_{3N}$ to replace $x_1, y_1, z_1 \ldots x_N, y_N, z_N$ so that

$$2T = \dot{q}_1^2 + \dot{q}_2^2 \ldots + \dot{q}_{3N}^2, \quad \dot{q}_i = \mathrm{d}q_i/\mathrm{d}t \tag{3.67}$$

and

$$2V = \lambda_1 q_1^2 + \lambda_2 q_2^2 + \ldots + \lambda_{3N}\dot{q}_{3N}^2 \tag{3.68}$$

then, on conversion, the Hamiltonian operator becomes

$$-\frac{h^2}{8\pi^2}\left[\frac{\partial^2}{\partial q_1^2} + \frac{\partial^2}{\partial q_2^2} + \ldots +\right] + \tfrac{1}{2}\left[\lambda_1 q_1^2 + \lambda_2 q_2^2 + \ldots\right] \tag{3.69}$$

and the total Schrödinger equation is then the sum of $3N$ separate equations in each of the $3N$ coordinates. Only if there are no cross terms can the total equation be separated. Properly chosen coordinates $q_1 \ldots q_{3N}$ are called *normal coordinates*.

We have discussed the diatomic molecule without specific reference to normal coordinates. Now, it will serve to illustrate the procedure which is required for polyatomic molecules by examining the six normal coordinates of a diatomic molecule with atomic mass m_1 and m_2. We will test the following coordinates as possible normal coordinates:

$$\begin{aligned}
q_1 &= \frac{\alpha_1\alpha_2}{\beta}(x_1 - x_2) \\
q_2 &= \frac{\alpha_1^2}{\beta}x_1 + \frac{\alpha_2^2}{\beta}x_2 \\
q_3 &= \frac{\alpha_1\alpha_2}{\beta}(y_1 - y_2) \\
q_4 &= \frac{\alpha_1^2}{\beta}y_1 + \frac{\alpha_2^2}{\beta}y_2 \\
q_5 &= \frac{\alpha_1\alpha_2}{\beta}(z_1 - z_2) \\
q_6 &= \frac{\alpha_1^2}{\beta}z_1 + \frac{\alpha_2^2}{\beta}z_2
\end{aligned} \tag{3.70}$$

where $\alpha_1 = \sqrt{m_1}$, $\alpha_2 = \sqrt{m_2}$, and $\beta = \sqrt{m_1 + m_2} - \sqrt{\alpha_1^2 + \alpha_2^2}$.

It can readily be shown that

$$\dot{q}_1^2 + \dot{q}_2^2 + \ldots + \dot{q}_6^2 = m_1(\dot{x}_1^2 + \dot{y}_1^2 + \dot{z}_1^2) + m_2(\dot{x}_2^2 + \dot{y}_2^2 + \dot{z}_2^2) = 2T \tag{3.71}$$

and that:

$$\begin{aligned}
&\lambda_1 q_1^2 + \lambda_2 q_2^2 + \ldots + \lambda_6 q_6^2 = k(x_1 - x_2)^2 - 2V \\
&\text{if } \lambda_1 = \frac{m_1 + m_2}{m_1 m_2}k \text{ and } \lambda_2 = \lambda_3 = \lambda_4 = \lambda_5 = \lambda_6 = 0
\end{aligned} \tag{3.72}$$

References p. 216–217

Thus the normal coordinates (3.70) satisfy the requirements of (3.67) and (3.68).

The significance of the coefficients λ may be seen if the equations of motion are written down for the two particles in the x direction, it being assumed that the molecule lies parallel to the x-axis and that therefore the restoring force acts in this direction. The value of the force acting on the first atom is $-\partial V/\partial x_1$: and on the second $-\partial V/\partial x_2$. The equations of motion are therefore

$$-k(x_1 - x_2) = m_1 \ddot{x}_1 = \alpha_1^2 \ddot{x}_1$$

and

$$+k(x_1 - x_2) = m_2 \ddot{x}_2 = \alpha_2^2 \ddot{x}_2 \tag{3.73}$$

Now from (3.70)

$$x_1 = \frac{\alpha_2 q_1 + \alpha_1 q_2}{\alpha_1 \beta} \quad \text{and} \quad x_2 = \frac{\alpha_2 q_2 - \alpha_1 q_1}{\alpha_2 \beta} \tag{3.74}$$

and therefore equations (3.73) become

$$-k(\alpha_1^2 + \alpha_2^2) q_1 - \alpha_1^2 \alpha_2^2 \ddot{q}_1 - \alpha_1^3 \alpha_2 \ddot{q}_2 = 0$$
$$+k(\alpha_1^2 + \alpha_2^2) q_1 + \alpha_1^2 \alpha_2^2 \ddot{q}_1 - \alpha_1 \alpha_2^3 \ddot{q}_2 = 0 \tag{3.75}$$

These separate into:

$$\ddot{q}_2 = 0 \tag{3.76}$$

and

$$k\beta^2 q_1 + \alpha_1^2 \alpha_2^2 \ddot{q}_1 = 0 \tag{3.77}$$

Equation (3.77) is the equation for simple harmonic motion and has a periodic solution:

$$q_1 = a_1 \cos(\lambda_1^{\frac{1}{2}} t + b) \tag{3.78}$$

where

$$\lambda_1 = \frac{\beta^2}{\alpha_1^2 \alpha_2^2} k = \frac{m_1 + m_2}{m_1 m_2} k \tag{3.79}$$

The vibration frequency ν_1 is given by

$$2\pi\nu_1 = \lambda_1^{\frac{1}{2}} = \sqrt{\frac{m_1 + m_2}{m_1 m_2}}\, k^{\frac{1}{2}} \tag{3.80}$$

a result which we have already used on p. 131 (equation 3.2).

The equation (3.76) does not have a periodic solution: *i.e.* $\lambda_2 = 0$. Similarly, since there is no restoring force in the x and y directions, the equations of motion in those directions lead to

$$\ddot{q}_3 = \ddot{q}_4 = \ddot{q}_5 = \ddot{q}_6 = 0$$

yielding four more non-periodic solutions. The conditions for λ given in equation (3.72) therefore imply that a diatomic molecule (linear) has $3N - 5 = 1$ vibrational motion and five non-vibrational. It can readily be seen from the form of the normal coordinates that q_2, q_4 and q_6 are concerned with translation of the molecule and that q_3 and q_5 represent rotation in the xy and xz planes respectively.

We will not attempt a proof here but it can be shown that there always exists a set of normal coordinates which (a) will transform a potential energy function, originally quadratic in cartesian coordinates but containing cross terms, into a function containing only square terms, and (b) will transform the kinetic energy function into one containing only square terms in momenta. It is important to know that such a set of normal coordinates does exist because further general conclusions follow which are quite independent of the task of finding the normal coordinates in any special case. The general theory[11,32] shows that there are $3N$ independent modes of nuclear motion, each one in terms of its own normal coordinate. Six of these (or five if the molecule is linear) are the non-periodic modes of translation and vibration. The remaining $3N-6$ (or $3N-5$) are independent simple harmonic variations of the several normal coordinates: they are referred to as the *normal modes* of vibration. In a normal vibration all the atoms move in simple harmonic motion with the same frequency but in general with different amplitudes. Any overall motion of the molecule can be represented by superposition of the normal modes and the five or six *non-genuine vibrations* of zero frequency.

Each normal mode is represented by an equation such as (3.78) which shows the periodic variation of a single coordinate q_1 with an amplitude a_1. For pictorial representations of the motion of the molecule when any one mode is activated, the normal coordinate must be resolved into the various displacement coordinates of which it is composed. Thus, for the diatomic molecule, equations (3.74) show that two normal modes (one non-genuine) contribute to the variation in x_1 and x_2. If q_2 is set equal to zero, equation (3.70) gives the relative variations in x_1 and x_2 when the vibrational mode q_1 alone is excited. The amplitudes are $\alpha_2 a_1/\alpha_1\beta$ and $-\alpha_1 a_1/\alpha_2\beta$ respectively. Thus the displacements shown for HF in Fig. 64 should be drawn in the ratio α_2^2: α_1^2 or 19:1. If q_1 is set equal to zero in equations (3.70) x_1 and x_2 are equal and of the same sign: this corresponds to the translatory mode q_2.

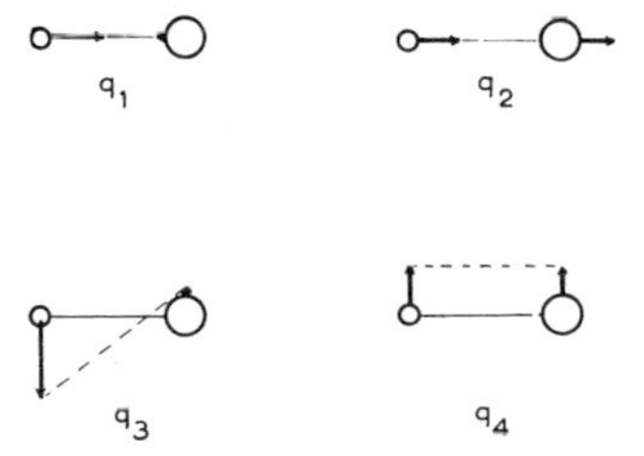

Fig. 64. One vibrational mode (q_1) and three non-periodic modes (q_2, q_3, q_4) for HF.

The same procedure for y_1 and y_2 in the normal coordinates q_3 and q_4 leads to the rotation and translation shown.

In the typical triatomic molecules CO_2 (linear) and SO_2 (non-linear) the normal modes and the rotational motions are illustrated in Fig. 65. For SO_2

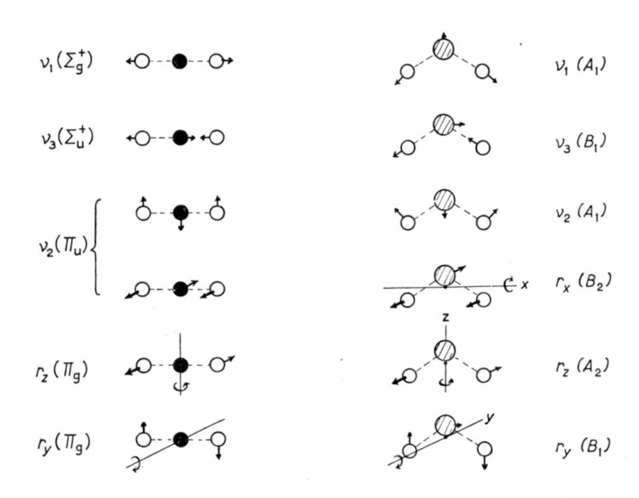

Fig. 65. Normal modes and rotation of CO_2 and SO_2.

there are the expected $3N - 6 = 3$ normal modes with three frequencies ν_1, ν_2, and ν_3. The linear molecule CO_2 has $3N - 5 = 4$ modes of vibration but it is noteworthy that two of the modes — the deformation or "flapping" modes — have the same frequency ν_2. Comparison with the adjacent representations for SO_2 shows that one of the deformation modes becomes rotation when the equilibrium configuration of the molecule is bent in the plane perpendicular to the normal coordinate. For any linear molecule there will always be one, and only one, deformation mode which would become a rotation in this way: hence the $3N - 5$ rule for linear molecules.

The two deformations of CO_2 have the same frequency by reason of the symmetry of the molecule. They are said to be *degenerate*. The two normal coordinates, say q_{2y} and q_{2z}, have the same λ_2. The q_{2y} is a function of displacements y_1, y_2 and y_3 of the three nuclei and q_{2z} is an identical function of z_1, z_2 and z_3. The equations of motion will accordingly also be satisfied by any linear combination of the displacements, *viz.* $ay_1 + bz_1$, $ay_2 + bz_2$, $ay_3 + bz_3$. The resulting motion is still simple harmonic, with frequency ν_2, but the nuclei execute circular or elliptical figures rather than straight-line displacements.

Quantum mechanical description

The existence, in principle, of normal coordinates for any molecule for which a quadratic potential function is valid enables the Schrödinger equations for all such molecules to be separated into $3N$ separate equations, one in each of the normal coordinates. Since for each normal coordinate, q_i, of a vibrational mode the contributions to the kinetic energy and potential energy are

$$T_1 = \tfrac{1}{2}\dot{q}_1^2 \text{ and } V_1 = \tfrac{1}{2}\lambda_1 q_1^2 \qquad (3.81)$$

the quantum mechanical solution for each mode is analogous to the solution for the diatomic harmonic oscillator, equation (3.3):

$$E_{v_1} = (v_1 + \tfrac{1}{2})\, h \frac{\lambda_1^{1/2}}{2\pi} = (v_1 + \tfrac{1}{2})\, h\nu_1$$

or

$$G(v_1) = (v_1 + \tfrac{1}{2})\omega_i \qquad (3.82)$$

The vibrational eigenfunctions for each mode are functions similar to (3.5) in which the normal coordinate q_1 replaces $\sqrt{mx}$ (the $\sqrt{m}$ is included because the normal coordinates are already mass-weighted).

A characteristic vibrational state of a molecule is only properly specified if each of the quantum numbers, $v_1, v_2, v_3 \ldots$ is given. The total energy is then the sum:

$$E_{v_1} + E_{v_2} + E_{v_3} + \ldots$$

or

$$G(v_1 v_2 v_3) = G(v_1) + G(v_2) + G(v_3) + \ldots \qquad (3.83)$$

and the total eigenfunction is the product of the eigenfunctions for each normal mode.

The various terms for SO_2 are illustrated in Fig. 66, the numbers in brackets giving the quantum numbers $(v_1 v_2 v_3)$. It will be noticed that the residual energy of the (000) level is 1515 cm^{-1} = 4.30 kcal/mole: in larger molecules the residual energy may be considerable, *e.g.* 27 kcal/mole for methane and 124 kcal/mole for benzene. Quanta corresponding to the transitions 000—100, 000—010, 000—001 have wave numbers ω_1, ω_2, and ω_3. Their frequencies, ν_1, ν_2, and ν_3 correspond (equation 3.82) to the frequencies of the normal modes in the mechanical model. It is thus possible, as in the diatomic case, to speak *as if* the absorption of a quantum of frequency ν_1 were exciting the molecule from a vibrationaless state to one in which it vibrates with frequency ν_1 in the normal mode q_1. The bands at ω_1, ω_2, ω_3 are the *fundamental* bands. The spectrum of SO_2 also provides examples of a *harmonic* or *overtone*, $2\omega_1$ (000—200), a *combination* band, $\omega_1 + \omega_3$ (000—101), and a *difference band* $\omega_1 - \omega_2$ (010—100). $2\omega_1$ = 2302 cm^{-1} and the band is observed at 2305 cm^{-1}: $\omega_1 + \omega_3$ = 2572 cm^{-1} compared with 2490

References p. 216–217

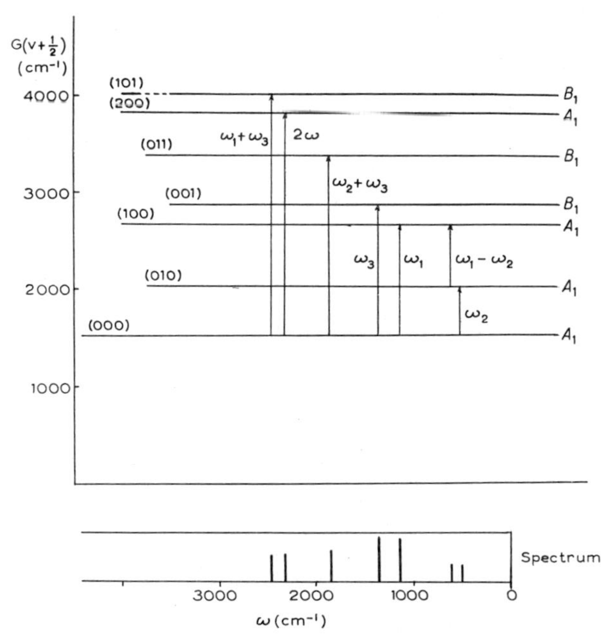

Fig. 66. Vibrational terms for SO_2 and observed transitions. The terms are labelled with values of quantum numbers $(v_1 v_2 v_3)$ and with species symbols.

cm^{-1} observed: $\omega_1 - \omega_2 = 632$ cm^{-1} compared with 606 cm^{-1} observed. Some discrepancies are to be expected on the grounds of uncertainty in the location of the origin of some weak bands and of anharmonicity. Recently much more accurate measurements of band origins have enabled a start to be made in the study of anharmonicity in polyatomic molecules.

Symmetry of the mechanical model

Vibrational analysis derives considerable assistance from arguments based upon the symmetry of the molecule and of its vibrational states. It is necessary first to recognise the symmetry elements of any conceivable molecular configuration. For example SO_2 has but three elements: there is a twofold axis of symmetry which is designated C_2 and two planes of symmetry, labelled $\sigma(xz)$ and $\sigma(yz)$. It is customary to show an axis of symmetry in the vertical position and consequently the two planes are both vertical. All this is evident from Fig. 67.

A set of such symmetry elements constitutes a *point group*. This particular set comprises the point group C_{2v}. The notation we have used is the co-salled Schoenflies notation. An alternative description (Hermann-Mauguin) denotes a twofold axis by the symbol 2 and a plane of symmetry (mirror plane)

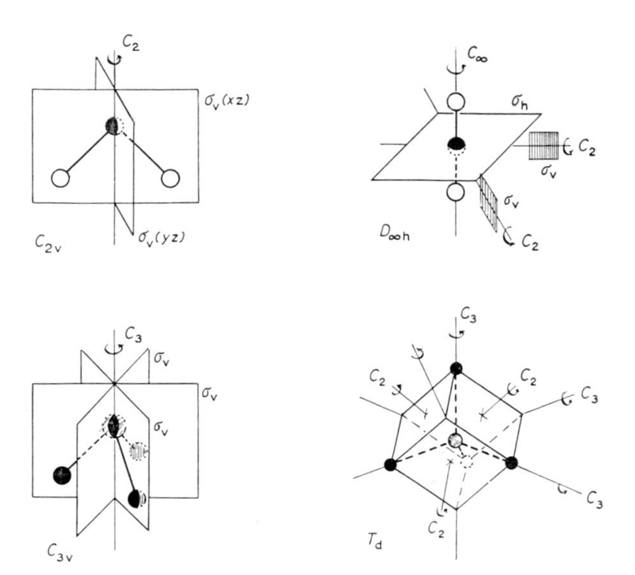

Fig. 67. Symmetry elements in SO_2, CO_2, NH_3 and CH_4. Rotation axes are denoted C_2, C_3 or C_∞: planes of symmetry are labelled σ.

by m. The point group is then denoted $2m$ or mm: only two elements of symmetry are indicated in the symbol because the third is a necessary consequence of the occurrence of the two.

Linear CO_2 has an infinite number of two-fold axes C_2, an ∞-fold axis C_∞, a plane σ_h perpendicular to the C_∞ axis, and an infinite number of planes parallel to that axis (Fig. 67). It also has a centre of symmetry. The Schoenflies notation for this point group to which CO_2 belongs is $D_{\infty h}$: the Hermann-Mauguin symbol is ∞/m. A homonuclear diatomic molecule belongs to the same point group. A heteronuclear diatomic and any unsymmetrical linear molecule has only the infinite number of σ_v and the ∞-fold axis, C_∞: it belongs to point group $C_{\infty v}$ or ∞m. Pyramidal NH_3 has a three-fold axis C_3, and three σ_v: point group, C_{3v}, or $3m$. Tetrahedral CH_4 has four C_3, three C_2, six σ, and three four-fold rotation-reflection axes, S_4 or $\bar{4}$: it does *not* have a centre of symmetry. The point group is T_d, or $\bar{4}3m$.

Having established the point group to which a chosen model belongs the next step is to examine the way in which the normal modes behave under the various symmetry operations which comprise the group. The potential energy (3.81) varies as the square of the normal coordinate and must remain unchanged when any symmetry operation is performed upon the molecule. Evidently two ways in which this can be achieved are for the operation either to leave the normal coordinate unchanged or to change its sign.

References p. 216–217

Comparison of Fig. 65 and 67 should make it clear that the displaced configurations in q_1 and q_2 (the ends of the arrows) have the same symmetry elements as the equilibrium position: that is to say, the displacements are symmetrical with respect to each of the operations, C_2, $\sigma_V(xz)$ and $\sigma_V(yz)$ of the point group. q_3, on the other hand, is affected by these three operations in the manner shown in Fig. 68: C_2 and $\sigma_V(yz)$ change the sign of the normal

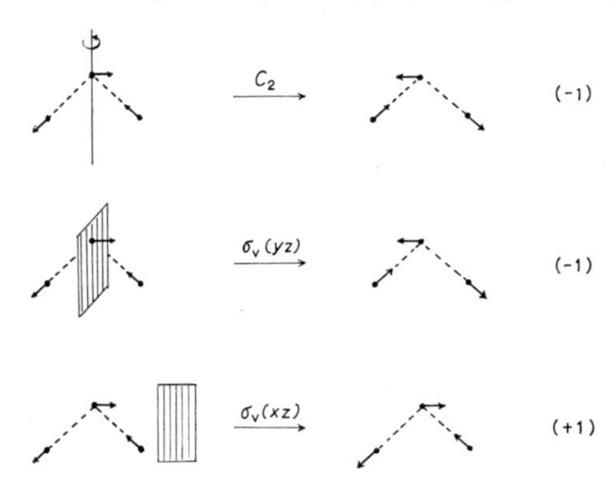

Fig. 68. Effect of symmetry operations C_2, $\sigma_V(xz)$ and $\sigma_V(yz)$ on the normal mode q_3 of SO_2.

coordinate (— 1) and $\sigma_V(xz)$ leaves it unchanged (+ 1). The reader should have no difficulty in verifying that the rotation r_Y of SO_2 in Fig. 65 is similarly affected by the three symmetry operations; that for r_Z the operations C_2, $\sigma_V(xz)$ and $\sigma_V(yz)$ have the characters + 1, — 1 and — 1; and that for r_X the operations have the characters — 1, — 1 and + 1.

The normal modes, the rotations, and also the translations, of the molecule can thus be classified. If they behave similarly to all the symmetry operations of the group they belong to the same *class* or *species*. The factors + 1 and — 1 which form a set showing how the modes in any class behave under the various symmetry operations are called the *characters* of the species. The species are labelled, according to convention with the letter A if they are symmetric, and B if they are antisymmetric with respect to a rotation axis. The subscript 1 is generally used if the species is symmetric (and 2 if it is antisymmetric) with respect to the plane of the molecule: some ambiguity arises here which, in practice, is unimportant. The species whose characters are + 1 throughout, *i.e.* A_1, is *totally symmetric*.

With these definitions we may now draw up a *character table* (Table 13), based on SO_2 but appropriate to any molecule belonging to the point group C_{2V}. The columns are headed by the symbols for the symmetry operations of the group. These include, for completeness, the identity operation I which

TABLE 13

CHARACTER TABLE FOR THE POINT GROUP C_{2v}

	I	C_2	$\sigma_v(XZ)$	$\sigma_v(YZ)$		for XY_2
A_1	+1	+1	+1	+1	t_z	ν_1, ν_2
A_2	+1	+1	−1	−1	r_z	
B_1	+1	−1	+1	−1	r_y, t_x	ν_3
B_2	+1	−1	−1	+1	r_x, t_y	

leaves all positions unchanged and therefore has the character +1. The characters for the four possible species (A_1, A_2, B_1 and B_2) of the group are given in rows. We have already seen that the normal modes for SO_2 and the rotations belong to the species indicated both in the table and in Fig. 65. The species of the translations are also given in Table 13.

The species for rotation and translation are, of course, the same for any molecule belonging to the same point group. The vibrational modes of all XY_2 molecules with C_{2v} symmetry are of the species given for SO_2. Other molecules of C_{2v} symmetry have more normal modes: for example the nine fundamentals of methylene dichloride, CH_2Cl_2, are distributed over the species as $4A_1 + A_2 + 2B_1 + 2B_2$.

A similar account could be given of the effect of the appropriate symmetry operations on unsymmetrical linear and diatomic molecules, point group $C_{\infty v}$, and on symmetrical linear and homonuclear diatomic molecules, $D_{\infty h}$. The one vibrational mode of a diatomic molecule will be seen to be totally symmetric. The symmetrical "breathing" mode of CO_2 (ν_1, Fig. 65) is also symmetric with respect to all operations of the group.

The general convention for labelling the species is not followed in the description of linear molecules. We shall see that the species symbols are essentially term symbols for vibrational states. By analogy with the symbols used for the electronic states of diatomic molecules (p. 118) the Greek capitals Σ, Π, Δ . . . are also used for the species of the vibrational states of diatomic and linear molecules. Thus the totally symmetric mode belongs to the species Σ. Being symmetric with respect to any of the σ_v planes this species is labelled Σ^+; being symmetric with respect to inversion through the centre of symmetry it is labelled Σ_g^+ (g for *gerade* = even). The antisymmetric mode ν_3 of CO_2 is antisymmetric with respect to the centre and is therefore of species Σ_u^+ (u for *ungerade* = odd). In unsymmetrical linear molecules, the g, u subscripts are no longer appropriate since there is no centre of symmetry.

The bending vibration ν_3 of CO_2 calls for comment since it provides an example of a degenerate mode. We have seen that it is doubly degenerate:

References p. 216 217

character tables give the character 2 for the identity operation, I, and -2 for inversion through the centre, i (Fig. 69). This merely signifies that the

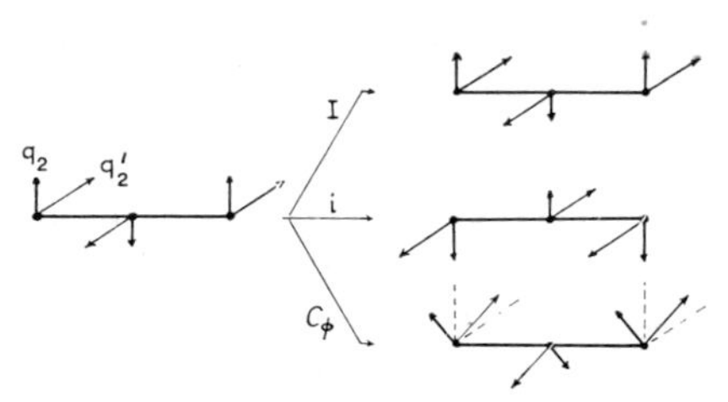

Fig. 69. Effect of I, i and C_ϕ on the degenerate bending mode q_2 of CO_2.

vibration is represented by a linear combination of two normal coordinates $q_{2y} + q_{2z}$. The degeneracy arises only in molecules having a more-than-two-fold symmetry axis, and in the following manner. The contribution to the potential energy of the two normal coordinates in question is $\frac{1}{2}\lambda_{2y}q^2{}_{2y} + \frac{1}{2}\lambda_{2z}q^2{}_{2z}$. The effect of the operation C_ϕ (rotate through any angle ϕ about the C_∞ axis, Fig. 69) is thus:

$$q_{2y} \rightarrow q^1{}_{2y} = q_{2y} \cos \phi + q_{2z} \sin \phi$$
$$q_{2z} \rightarrow q^1{}_{2z} = q_{2z} \cos \phi - q_{2y} \sin \phi \qquad (3.84)$$

Only if $\lambda_{2y} = \lambda_{2z} = \lambda_2$ does the potential energy after the operation, $\frac{1}{2}\lambda_2(q'_{2y}{}^2 + q'_{2z}{}^2)$ equal $\frac{1}{2}\lambda_2(q^2_{2y} + q^2_{2z})$. Hence the frequencies of the two modes must be equal. The character of the operation is given in the character tables as $2 \cos \phi$.

An appropriate linear combination of q_{2y} and q_{2z} results in a rotary motion of each atom about the C_∞ axis. The mode can thus give rise to *vibrational angular momentum*, which enters into the discussion of rotational structure. The species symbol Π is therefore appropriate. The species of doubly degenerate modes in molecules other than linear is always denoted by E (with or without subscripts). Two normal frequencies of ammonia are doubly-degenerate. Fig. 70 shows the displacements for one pair of degenerate vibrations. Linear combinations can here lead to circular motion of the nitrogen atom: the hydrogen atoms remain in more-or-less straight line oscillation.

In molecules of higher symmetry (tetrahedral and octahedral) triple degeneracy may arise: species symbol F. Methane has its normal modes distributed as $A_1 + E + 2F_2$: if account is taken of the fact that E represents two modes and each F represents 3 the total is 9 $(= 3 \times 5 - 6)$.

Symmetry of vibrational states

The preceding discussion has been concerned with the symmetry of molecular models at rest in the equilibrium configuration and with the symmetry

Fig. 70. One pair of degenerate (E) vibrations of NH_3 and the resultant combination.

of the mechanical vibration of that model. The symmetry of the characteristic vibrational states themselves is to be deduced from the form of the eigenfunction. We have seen (p. 163) that the total eigenfunction is the product of the eigenfunctions for each normal mode and these have the form of equations (3.5) on p. 132. This the ground state of SO_2 is described by:

$$\begin{aligned}\psi_{000} &= \psi_0(q_1) \cdot \psi_0(q_2) \cdot \psi_0(q_3)\\ &= C \exp\{-\tfrac{1}{2}\lambda_1^{\frac{1}{2}} q_1^2 - \tfrac{1}{2}\lambda_2^{\frac{1}{2}} q_2^2 - \tfrac{1}{2}\lambda_3^{\frac{1}{2}} q_3^2\}\end{aligned} \tag{3.85}$$

C being a constant.

The symmetry of the mechanical model itself is determined from the equilibrium configuration: the equilibrium configuration is therefore, by definition, totally symmetric. From (3.85) it is evident that any operation which has the character $+1$ or -1 leaves ψ_{000} unchanged since the normal coordinates occur as square terms. If there are degenerate modes, as q_{2y} and q_{2z} for CO_2, they occur in an equation such as (3.85) in the form $-\lambda_2^{\frac{1}{2}}(q_{2Y}^2 + q_{2Z}^2)$. Since the degeneracy in the mechanical model arises because two modes have the same potential energy, which contains the term $\lambda_2^{\frac{1}{2}}(q_{2Y}^2 + q_{2Z}^2)$ and must be unchanged by any symmetry operation, the term $-\lambda^{\frac{1}{2}}{}_2(q_{2Y}^2 + q_{2Z}^2)$ must also be totally symmetric. Thus the quantum mechanical analogue of the equilibrium configuration, ψ_{000}, is always totally symmetric, of species A, A_1, Σ^+ or $\Sigma_g{}^+$, depending upon the point group.

Excitation of one mode, say $v_1 = 1$, leads to (comparing equations 3.5)

$$\psi_{100} = c' \cdot q_1 \cdot \exp\{-\tfrac{1}{2}\lambda_1^{\frac{1}{2}} q_1^2 - \tfrac{1}{2}\lambda_2^{\frac{1}{2}} q_2^2 - \tfrac{1}{2}\lambda_3^{\frac{1}{2}} q_3^2\} \tag{3.86}$$

Since the exponential term is totally symmetric, ψ_{100}, behaves towards all operations of the group in the same way as q_1. The characteristic state or fundamental level corresponding to the excitation of one normal mode is of the same species as the normal coordinate. Not only does the quantum of absorbed radiation *for a fundamental mode* have the same frequency as the classical frequency of vibration of the molecule, but it is associated with a transition from a totally symmetric ground state to an excited state which has the same symmetry as the classical vibration.

For an overtone level, say $v_1 = 2$, equation (3.5) indicates:

$$\psi_{200} = (Cq_1^2 - C') \exp\{-\tfrac{1}{2}\lambda_1^{\frac{1}{2}}q_1^2 - \tfrac{1}{2}\lambda_2^{\frac{1}{2}}q_2^2 - \tfrac{1}{2}\lambda_3^{\frac{1}{2}}q_3^2\} \tag{3.87}$$

and thus any first overtone of a non-degenerate vibration state is totally symmetric. By similar reasoning a non-degenerate second overtone ($v = 3$) has the same symmetry as the fundamental. These results are less easily seen in classical mechanical terms. Multiple excitation of degenerate modes is more complex.

Single excitation of two modes, say $v_1 = 1$ and $v_2 = 1$, gives:

$$\psi_{110} = C \cdot q_1 \cdot q_2 \exp\{-\tfrac{1}{2}\lambda_1^{\frac{1}{2}}q_1^2 - \tfrac{1}{2}\lambda_2^{\frac{1}{2}}q_2^2 - \tfrac{1}{2}\lambda_3^{\frac{1}{2}}q_3^2\} \tag{3.88}$$

and the symmetry depends upon the symmetry of the product $q_1 \cdot q_2$. This can be obtained by multiplication of the characters of the species of q_1 and q_2. If one of the modes, say q_1, is totally symmetric the combination level will have the same species as q_2. Thus for SO_2, in which states (100), (010) are of species A_1 and (001) is of species B_1, the state (110) is A_1 and (101) and (011) are of species B_1. Where a molecule belongs to point group C_{2v} and has normal modes of species A_2 and B_2 as well as A_1 and B_1, the possibility arises of the combination modes $A_2B_1 = B_2$, $A_2B_2 = B_1$ and $B_1B_2 = A_2$. It can be seen from Table 13 that multiplication of the characters of each species for each symmetry operation gives the characters of the combination species. Some authors denote the species of fundamentals with lower case letters and combination modes with upper case: thus $a_2b_1 = B_2$, etc. . . .

Selection rule for infra-red spectra

In considering the selection rule for the spectroscopic observation of vibrational transitions in a diatomic molecule (p. 134) the dipole moment of the molecule was assumed to be a linear function of the displacement from equilibrium. In a vibrating polyatomic molecule the dipole moment $\boldsymbol{\mu}$ will, in general, have components μ_x, μ_y, and μ_z, and each of these components will be a linear function of the normal coordinates (compare equation 3.7):

$$\mu_x = \mu^0{}_x + \frac{\partial \mu_x}{\partial q_1} \cdot q_1 + \frac{\partial \mu_x}{\partial q_2} \cdot q_2 + \ldots \tag{3.89}$$

and similarly for μ_y and μ_z. μ_x^0, μ_y^0 and μ_2^0 are the components of the permanent dipole of the molecule.

The intensity of a dipole transition depends again upon integrals such as (3.8) which vanish unless the transition is one for which

$$\Delta v_1 = \pm 1 \tag{3.90}$$

Furthermore if the normal coordinates are independent there cannot be simultaneous changes of two different quantum numbers. Thus, to the degree of approximation assumed, the allowed transitions are restricted to those in which one vibrational quantum number changes by one unit.

According to this rule only fundamental frequencies should appear in the infra-red and Raman spectra. The occurrence of overtones and combinations depends upon mechanical and electrical anharmonicity and the transition probabilities are usually considerably smaller than for fundamentals. Difference transitions, such as $\omega_1 - \omega_2$ for SO_2 (Fig. 66), not only involve simultaneous changes in two quantum numbers, *viz.* $\Delta v_1 = +1$ and $\Delta v_2 = -1$, but depend upon the population of an excited lower state. They are therefore expected to be weak.

The selection rule (3.90) imposes a considerable restriction on the number of transitions which might otherwise occur. Further restrictions are imposed by the symmetry of the molecule. The inactivity of the vibration of homonuclear diatomic molecules in absorption is essentially a symmetry restriction. In a diatomic molecule $\mathrm{d}\mu/\mathrm{d}x = 0$ if $\mu_x{}^0 = 0$. However, in a polyatomic molecule it does not necessarily follow that all the $\partial\mu/\partial q$ vanish when $\mu_x{}^0 = \mu_y{}^0 = \mu_z{}^0 = 0$. The symmetry restrictions are essentially those which limit the $\partial\mu/\partial q$: if the symmetry of the molecule is such that $\partial\mu_x/\partial q_i = \partial\mu_y/\partial q_i = \partial\mu_z/\partial q_i = 0$ a transition involving only the ith mode is forbidden.

It can be shown that, for absorptions (dipole transitions) originating from the totally symmetric ground state, the upper states to which a transition is allowed on symmetry grounds are those which are of the same symmetry species as one or more of the components $\partial\mu_x/\partial q$, $\partial\mu_y/\partial q$, $\partial\mu_z/\partial q$ (and therefore of the same species as one or more of the axes x, y or z or the translatory modes t_x, t_y, t_z). It follows that no more than three species can be represented among the infra-red active fundamentals. This selection rule is independent of mechanical or electrical harmonicity: it is rigorous so long as interaction with rotation and electronic motion can be neglected.

For example, the levels (100) and (010) of SO_2 are of class A_1: one of the translations, t_z, is also of that class and therefore the transition from the ground state is allowed. Likewise the level (001) is of species B_1, as also is the translation t_x: that fundamental transition is also allowed. In CO_2 on the other hand, none of the translations is symmetric with respect to all the operations of the point group $D_{\infty h}$ and hence the totally symmetric breathing mode, $\nu_1(\Sigma_g{}^+)$, is forbidden in the infra-red. In fact the totally symmetric modes of most of the symmetrical molecules are forbidden in the infra-red. This can be readily seen: the symmetrical stretching of linear CO_2, of the tetrahedral CH_4, and of octahedral SF_6 will not alter the dipole moments from their equilibrium (zero) values.

References p. 216–217

The symmetry selection rules for combination or difference bands, or generally for transitions which originate from states other than the totally symmetric ground state, are similar. In this case it is the species of the product of the upper and lower states, obtained as on p. 170, which must be the same as the species of t_x, t_y or t_z for the infra-red activity. We have seen (p. 170) that the $v = 2$ level for any mode is totally symmetric and that consequently any first overtone is totally symmetric. Thus it is that in molecules of high symmetry like CO_2 or SF_6, where totally symmetric fundamentals are forbidden in the infra-red, so also are the first overtones of all modes. The second overtones, however, again have the same activity as the fundamentals.

Selection rules for Raman spectra

To extend the discussion[12] of selection rules for Raman activity from diatomic (p. 136) to polyatomic molecules we must take into account the fact that the induced polarization μ is, in general, a vector whose direction differs from that of the applied field F. In consequence the polarizability α in $\mu = \alpha F$ (3.13) is a tensor. Each component of F contributes to each component of μ and thus:

$$\begin{aligned} \mu_x &= \alpha_{xx} F_x + \alpha_{xy} F_y + \alpha_{xz} F_z \\ \mu_y &= \alpha_{yx} F_x + \alpha_{yy} F_y + \alpha_{yz} F_z \\ \mu_z &= \alpha_{zx} F_x + \alpha_{zy} F_y + \alpha_{zz} F_z \end{aligned} \tag{3.91}$$

It can be shown that

$$\alpha_{xy} = \alpha_{yx}, \qquad \alpha_{xz} = \alpha_{zx}, \qquad \alpha_{yz} = \alpha_{zy}$$

so that there remain six components, α_{xx}, α_{xy}, α_{xz}, α_{yy}, α_{yz}, α_{zz} of the polarizability tensor, α. These can be used as coefficients of the equation of an ellipsoid:

$$\alpha_{xx}x^2 + \alpha_{yy}y^2 + \alpha_{zz}z^2 + 2\alpha_{xy}xy + 2\alpha_{yz}yz + 2\alpha_{zx}zx = 1 \tag{3.92}$$

and equation (3.92) defines the *polarizability ellipsoid* which serves as a pictorial representation of the tensor. The principal axes of this ellipsoid coincide with the axes of a new coordinate system (X, Y, Z, obtained by rotation of the old about the origin) in which the equation (3.92) becomes

$$\alpha_{XX}X^2 + \alpha_{YY}Y^2 + \alpha_{ZZ}Z^2 = 1 \tag{3.93}$$

The new α_{XX}, α_{YY} and α_{ZZ} are the *principal values* of α. The ellipsoid intercepts the principal axes at values $X = \alpha_{XX}^{-\frac{1}{2}}$, $Y = \alpha_{YY}^{-\frac{1}{2}}$, $Z = \alpha_{ZZ}^{-\frac{1}{2}}$.

Vibration of the nuclei leads to a change in polarizability and we may write

$$\alpha = \alpha^0 + \frac{\partial\alpha}{\partial q_1} \cdot q_1 + \frac{\partial\alpha}{\partial q_2} \cdot q_2 + \ldots \tag{3.94}$$

$\partial\alpha/\partial q_i$ is a tensor, as is α, and is called the *derived polarizability tensor*. It contains terms $\partial\alpha_{xx}/\partial q_i$, $\partial\alpha_{xy}/\partial q_i$ . . ., for the normal vibration defined by q_i. As for diatomic molecules (equation (3.15)) scattering involving states v' and v'' depends upon integrals such as:

$$[\alpha_{xx}]^{v'v''} = \int \psi^*{}_{v'_i}\,\alpha_{xx}\,\psi_{v''_i}\,d\tau \qquad (3.95)$$

There are, in view of the composition of α, six such integrals. For $v' = v''$ the parts of the integrals which contain the change of polarizability ($\partial\alpha_{xx}/\partial q_i$, . . .) vanish and leave terms involving only the equilibrium polarizability, α^0. This accounts for Rayleigh scattering. For $v'_i \neq v''_i$, the integrals involving the derived polarizabilities do not vanish so long as v'_i and v''_i differ by one unit, *i.e.*

$$\Delta v_i = \pm 1 \qquad (3.90)$$

Again a symmetry selection rule operates. This time it depends upon the symmetry behaviour of the six components of α and their change with vibration. It can be shown that a Raman transition involving the totally symmetric ground state is allowed if the upper state is of the same species as at least one of the quantities α_{xx}, α_{xy}, α_{xz}, α_{yy}, α_{yz}, α_{zz}. Since these coincide with the species of the products t_xt_x, t_xt_y, t_xt_z, t_yt_y, t_yt_z, t_zt_z the symmetry selection rules can again be derived from the character tables.

For point group C_{2v} the character table (Table 13) enables the species of the six products to be written down, thus: A_1, A_2, B_1, A_1, B_2, A_1. In this case all possible species are represented so that all modes of a C_{2v} molecule will be allowed in the Raman spectrum. For CO_2, on the other hand, the species of the translations t_x, t_y and t_z are Σ_n^+, Π_n and Π_n. The six products are Σ_g^+ (and Δ_g), Δ_g, Π_g, Σ_g^+ (and Δ_g), Π_g, Σ_g^+ among which Σ_u^+ and Π_u are *not* represented. Accordingly the two fundamentals ν_3 and ν_2 (Fig. 65) do not appear in the Raman spectrum of CO_2, but ν_1 does. So far as ν_2 is concerned it appears reasonable that the polarizability does not change with a vibration in which, to a first approximation, the bond lengths do not change.

Centrosymmetric molecules

We have now shown that, for CO_2, ν_2 and ν_3, but not ν_1, should appear in the infra-red spectrum, and that ν_1, but not ν_2 and ν_3, should appear in the Raman. This is as found (but see p. 178) and provides an example of the general rule that for a molecule with a centre of symmetry transitions that are allowed in the infra-red are forbidden in the Raman spectrum, and vice versa. This is the *rule of mutual exclusion*. If the two spectra of a molecule are obtained in sufficient detail and do not contain any coincidences it is

References p. 216–217

reasonable to deduce that the molecule has a centre of symmetry. This conclusion must always be subject to the proviso that a transition may not have been observed because it was too weak rather than because it was altogether absent.

Polarization of Raman lines

When a sample of liquid or gas is irradiated in a Raman tube by natural light it is effectively subjected to two oscillatory electric vectors of equal magnitude. Suppose the direction of irradiation to be along the axis or perpendicular to the xy plane (Fig. 71). The electric vectors are F_x and F_y and

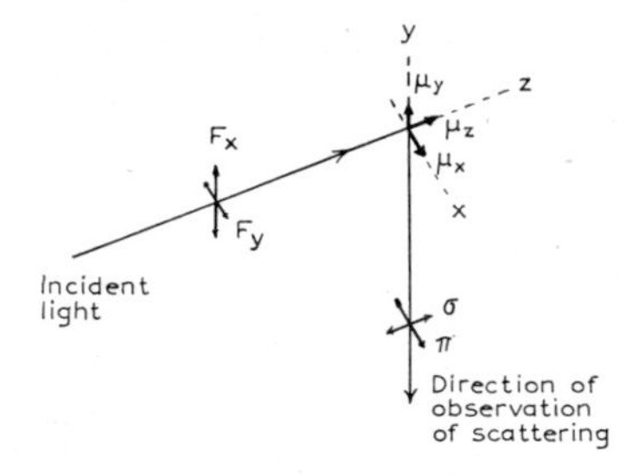

Fig. 71. Polarization of scattered light. σ and π refer to components of the scattered light plane-polarized in the yz plane and in the xy plane respectively.

will induce component moments μ_x, μ_y and μ_z. If observation is along the y-axis only the components μ_x and μ_z will contribute to the observed scattering since an oscillating dipole does not radiate in the direction of its oscillation. The scattered light may be analysed into two plane-polarized components of intensity $I(\sigma)$ in the yz plane and $I(\pi)$ in the xy plane (Kohlrausch[31] uses $I(\sigma)$ and $I(\pi)$; Herzberg[30] calls them $I(\perp)$ and $I(||)$, referring to perpendicular and parallel to the xy plane). The measured ratio $\rho = I(\sigma)/I(\pi)$ is called the *degree of depolarization.*

Now the polarizability ellipsoid, representing α, has two characteristics which are *invariant* with respect to orientation of the ellipsoid. The *mean-value invariant a* is the average of the three principal values of α:

$$a = \tfrac{1}{3}\,(\alpha_{XX} + \alpha_{YY} + \alpha_{ZZ}) \tag{3.96}$$

and is a rough estimate of the size of this ellipsoid. The *anisotropy* γ is defined by:

$$\begin{aligned}\gamma &= \tfrac{1}{2}(\alpha_{xx} - \alpha_{yy})^2 + (\alpha_{yy} - \alpha_{zz})^2 + (\alpha_{zz} - \alpha_{xx})^2 + 6(\alpha_{xy}{}^2 + \alpha_{yz}{}^2 + \alpha_{zx}{}^2)\\ &= \tfrac{1}{2}(\alpha_{XX} - \alpha_{YY})^2 + (\alpha_{YY} - \alpha_{ZZ})^2 + (\alpha_{ZZ} - \alpha_{XX})^2\end{aligned} \tag{3.97}$$

If $\alpha_{XX} = \alpha_{YY} = \alpha_{ZZ}$ the ellipsoid is a sphere, the molecule is isotropic, and $\gamma = 0$. The anisotropy thus measures the departure of the ellipsoid from the spherical.

The derived polarizability tensor, although it cannot be represented by an ellipsoid, nevertheless has the analogous invariants which we may denote by a' and γ'. It can be shown that the degree of polarization of Raman scattering, when averaged over all orientations of the molecule is given by:

$$\rho = \frac{6\gamma'^2}{45a'^2 + 7\gamma'^2} \tag{3.98}$$

The deformation of a molecule will in general affect the polarizability ellipsoid and symmetry considerations may be invoked to distinguish two possibilities for ρ. The ellipsoid will conform to the symmetry of the molecule: an axis of symmetry will coincide with one of the principal axes, X, Y, Z of the ellipsoid: a three-fold axis of symmetry requires that the ellipsoid be an ellipsoid of revolution: a molecule with tetrahedral or octahedral symmetry has a spherical ellipsoid (isotropic).

If the normal vibration is *not* totally symmetric the effect of some symmetry operation is to change the sign of the normal coordinate q. The effect on the polarizability ellipsoid is to move its axes but not change its size. The principal values remain unchanged, *i.e.* $\partial\alpha_{XX}/\partial q = \partial\alpha_{YY}/\partial q = \partial\alpha_{ZZ}/\partial q = 0$, but since the axes x, y, z no longer coincide with X, Y, Z the derivatives $\partial\alpha_{xy}/\partial q$, $\partial\alpha_{yz}/\partial q$, $\partial\alpha_{xz}/\partial q$ do not necessarily differ from zero. Consequently in the derived invariants corresponding to (3.96) and (3.97), $a' = 0$ but in general $\gamma' \neq 0$. Hence by (3.98), $\rho = 6/7$ and the Raman line is said to be depolarized. If symmetry also requires that $\gamma' = 0$ then the Raman line is forbidden.

If a normal vibration is totally symmetric then all symmetry operations leave q unchanged. The axes of the polarizability ellipsoid remain unmoved but the principal values change. Now $\partial\alpha_{XX}/\partial q$, $\partial\alpha_{YY}/\partial q$, $\partial\alpha_{ZZ}/\partial q$ are no longer zero but $\partial\alpha_{xy}/\delta q = \partial\alpha_{yz}/\partial q = \partial\alpha_{zx}/\partial q = 0$. Consequently $a' \neq 0$, and $\gamma' \neq 0$. Therefore $\rho < 6/7$ and the line is polarized. An isotropic molecule under a totally symmetrical vibration will have $\partial\alpha_{XX}/\partial q = \partial\alpha_{YY}/\partial q = \partial\alpha_{ZZ}/\partial q = 0$ and then $\gamma' = 0$. In that case $\rho = 0$ and the line is *completely polarized*.

Evidently the observation of the degree of depolarization is a valuable aid to the characterization of vibrational states. The value of ρ is not easily measured with accuracy so that it is not always possible to say with certainty that a line is depolarized. On the other hand it is normally possible to identify a polarized line.

Use of character tables

The symmetry arguments which we have illustrated mainly with reference to triatomic molecules become very involved if the normal coordinates have

References p. 216–217

to be deduced first and then symmetry behaviour examined under each operation. The whole problem is handled with much greater generality by *group theory*. A group, in the mathematical sense, is a set of quantities satisfying certain conditions about their interaction. Group theory develops an algebra based on those conditions and has its own notation. In particular the groups may be represented as matrices and their properties derived with the aid of matrix algebra. The symmetry point groups comply with the conditions and are therefore subjects for group theory[13, 14].

The interested reader has a choice of two monographs in particular in which molecular vibrations are very fully discussed. Wilson, Decius and Cross[32] employ group theory extensively whereas Herzberg[29] avoids its formalism and adopts a more physical approach. Fortunately it is not necessary to understand group theory in order to make use of its results.

In the elucidation of molecular structure from vibrational spectra the problem is to reconcile with possible molecular models the number of fundamentals observed in the infra-red and Raman spectra, the number of coincidences between the two spectra, and the states of polarization of the Raman lines. For any eligible model, therefore, it is necessary first to determine the symmetry and hence the point group to which it belongs. A useful guide for this step is either Herzberg or the introduction to the Chemical Society publication *Interatomic Distances*[38] (see also Refs. 36, 37).

Next the character table is required for the appropriate point group (Herzberg[29], Tables 12—30, pp. 105—123 or Wilson, Decius and Cross[32], Appendix X). This table will indicate the number of classes or species into which the normal modes and the vibrational levels may fall. It will not show how many normal modes occur in each class: this, in principle, is done by setting up the equations of motion (the *secular equation*) and noting the number of elements in each non-mixing block of the secular determinant. Again, however, Herzberg[29] gives tables (Tables 35 and 36, pp. 134—139) from which the number of modes in each species is obtainable in terms of the numbers of nuclei and sets of nuclei in the model which lie on some or all of the symmetry elements.

The character table also gives the species of the translatory modes, and their products, from which the selection rules for each species may be deduced. Herzberg[29] (Table 55, p. 252) conveniently lists the symmetry types of the components of dipole moment and of polarizability for the most important point groups. Those vibrations belonging to species which are represented among the components of dipole moment are infra-red active and those whose species are represented among the components of polarizability are Raman active. Those not represented in either are forbidden.

Determination of molecular symmetry

The above process may now be illustrated with reference to various models for triatomic molecules XYZ and XY_2 and tetratomic XY_3. Fig. 72 shows the models considered. Some of them may not be very plausible on chemical

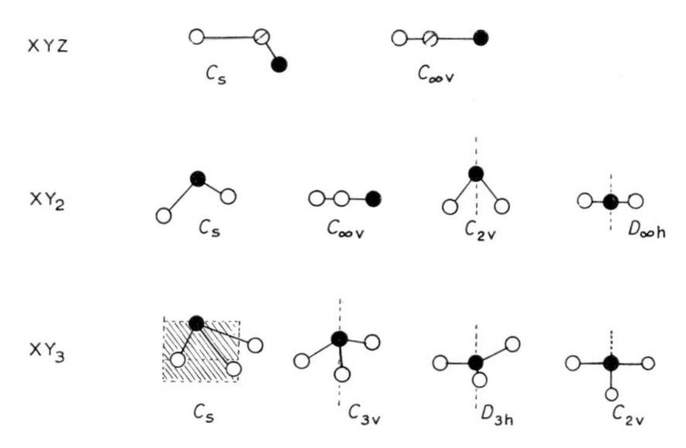

Fig. 72. Models for molecules of formulae XYZ, XY_2 and XY_3.

grounds. All triatomic molecules must have at least a plane of symmetry (the plane of the molecule): XYZ may either be bent or linear: XY_2 may be bent and unsymmetrical, bent and symmetrical, linear and unsymmetrical, linear and symmetrical. The XY_3 models do not include all possibilities: the wholly unsymmetrical point group C_1 is not represented: the C_s model has a plane of symmetry including the shorter bond and bisecting the angle between the two longer bonds. The point groups to which these models belong are given in the figure.

We now find from the character tables that the possible species for each point group are as follows:

C_{s} : A', A''
$C_{2\mathrm{v}}$: A_1, A_2, B_1, B_2
$C_{3\mathrm{v}}$: A_1, A_2, E
C_{sv} : $\Sigma^+, \Sigma^-, \Pi, \Delta, \ldots$
$D_{3\mathrm{h}}$: $A_1', A_2', E', A_1'', A_2'', E''$
$D_{\infty\mathrm{h}}$: $\Sigma^+{}_{\mathrm{g}}, \Sigma^-{}_{\mathrm{g}}, \Pi_{\mathrm{g}}, \Delta_{\mathrm{g}} \ldots \Sigma^+{}_{\mathrm{u}}, \Sigma^-{}_{\mathrm{u}}, \Pi_{\mathrm{u}} \ldots$

The distributions of normal modes among the species are now found:

XYZ:	C_{s} :	$3A'$	no A''
	$C_{\infty\mathrm{v}}$:	$2\Sigma^+ + \Pi$	no $\Sigma^-, \Pi, \Delta \ldots$
XY_2:	C_{s} :	$3A'$	no A''
	$C_{2\mathrm{v}}$:	$2A_1 + B_1$	no A_2, B_2
	$C_{\infty\mathrm{v}}$:	$2\Sigma^+ + \Pi$	no $\Sigma^-{}_{\mathrm{g}}, \Sigma^-{}_{\mathrm{u}}, \Delta \ldots$
	$D_{\infty\mathrm{h}}$:	$\Sigma^+{}_{\mathrm{g}} + \Sigma^+{}_{\mathrm{u}} + \Pi_{\mathrm{u}}$	no $\Sigma^-{}_{\mathrm{g}}, \Sigma^-{}_{\mathrm{u}}, \Pi_{\mathrm{g}} \ldots$
XY_3	C_{s} :	$4A' + 2A''$	
	$C_{3\mathrm{v}}$:	$2A_1 + 2E$	no A_2
	$D_{3\mathrm{h}}$:	$A_1' + A_2'' + 2E'$	no A_1'', A_2', E''
	$C_{2\mathrm{v}}$:	$3A_1 + 2B_1 + B_2$	no A_2

References p. 216–217

The activity of these models is indicated in Table 14.

TABLE 14

SPECTRAL ACTIVITY OF VARIOUS CONFIGURATIONS OF XYZ, XY_2, XY_3

		Numbers of:				
		fundamentals	infra-red active	Raman active	coincidences	polarized lines
XYZ:	C_s	3	3	3	3	3
	$C_{\infty v}$	3	3	3	3	2
XY_2:	C_s	3	3	3	3	3
	$C_{\infty v}$	3	3	3	3	2
	C_{2v}	3	3	3	3	2
	$D_{\infty h}$	3	2	1	0	1
XY_3:	C_s	6	6	6	6	4
	C_{2v}	6	6	6	6	3
	C_{3v}	4	4	4	4	2
	D_{3h}	4	3	3	$2(E')$	1

From this table we may draw a number of conclusions. For XYZ molecules the vibrational spectrum will evidently give little help in distinguishing between bent and linear, since the only difference lies in the number of polarized Raman lines. In HCN for example all three fundamentals have been observed in the infra-red but only two in the Raman spectrum. Degrees of polarization have not been measured but they would be of little assistance without the third line.

Among XY_2 molecules we may note that the numbers given for C_{2v} and $D_{\infty h}$ agree with our earlier discussion of CO_2 and SO_2. In the case of CO_2 the situation is complicated by the appearance of a doublet in the Raman spectrum where only one line is expected. Neither frequency coincides with the two infra-red bands which are clearly fundamentals. The explanation of this phenomenon is illustrated in Fig. 73 and, far from weakening the argument for linearity, is conclusive evidence for the $D_{\infty h}$ symmetry. ω_2 (667.3 cm^{-1}) and ω_3 (2349.3 cm^{-1}) are the strong infra-red fundamentals. Now $2 \times 667.3 = 1334.6$ cm^{-1} giving the expected value of the (020) level which has the same species (Σ_g^+) as the (100) level. If the (100) level was expected to occur at about 1340 cm^{-1} the effect on two such closely lying levels would be a mutual perturbation. The two energy levels in effect repel each other and the eigenfunctions are mixed, both states taking some of the character of each vibration. The deviation of energy and the mixing of eigenfunctions is greater the smaller the difference in the unperturbed energy levels. It can be shown that the interaction of the vibrations occurs mainly through anharmonicity in the potential energy and, from symmetry considerations, can

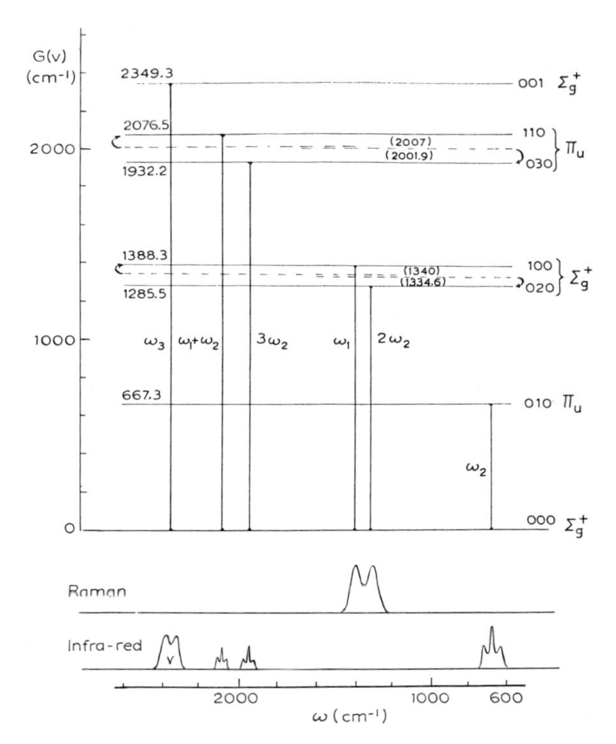

Fig. 73. Part of the vibrational term manifold for linear CO_2. Note that, in contrast to Fig. 66, energy is shown measured from the lowest level (000). The bands (of exaggerated width) in the infra-red spectrum show rotational contour.

only occur between states of the same symmetry type. This condition is satisfied by the (100) and (020) levels in CO_2 and the resultant perturbation is about 50 cm^{-1}. A further consequence of mixing (both eigenfunctions having some 100 components) is that the transition $2\omega_2$, which would normally be weak, "borrows" intensity from ω_1 and hence a doublet appears where one line is expected.

This is an example (and the first to be recognised, by Fermi in 1931) of *Fermi resonance*. Despite its restriction to states of the same species, the occurrence is widespread, especially among large molecules. The CO_2 spectrum contains evidence of many more perturbed levels. Fig. 73 shows the pair of transitions to the mixed (110, 030) levels: since the species is Π_u these occur in the infra-red.

With SO_2 the observation of three frequencies in both spectra immediately rules out the linear symmetrical arrangement and since one Raman line is depolarized the unsymmetrical C_s structure is also eliminated. The vibration spectrum does not, however, enable a decision to be made between $C_{\infty v}$ an C_{2v}. N_2O has three infra-red active fundamentals, two of which have also been found in the Raman spectrum. This eliminates the linear symmetrical structure N—O—N. From the vibration spectrum alone no further decision can be made but it can be said that the evidence is compatible with linear N—N—O. Similarly, although the less symmetrical structures are not conclusively rejected, the vibrational spectra distinguish between bent and linear structures in the following: linear; CS_2, $ZnCl_2$ in alcoholic solution, HF_2^-, N_3^-, NO_2^+, BO_2^- (if the metaborate ion is triatomic at all): bent; H_2O, H_2S, H_2Se, OF_2, OCl_2, NO_2, NO_2^-, UO_2^{2+}.

Examples of XY_3 molecules are BF_3, NH_3 and ClF_3. For BF_3 four fundamentals have been found, three in the infra-red, and two in the Raman, one frequency being common. The argument for the planarity of BF_3 (D_{3H}) would thus rest upon the absence of bands and is therefore not conclusive. In this case the fact that the doubling of infra-red bands, due to the presence of isotopes ^{10}B and ^{11}B, is not found in one of the Raman lines shows that in that mode the boron atom is not displaced. This can only be so if the molecule is planar, and then only in the totally symmetric A_1' mode. In NH_3 three fundamental bands are definitely observed in the infra-red: two of these also appear in the Raman spectrum, one being polarized. A further infra-red band is probably another fundamental but if this is uncertain the planar structure cannot be eliminated. On the other hand no more than four fundamentals are required to account for a large number of overtone and combination bands, which makes the less symmetrical modes improbable. In the Raman spectrum of ClF_3 both C_{3v} and D_{3H} are ruled out by the observation of five lines. The result is compatible with the C_{2v} structure, which has considerable interest in relation to valence.

Many of the above examples, and others which could be cited, refer to molecules whose rotational spectra have been observed, and whose symmetries are in fact better established by that means. On the other hand for larger and heavier molecules the rotational spectrum may be unresolved, especially if the molecule is non-polar and therefore inaccessible to the high resolution of a microwave spectrometer. Then the vibrational spectrum may make important contributions.

Among the six hexafluorides SF_6, SeF_6, TeF_6, MoF_6, WF_6 and UF_6, the regular octahedral structure of the lighter molecules has hardly been in doubt but electron diffraction measurements have suggested that MoF_6, WF_6 and UF_6 were unsymmetrical distorted octahedra. The selection rules for the regular octahedral model (point group O_H) are fairly strict. The $3N - 6 = 15$ modes are distributed $A_{1g} + E_g + 2F_{1u} + F_{2g} + F_{2u}$: six fundamental frequencies of which the last (F_{2u}) is inactive, except in combination, in both spectra. Two should appear in the infra-red and three in the Raman, with no coincidences. One Raman line should be polarized, completely in the isotropic molecule. In fact in all except UF_6 three Raman lines only are found, one being very strongly polarized. One infra-red fundamental is observed in all the fluorides but the second is only found in SF_6 and SeF_6. The overtones and combinations for an octahedral molecule are limited: only eight binary combinations are allowed in the infra-red, no first overtones (since they have the species A_{1g} — totally symmetric) and three second overtones. It nevertheless proves possible to account for the remainder of the infra-red spectra in terms of the observed frequencies, an assigned value for the inactive fundamental, and an assigned value for the other fundamental which should be active in the infra-red. The values assigned indicate that the band occurs at frequencies too low for observation. The entire consistency of the data for the six fluorides thus confirms the octahedral symmetry[15].

Osmium and ruthenium tetroxides are two other molecules in which uncertainty had arisen in the interpretation of the electron diffraction data. Here tetrahedral symmetry requires four frequencies ($A_1 + E + 2F_2$) of which all four are Raman-active (one polarized) and two should appear in the infra-red. This is found to be the case for OsO_4 except that only three Raman lines appear. It is possible that one broad Raman line in fact comprises two transitions of about the same frequency (as also in the Raman spectrum of ReO_4^- and WO_4^{2-} ions). In any case the simplicity of the spectra is strong evidence for the tetrahedral configuration[16, 17] since any lower symmetry requires at least six fundamental frequencies. Indeed in SF_4 the spectra do contain at least six distinct fundamental frequencies and must be interpreted[18] in terms of a structure in which the five pairs of valence electrons are all stereochemically active although only four are involved in bonds.

Limitations

Vibration spectra do not provide values of bond lengths nor of bond angles when these are not defined by symmetry. It might be asked: "By how much may an angle vary from the symmetry requirement (60°, 90°, 180°) before breakdown of selection rules becomes apparent?" This is a valid question if it may be argued that small departures from a particular configuration to one of lesser symmetry do not immediately release the full activity permitted by the new point group. It may be imagined that the intensity of newly permitted transitions will be small. If this is so small departures from a particular configuration will go undetected. An interesting example of this is provided by disilylether, SiH_3—O—SiH_3. The infra-red and Raman spectra were interpreted[19] on the basis of the selection rules, as indicating a linear Si—O—Si

References p. 216–217

skeleton, suggesting d—pπ bonding similar to that which is involved in the planar structure of trisilylamine, $N(SiH_3)_3$. If this were the case the frequency corresponding to the symmetrical stretching of the two Si—O bonds would remain unchanged on substituting ^{18}O for ^{16}O. That a small displacement of frequency is observed[20] shows that the oxygen atom must move in this normal mode and hence the Si—O—Si angle must be less than 180°. The angle is probably about 150° and apparently not sufficiently different from 180° for the selection rules for a linear Si—O—Si skeleton to be observably violated.

Assignment of frequencies

If the molecular symmetry has been established it is known how many frequencies belong to each species. It is customary, in numbering the frequencies, to start with ω_1 as the highest frequency of the totally symmetric species, to continue enumeration in the same species and in order of decreasing frequency; then to take the next species as it appears in the character tables, and again number in order of decreasing frequency. Thus the six frequencies of $CHCl_3$ (point group C_{3v}) are distributed $3A_1 + 3E$ and are accordingly numbered: $\omega_1(3019)$, $\omega_2(668)$, $\omega_3(366)$ in class A_1; $\omega_4(1216)$, $\omega_5(761)$, $\omega_6(262\ \text{cm}^{-1})$ in class E. This convention is not followed in triatomic molecules where the bending mode is always labelled ω_2.

If the molecule is of fairly high symmetry there will be few totally symmetric modes and their frequencies will, in principle, be identified by polarization. Other frequencies may be assigned to other classes by virtue of other selection rules which permit or forbid appearance in the infra-red and Raman or which govern participation in overtone or combination frequencies. The shape of the rotational structure of infra-red bands depends upon the orientation of the rotational axes (p. 185) with respect to the direction of oscillation of the dipole moment (which can be deduced for each species from the character tables). These are various means which may in favourable circum stances, enable all the frequencies to be assigned to the symmetry classes.

Beyond this it may be possible to carry out a normal coordinate analysis and insert reasonable values of force constants in order to show the regions in which to expect the members of each class. This procedure involves the assumptions that molecules are less resistant to deformation by bending than by bond stretching and that force constants for similar bonds and similar bond angles are transferable from one molecule to another. Without a mathematical analysis it is often possible (with the aid of intuition or experience and a knowledge of the direction of dipole oscillation) to obtain a fair idea of what the vibrations look like, if there are only a few modes in any one species.

The effect of isotopic substitution provides additional evidence: Fig. 74

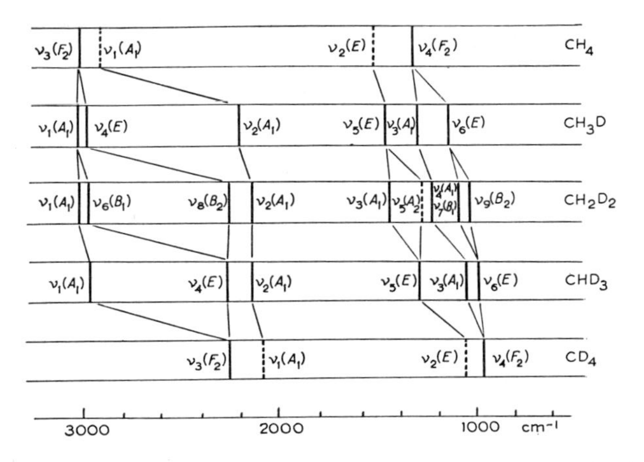

Fig. 74. Correlation of spectra of deuterated methanes. A dotted line indicates that the transition is active only in the Raman.

shows the correlation of fundamentals in the spectra of methane and deuterated methanes. Similar correlations may be made between the spectra of compounds in which one or more atoms are successively replaced by chemically similar atoms of different mass, *e.g.* the series CH_4, CH_3Cl, CH_2Cl_2, $CHCl_3$, CCl_4; or CH_3F, CH_3Cl, CH_3Br, CH_3I; or FCN, ClCN, BrCN, ICN; or OsO_4, OsO_3N^-.

To be rigorous the process of assignment is complete when the correct number of frequencies has been assinged to a symmetry class. Often, however, a mode is described in terms such as C—F stretching, S=O stretching, NO_2 symmetrical stretch, CH out-of-plane deformation or C—O—H bending. It is important to remember that this terminology is only an approximation. In a few cases only is it strictly legitimate. The symmetrical breathing modes of CO_2, BF_3, CH_4, SF_6 are indeed solely bond stretching modes. The out-of-line deformation of CO_2 and out-of-plane deformation of BF_3 are certainly pure bending modes. In general, however, every nucleus moves and every bond and angle is affected. It is nevertheless a convenience to describe a mode in accordance with the largest change in the configuration and the description is the more meaningful the further the frequency in question is removed from other frequencies of the same class. The empirical correlations between frequency and structure, to which we return in the next chapter, depend largely upon this consideration.

Force constants

For an N-atomic molecule, the most general potential energy function, in terms of internal coordinates and employing only quadratic terms, requires $\frac{1}{2}(3N - 6)(3N - 5)$ constants (p. 157). This number is reduced by any elements of symmetry in the molecule, but so also is the number of frequencies. To determine n force constants it is necessary to set up an equal number of simultaneous equations with n frequency values. For most molecules there are insufficient frequencies to enable this to be done. Isotopic substitution, which is assumed not to affect the force constant, can sometimes give a sufficient number of frequencies, but it generally becomes necessary to make simplifications in the general force field.

Although the whole subject is of chemical interest, since in principle, the quantities obtained are related to chemical binding and the stability of molecules, it is not one which can be usefully pursued without discussion of problems of computation. We shall therefore merely note the two earliest approximate potential energy functions to be used.

The central force field assumes the force on any atom to be the sum of all attractions or repulsions by all other atoms. With reference to the general triatomic molecule ABC (Fig. 75) this is the equivalent to writing:

$$2V = a\Delta r_{AB}^2 + b\Delta r_{BC}^2 + c\Delta r_{CA}^2 \tag{3.99}$$

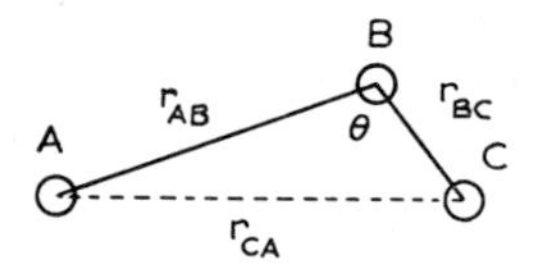

Fig. 75. Bond lengths and angles in a triatomic molecule.

The function (3.99) has three constants and the molecule has three normal frequencies. For a symmetrical non-linear XY_2 molecule the number of constants reduces to two and therefore the three frequencies allow an internal check on the validity of the force field. For tetratomic molecules the force field contains six constants for the unsymmetrical model (six frequencies) but only two (four frequencies) for the pyramidal XY_3. The unsymmetrical tetrahedral pentatomic molecule has one too many constants (10) for the total number of frequencies (9) but in the regular tetrahedral model only two constants are required for four frequencies. The field appears most satisfactory with such tetrahedral molecules. It does not take into account any restoring force restricting the bending of CO_2 or the out-of-plane deformation of BF_3: these would therefore have zero frequency.

The simple *valence force field* assumes (a) a strong restoring force along the line of each valence, resisting bond stretching, and (b) a restoring force resisting any change in angle between adjacent bonds. The appropriate equation for the general triatomic now becomes

$$2V = a\Delta r_{AB}^2 + b\Delta r_{BC}^2 + c\Delta\theta^2 \qquad (3.100)$$

There are the same number of constants, reducing to two for symmetrical XY_2. Six constants are needed for the general tetratomic, reducing to three for pyramidal XY_3. In regular tetrahedra the number of constants is again two.

Other and more general potential fields have been tested and shown to give more satisfactory accounts of vibrational frequencies. It is by no means certain which is the best: either for a comparison with other criteria of bond strength or for correlation between similar molecules. For further details see Linnett[21] (see also Ref. 32).

Rotational States

Of the $3N$ degrees of freedom of an assembly of N particles, three belong to the rotation of the whole assembly about three mutually perpendicular axes. To a first approximation interaction between rotation and vibration is assumed to be negligible and thus we may assume for the present purpose that the molecule is rigid.

The *moment of inertia* of the assembly about any axis is given by:

$$I = m_1 r_1^2 + m_2 r_2^2 + \ldots \qquad (3.10)$$

where m_1 is the mass of the particle whose perpendicular distance from the axis is r_1. It can be shown that there is one orientation of three mutually perpendicular axes for which the corresponding moments of inertia are maxima or minima, and that the origin of these axes is the centre of mass. The maximum and minimum values are called the *principal moments of inertia* (usually implied by the unqualified term, moments of inertia) and their axes are the *principal axes*.

Thus a diatomic molecule, masses m_1 and m_2 distance r_{12} apart, has its centre of mass a distance $m_1 r/(m_1 + m_2)$ from m_2 along the internuclear line. The moment of inertia about any axis, at angle θ to the internuclear line and intercepting that line at distance x from the centre of mass (Fig. 76), is given by

$$I = m_1 \left(\frac{m_2}{m_1 + m_2} r_{12} - x\right)^2 \sin^2\theta + m_2 \left(\frac{m_1}{m_1 + m_2} r_{12} + x\right)^2 \sin^2\theta$$

Setting $\partial I/\partial\theta = \partial I/\partial x = 0$ we find that $x = 0$ and $\theta = 0$ or $90°$. The

References p. 216–217

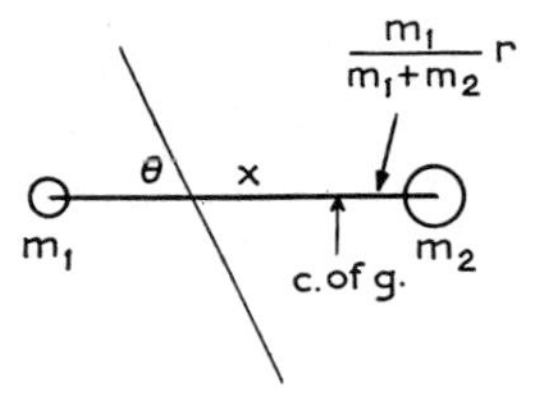

Fig. 76. Rotation of a diatomic molecule about an axis inclined at angle θ to the internuclear line at distance x along that line from the centre of gravity.

principal axes thus lie along the molecular axis and perpendicular to it and their origin is at the centre of mass. Furthermore when $\theta = 0$, $I = 0$ and when $\theta = 90°$ and $x = 0$

$$I = \frac{m_1 m_2}{m_1 + m_2} \cdot r^2{}_{12} = m_{12} r^2{}_{12} \tag{3.102}$$

where m_{12} is the reduced mass.

It is conventional to label the principal moments of inertia I^c, I^b, I^a, in order of decreasing magnitude: thus $I^c \geqq I^b \geqq I^a$. We can distinguish four cases:

i) asymmetric tops: three distinct moments of inertia, $I^c \neq I^b \neq I^a$,

ii) symmetric tops: two moments of inertia equal: if $I^c > I^b = I^a$ we refer to an *oblate* symmetric top (disc) and if $I^c = I^b > I^a$ to a *prolate* symmetric top (rugby football). The two equal moments may be designated I^b but it is not always known whether the other is larger or smaller. Herzberg labels the unequal moment I^a in either case: it should be remembered that this moment might turn out to be I^c.

iii) spherical tops: all moments equal, $I^a = I^b = I^c = I$,

iv) linear and diatomic molecules: two equal moments and one zero, $I^a = I^b = I$, $I^c = 0$.

The rotational energy levels are determined by the magnitudes of the moments of inertia so that the characterization of rotational states should offer a means of obtaining molecular dimensions with some precision. We shall also see that it gives information about molecular symmetry and contributes to the assignment of vibrational states.

Linear and diatomic molecules

Quantum mechanics treats the rigid rotator by replacing it with a single mass, rigidly tied to a centre of rotation at such a distance that it has the same moment of inertia about that centre as the molecule about its centre of mass. The justification for this in the case of the diatomic molecule, can be seen in equation (3.102). A particle of mass m_{12} rotating at the fixed distance r_{12} about a point has the same moment of inertia as the molecule, of reduced mass m_{12}, with an internuclear distance r_{12}.

The Schrödinger equation is set up for such a system with zero potential energy, $V = 0$, there being no force resisting rotation. The equation is best solved after conversion from Cartesian to spherical polar coordinates. These define the position of a particle in space in terms of a distance r from the origin and two angles θ and ϕ (Fig. 77). Keeping r constant the Schrödinger equation becomes one involving only the two angles and its solutions give the probability distribution in terms of other angles. The allowed solutions are:

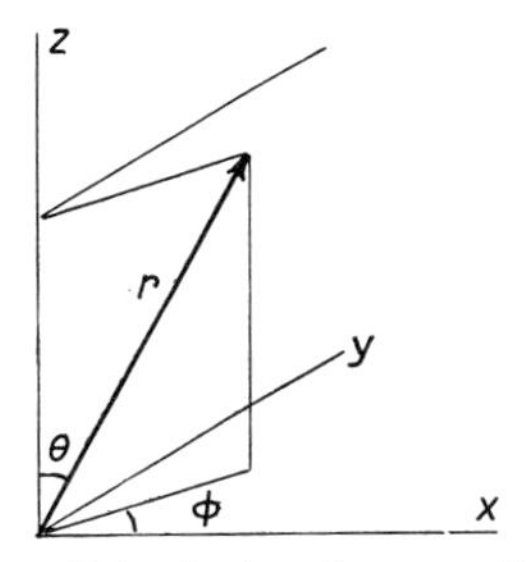

Fig. 77. Spherical polar coordinates.

$$\psi(\theta, \phi) = \Theta(\theta) \cdot \Phi(\phi)$$

where

$$\Phi_M(\phi) = \frac{1}{\sqrt{2\pi}} e^{iM\phi} \tag{3.103}$$

and $\Theta(\theta)$ is a function of θ containing quantum number M and J. The first few Θ functions are:

$$\Theta_{00} = 1/\sqrt{2}$$

$$\Theta_{1,0} = \sqrt{\frac{3}{2}} \cos\theta \quad \Theta_{1,1} = \tfrac{1}{2}\sqrt{3} \sin\theta$$

$$\Theta_{2,0} = \tfrac{1}{2}\sqrt{\frac{5}{2}}(3\cos^2\theta\pi 1) \quad \Theta_{2,1} = \tfrac{1}{2}\sqrt{15}\sin\theta\cos\theta \quad \Theta_{2,2} = \tfrac{1}{4}\sqrt{15}\sin^2\theta \tag{3.104}$$

The corresponding eigenvalues of energy are found to be

$$E_J = \frac{h^2}{8\pi^2 I} J(J+1) \tag{3.105}$$

The rotational term values are:

$$F(J) = BJ(J+1) \tag{3.106}$$

where B is the *rotational constant*

$$B = \frac{h}{8\pi^2 cI} \text{ (in cm}^{-1}) = \frac{h}{8\pi^2 I} \text{ (in sec}^{-1}) \tag{3.107}$$

The total angular momentum $\boldsymbol{J}$ is quantized:

$$\text{magnitude of } \boldsymbol{J} = \sqrt{J(J+1)}\,\hbar \tag{3.108}$$

and its direction is also restricted so that its component $\boldsymbol{M}_z$ in one direction (the z-direction) is quantized:

References p. 216–217

$$\text{magnitude of } \boldsymbol{M}_Z = M\hbar \tag{3.109}$$

$\hbar = h/2\pi$ is the natural unit of angular momentum.

The quantum numbers may take the values:

$$J = 0, 1, 2, 3 \ldots$$
$$M = 0, \pm 1, \pm 2 \ldots \pm J \tag{3.110}$$

The quantization of angular momentum (3.108 and 109) is parallel to the results for electronic motion (p. 92) and indeed has the same origin. So far as molecules are concerned we will need these results later.

Transitions solely between rotational states are subject to the same overall considerations as vibrational states (p. 133), namely that dipole transitions (emission and absorption) require an oscillating dipole and that Raman transitions require an oscillating polarizability. For dipole transitions the molecule must possess a dipole μ^0, whose magnitude in the x and z directions, will be

$$\mu_X = \mu^0 \cos\phi \text{ and } \mu_Z = \mu^0 \cos\theta \tag{3.111}$$

Transition probabilities depend upon integrals such as

$$\int_0^{2\pi} \Phi^*_{M'}\, \mu^0 \cos\phi \Phi_{M''}\, d\phi \text{ and } \int_0^{2\pi} \Theta^*_{J'M'}\, \mu^0 \cos\theta\, \Theta_{J''M''} d\theta \tag{3.112}$$

of which the first can be shown to be generally zero unless $M' = M'' \pm 1$ (using equation 3.103 and remembering the meaning of the conjugate complex) and the second can be shown from equations (3.104) to be zero except for the transitions in which $J' = J'' \pm 1$. Hence the selection rules:

$$\Delta M = \pm 1 : \Delta J = \pm 1 \tag{3.113}$$

and, since the integrals (3.112) would vanish if $\mu^0 = 0$, the molecule must be polar. (Non-polar molecules may show rotational structure in association with vibrational changes and then, in certain cases, $\Delta J = 0$ is also allowed. In pure rotation, of course, $\Delta J = 0$ implies no change).

Raman transitions in rotation have different selection rules (in contrast to vibrational transitions where the selection rule is the same for both Raman and dipole transitions) which may first be illustrated in classical mechanical terms. The polarizability of a molecule has direction but not sign. When a linear molecule rotates about the z-axis perpendicular to the line of nuclei the dipole moment component μ_x returns to its original value after a whole revolution, but the polarizability component does so after half a revolution (Fig. 78). Thus if the molecule rotates, in a classical sense, at a frequency ν_2 the polarizability changes at a frequency $2\nu_2$. The induced dipole is given by αE, where E oscillates with the frequency of the incident light ν^*. αE therefore oscillates with frequency $\nu^* \pm 2\nu_2$. This argument hardly proves the selection rule but illustrates the correspondence. From Fig. 78 it can be seen that while the dipole moment changes with rotation as $\mu_x \cos\phi$ the polarizability changes as $\alpha \cos^2 \phi$. Thus the term which must be used in the first of the integrals for Raman transition (compare equation 3.95, p. 173) is of the form $\alpha \cos^2 \phi$ whereas (3.112) contains $\mu^0 \cos \phi$. Thus the same arguments which lead to (3.113) for the dipole transition give the Raman selection rule:

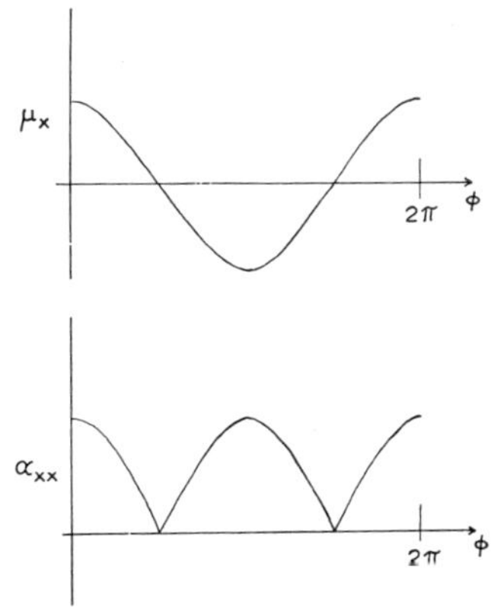

Fig. 78. Illustrating the classical explanation of the rotational selection rules for dipole transitions (μ) and Raman transitions (α).

$$\Delta J = \pm 2 \tag{3.114}$$

It should also be noted that the molecule need not have a permanent dipole moment.

From (3.106) it follows that the dipole transitions $J \rightarrow J + 1$ occur at

$$\nu_J (\omega_J) = F(J + 1) - F(J) = 2B(J + 1) \tag{3.115}$$

and for Raman transitions $J \rightarrow J + 2$

$$\omega_J = F(J + 2) - F(J) = 4B\left(J + \tfrac{3}{2}\right) \tag{3.116}$$

Since the rotational terms do not depend upon M (except in a magnetic field) the transitions, illustrated in Fig. 79, are also independent of M. Notice that the transitions between successive levels occur at different frequencies since the levels themselves diverge. This is by contrast with vibrational transitions and confirms the assignment of the set of low frequency lines in Fig. 56 to rotational changes.

The assignment is further confirmed by the measured wave numbers for HCl both in the far infra-red and the Raman. They are:

infra-red	83	104	124	145	165	186	206	227	cm^{-1}
Raman		101		143		188		229	cm^{-1}

These values are in accordance with Fig. 79 if it is assumed that the first three transitions in absorption, and one on the Raman effect have gone un-

References p. 216–217

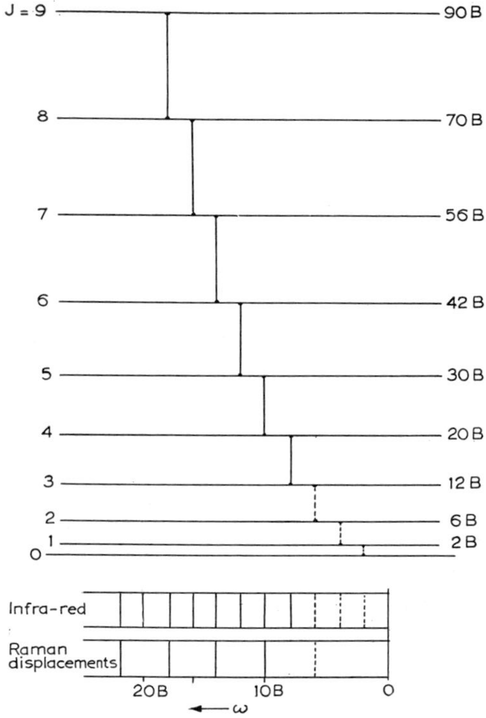

Fig. 79. Rotational term scheme for a diatomic or linear molecule with rotational constant B. Origin of pure rotational infra-red and Raman spectra.

observed. They are also, in accordance with (3.115) and (3.116) in that the spacing between the absorptions lines should be $2B$ and between Raman lines $4B$. The mean value of B from the above figures is

$$B(\text{HCl}) = 10.32 \text{ cm}^{-1} = 309{,}600 \text{ Mc/s}$$

The resultant value of I is 2.71×10^{-40} g cm^2 and hence r (assuming $H^{35}Cl$) is 1.29×10^{-8} cm or 1.29 Å. This is a result of the expected order and further confirms the interpretation.

The B value for HCl in Mc/s indicates that the pure rotational transitions fall at frequencies too high for the usual microwave region, which goes up to about 60,000 Mc/s. The first $J = 0 \rightarrow 1$ transition should occur at 20 cm^{-1} or 620,000 Mc/s, *i.e.* at wavelength 500 μ or 0.5 mm. This is just on the limit of optical spectroscopy (for infra-red) and falls in the more difficult "millimetre" range of microwave techniques. Heavier molecules than HCl have larger moments of inertia, small B values, and therefore pure rotational transitions at lower frequency (longer wavelength). They thus fall into the readily accessible region of microwave spectroscopy and can be measured with great accuracy. Recently high resolution Raman spectroscopy[22] has made it possible to observe pure rotational transitions from about 5 cm^{-1} to high frequencies and thus to cover the difficult gap between microwave and infra-red absorption. It also permits the measurement of pure rotational transitions of non-polar molecules. Fig. 80 shows such spectra for CO_2, N_2O and C_2H_2.

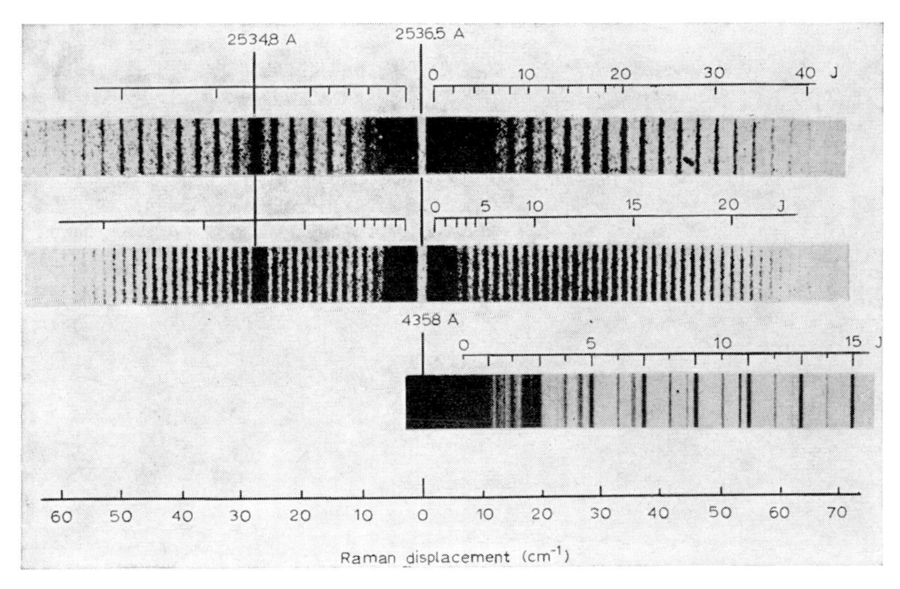

Fig. 80. Rotational Raman spectra of CO_2, N_2O and C_2H_2. The first two are excited by 2537 Å radiation which has been reabsorbed, after scattering, by a mercury vapour filter. The third is excited by 4358 Å radiation, shows only the Stokes lines and exhibits the expected alternation of intensity. The wave number scale is the same for all.

Although other factors must enter into a complete discussion (p. 211), to a first approximation linear polyatomic molecules give rotational spectra which conform to (3.115) and (3.116). Some examples of rotational constants obtained from pure rotational spectra in the microwave region and in the Raman effect are given in Table 15. Refinements in measurement enable B values to be obtained for upper vibrational states, and thus permit extrapolation to an "equilibrium value", B_e, corresponding to r_e, the internuclear distance at the unattainable minimum in the potential energy curve (p. 132). In the absence of information about higher states it is only possible to give the B_0 value for the ground vibrational state. Table 15 gives an indication of the precision with which those constants can be measured. The precision is so high in the microwave region that the effect of electron distribution on the moment of inertia has been detected and the gaseous molecules LiBr and LiI have to be treated as Li^+Br^- and Li^+I^-.

Internuclear distances in polyatomic linear molecules are not uniquely determined by measurement of one rotational constant. For this purpose use has to be made of isotopes plus the assumption that internuclear distances are not changed by isotopic replacement. Thus B_0 for $H^{12}C^{14}N$ is 44,315.97 and for $D^{12}C^{14}N$ is 35,207.5 Mc/s. The reader may confirm for himself that the distances are then r(C—H)1.066 and r(C—N)1.156.

TABLE 15

ROTATIONAL CONSTANTS FOR SOME DIATOMIC AND LINEAR MOLECULES FROM PURE ROTATIONAL TRANSITIONS IN THE RAMAN EFFECT AND MICROWAVE ABSORPTION

(B_0 and r_0 in Roman type; B_e and r_e in italic)

Raman molecule	B (cm^{-1})		r (Å)	
H_2	59.339 *60.841*		0.75105 *0.74173*	
HD	44.668		0.74973	
N_2	1.9897		1.1001 *1.09758*	
F_2	0.8828		1.4177	
CO_2	0.3904	r_{CO}	1.162	
C_2N_2	0.1575	r_{CC} r_{CN}	1.380 1.157	 assumed

Microwave molecule	B (Mc/s)		r (Å)	
$^{12}C^{16}O$	*57897.5*		*1.1282*	
$^{14}N^{16}O$	*51084.5*		*1.1510*	
$H^{12}C^{14}N$	44315.80	r_{CH}	1.066	
$D^{12}C^{14}N$	36207.5	r_{CN}	1.156	
$F^{35}Cl$	*15483.69*		*1.6281*	
$^6Li^{79}Br$	*19161.51*		*2.170*	
^6LiI	*15381.45*		*2.3919*	
$^{14}N^{14}N^{16}O$	12561.64	r_{NN}	1.126	
$^{15}N^{14}N^{16}O$	12137.30	r_{NO}	1.191	assumed
$F^{12}C^{14}N$	10554.2	r_{CF} r_{CN}	1.260 1.165	 assumed
$^{35}Cl^{12}C^{14}N$	5970.821	r_{CCl}	1.630	
$^{37}Cl^{12}C^{14}N$	5847.252	r_{CN}	1.163	

Spherical top molecules

There is again only one value of the moment of inertia in a spherical top molecule and the levels are, as in (3.106), given by

$$F(J) = BJ(J + 1) \qquad J = 0, 1, 2 \ldots \qquad (3.117)$$

Since a spherical top molecule is, by symmetry, non-polar, its rotation produces no change in moment and no dipole transitions occur. Rotational states can, however, be observed in combination with vibration. Since a spherical top molecule is also isotropic (polarizability ellipsoid a sphere) rotation produces no change in polarizability and therefore no rotational Raman spectrum is observed. Again rotational structure of Raman-active vibrations can be seen. It is conceivable that transitions between rotational levels in an upper vibrational state could be active, either in the Raman or in absorption, but none seems to have been observed.

Symmetric top molecules

Two moments of inertia are equal in symmetric top molecules and are labelled I^b. The other is not zero and differs from I^b. It is labelled I^a although it may be larger than I^b. Symmetry considerations show that any molecule with a three-fold rotation axis must be a symmetric top and that the I^a-axis (the top axis) coincides with the symmetry axis (C_3). The I^b-axes are then perpendicular to the top axis. NH_3 and CH_3Cl are two examples (Fig. 81) and indeed in NH_3 $I^b < I^a$.

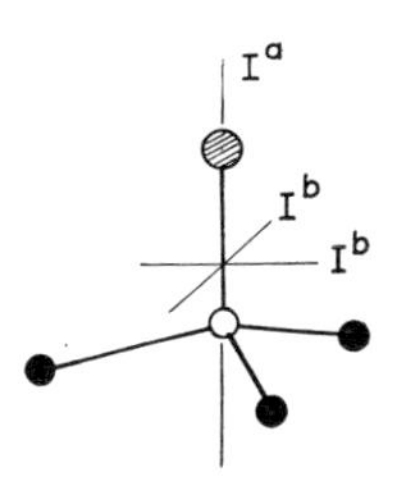

Fig. 81. Principal axes of methyl chloride (with ◍) or ammonia (without ◍).

The rotational energy in a symmetric top now depends upon two quantum numbers J and K. This is connected with the fact that, in such a molecule, the classical total angular momentum vector $\boldsymbol{J}$ is not necessarily directed perpendicular to the top axis. The preservation of the magnitude and direction of $\boldsymbol{J}$ is maintained in the classical motion by the rotation of the molecule about the I^a-axis and a *precession* or *nutation* of that axis about the direction of $\boldsymbol{J}$ (Fig. 82).

In quantum mechanical terms $\boldsymbol{J}$ is quantized and has magnitude, as before

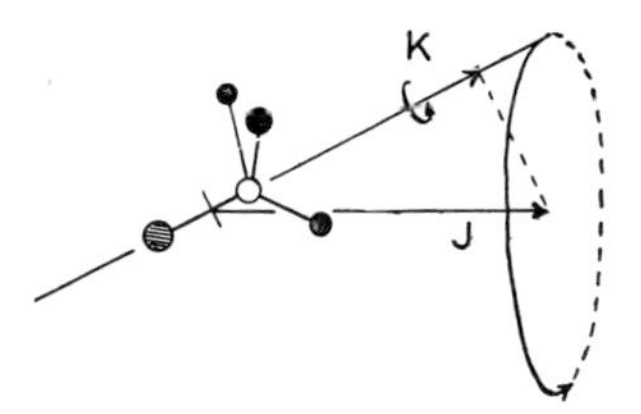

Fig. 82. Classical motion of a symmetric top, comprising rotation about the top axis and precession of that axis about the total angular momentum.

$$\text{magnitude of } \boldsymbol{J} = \sqrt{J(J+1)}\hbar \qquad (3.118)$$

In addition, its component $\boldsymbol{K}$ in the direction of the top axis is also quantized:

$$\text{magnitude of } \boldsymbol{K} = \hbar K, \qquad \hbar = h/2\pi \qquad (3.119)$$

There is, as before, quantization of the direction of $\boldsymbol{J}$ with respect to an axis fixed in space. (3.109) applies, and there is still a magnetic quantum number M. The term values of the symmetric top then become

$$F(J, K) = BJ(J+1) + (A - B)K^2 \qquad (3.120)$$

in which

$$B = \frac{h}{8\pi^2 cI^b} \text{ and } A = \frac{h}{8\pi^2 cI^a} \qquad (3.121)$$

(in cm^{-1}: omit c for sec^{-1}). The quantum numbers may take values

$$J = 0, 1, 2 \ldots$$
$$K = 0, \pm 1, \pm 2 \ldots \pm J \qquad (3.122)$$

and thus all rotational states with $K > 0$ are doubly degenerate.

The selection rules are:

$$\text{dipole transition: } \Delta J = 0, \pm 1; \Delta K = 0$$
$$\text{Raman transition: } \Delta J = 0, \pm 1, \pm 2; \Delta K = 0 \qquad (3.123)$$

The limitation of ΔK, which is relaxed in combination with vibration, means that in pure rotational absorption spectra the frequencies are again given by

$$\nu_J(\omega_J) = F(J+1, K) - F(J, K) = 2B(J+1) \qquad (3.124)$$

and do not contain the rotational constant, A. The explanation of the selection rule is that in a symmetric top the dipole moment lies in the direction of the top axis and therefore no rotation of the molecule about the axis can change any component of the dipole moment. The Raman selection rule leads to two series of lines, which are distinguished as S and R branches:

$$S;\ \Delta J = 2 \quad \omega = 4B\left(J + \tfrac{3}{2}\right) \quad J = 0, 1, 2 \ldots$$
$$R;\ \Delta J = 1 \quad \omega = 2B(J+1) \quad J = 1, 2, 3 \ldots \qquad (3.125)$$

The two branches will be distinguished by different intensities since every other R-line coincides with an S-line, as is shown in the rotational Raman spectrum of NH_3 (Fig. 83).

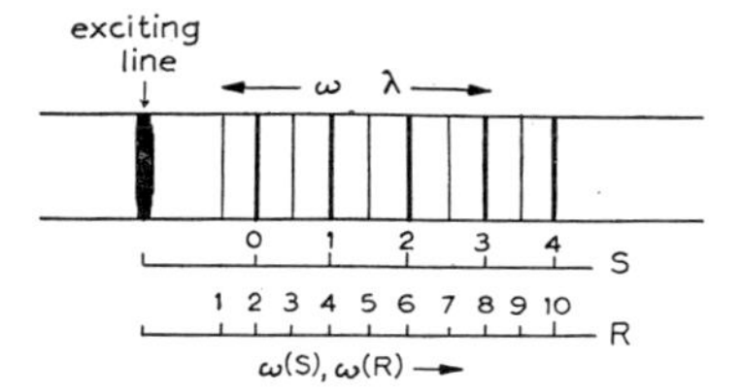

Fig. 83. Rotational Raman spectrum of a symmetric top, such as NH_3.

The appearance of an absorption spectrum still depends upon the molecule having a permanent dipole. Consequently planar symmetric tops such as BF_3 and SO_3 have no microwave spectrum, and its absence is the strongest argument for the planar structures. There seems no reason why these molecules should not have rotational Raman spectra, but none has been found.

TABLE 16

ROTATIONAL CONSTANTS FOR SOME SYMMETRIC TOP MOLECULES BY THE RAMAN AND MICROWAVE METHODS

Molecule	Method	B_0 (cm^{-1})	B_0 (Mc/s)	Parameters	
NH_3	R	9.92	(297,000)		
$^{14}NH_3$	M	(9.94)	298,000	r_{NH} = 1.008 Å (both)	
$^{14}ND_3$	M	(5.14)	154,000	$H\hat{N}H$ = 107.3° (both)	
NF_3	M	(0.35628)	10680.96	r_{NF} = 1.371,	$F\hat{N}F$ = 102.1°
$P^{35}Cl_3$	M	(0.08730)	2617.1	r_{PCl} = 2.043,	$Cl\hat{P}Cl$ = 100.1°
$CH_3{}^{35}Cl$	M	(0.44341)	13292.95	r_{CH} = 1.113, $H\hat{C}H$ = 110.5°	r_{CCl} = 1.781
$C^{35}Cl_3H$	M	(0.11014)	3301.94	r_{CH} = 1.073, $Cl\hat{C}Cl$ = 110.4°	r_{CCl} = 1.767
PF_3O	M	(0.15325)	4594.25	r_{PF} = 1.52, $F\hat{P}F$ = 102.5°	r_{PO} = 1.45
ReO_3Cl	M	(0.06986)	2094.20	r_{ReO} = 1.761, $O\hat{Re}O$ = 108.3°	r_{ReCl} = 2.230
$CH_2=C=CH_2$	R	0.29653	(8889.7)	r_{CC} = 1.309 (all three)	
$CD_2=C=CD_2$	R	0.23230	(6964.1)	r_{CH} = 1.07 (all three)	
$CH_2=C=CD_2$	R	0.26190	(7851.5)	$H\hat{C}H$ = 117° (all three)	
$^{12}CH_3{}^{12}C\equiv{}^{12}CH$	M	(0.28506)	8545.84	$r_{CH}(CH_3)$ = 1.10 (all five)	
$^{13}CH_3{}^{12}C\equiv{}^{12}CH$	M	(0.27730)	8313.23	$H\hat{C}H$ = 108° (all five)	
$^{12}CH_3{}^{12}C\equiv{}^{12}CD$	M	(0.25979)	7788.14	r_{C-C} = 1.46 (all five)	
$^{12}CD_3{}^{12}C\equiv{}^{12}CH$	M	(0.24536)	7355.75	$r_{C\equiv C}$ = 1.21 (all five)	
$^{12}CD_3{}^{12}C\equiv{}^{12}CD$	M	(0.22463)	6734.31	r_{CH} = 1.06 (all five)	
CF_3SF_5	M		1097.6		
C_6H_6	R	0.18960		r_{C-C} = 1.3973 (all three)	
C_6D_6	R	0.15681			
$C_6D_3H_3$	R	0.17165		r_{CH} = 1.084 (all three)	

References p. 216–217

Rotational constants obtained from symmetric top molecules are measured with great accuracy in both the microwave region and the Raman effect. Some values are given in Table 16. The parameters cannot be uniquely determined from one species since only B is determined, not A, and at least two parameters are involved, *e.g.* the N—H bond length and the HNH angle in NH_3. If isotopic molecules can be measured (and the intensity of the microwave absorption is often such that isotopic species can be observed in their natural abundance) values of distances and angles can be obtained. The moment of inertia I^b of a symmetric top molecule XY_3 in terms of the distance r_{XY} and YXY angle θ (Fig. 81) is

$$I^b = m_y r^2_{xy}(1 - \cos\theta) + \frac{m_x m_y r^2_{xy}}{3m_y + m_x}(1 + 2\cos\theta) \quad (3.126)$$

and it is again left to the reader to verify that the B values for NH_3 and ND_3 lead to the r_{NH} and HNH values given in Table 16.

Asymmetric top molecules

In this case the three moments of inertia are all different, the classical motion is complex and no simple formula can be given for the energy levels. We will not discuss here the characterization of these rotational states but merely note that they give rise to much more complex rotational spectra whose analysis is more difficult than in the previous groups. Nevertheless successful analysis yields three rotational constants. Among the asymmetric top molecules there are not only the large number of (principally organic) molecules with no symmetry at all but also the simpler molecules in the point groups C_s (one plane of symmetry; NOCl, HN_3, HNCO, $CH_2 = CH \cdot CN$) and C_{2v} (two planes of symmetry and a two-fold axis; non-linear XY_2, CX_2Y_2, $X_2C = CY_2$, C_6H_5X).

It is not possible in an unsymmetrical asymmetric top to judge the orientation of the principal axes of rotation. Where there is a plane of symmetry two of the axes lie in the plane and in the point group C_{2v} two axes lie in the plane and one coincides with C_2-axis. It is still not possible to say which principal axes take these positions but in planar molecules (C_s or C_{2v}) it is at least possible to assert that the I_c-axis is perpendicular to the plane and that

$$I^c = I^a + I^b \quad (3.127)$$

In non-linear symmetrical XY_2 the moments are

$$I^a \text{ (or } I^b) = 2m_y r^2_{xy} \sin^2\frac{\theta}{2}$$

$$I_b \text{ (or } I_a) = 2m_y r^2_{xy} \cos^2\frac{\theta}{2}$$

$$I^c = I^a + I^b = 2m_y r^2_{xy} \quad (3.128)$$

where θ is the angle YXY. The equations lead to ambiguity in θ. The reader may verify that the constants given in Table 17 for SO_2 give $\theta = 110°$ or $44°$. The larger value is chosen because it is more reasonable in comparison with other bond angles which are known unambiguously. Furthermore the smaller angle would imply an O—O distance of about 1.08 Å which is considerably smaller than the internuclear distance in O_2.

Table 17 gives some other molecules whose shapes and dimensions have been measured by microwave absorption. Again the great precision of the method is demonstrated, requiring incidentally that accurate isotopic masses

TABLE 17

ROTATIONAL CONSTANTS FOR SOME ASYMMETRIC TOP MOLECULES (MICROWAVE ABSORPTION)

Molecule	Rotational constants Mc/s	Structure	(r in Å)
$H^{12}CN^{32}S$	$B_0 = 5903.0$	H\N—C—S	$H\hat{N}C = 130.2°$
	$C_0 = 5828.0$		$r_{NH} = 1.013$
			$r_{NC} = 1.216$
			$r_{CS} = 1.561$
H_2CO	$A_0 = 282{,}106$	non planar	$H\hat{C}H = 118°$
	$B_0 = 38{,}834$		$r_{CH} = 1.12$
	$C_0 = 34{,}004$		$r_{CO} = 1.21$
C_5H_5N	$A_0 = 6039.13$	planar	$r_{C_2N} = 1.342$, $r_{C_2H} = 1.085$
	$B_0 = 5804.72$		$r_{C_2C_3} = 1.391$, $r_{C_3H} = 1.080$
	$C_0 = 2959.25$		$r_{C_3C_4} = 1.398$, $r_{C_4H} = 1.075$
NO_2F	$A_0 = 13203$	planar	$O\hat{N}O = 125°$
	$B_0 = 11447$		$r_{NO} = 1.23$
	$C_0 = 6120$		$r_{NF} = 1.35$
$^{35}ClF_3$	$A_0 = 13747.7$	planar	$F\hat{Cl}F = 87.5°$
	$B_0 = 4611.7$		$r_{ClF} = 1.698$ and 1.598
	$C_0 = 3448.7$		
$^{32}SO_2$	$A_0 = 60778.79$		$O\hat{S}O = 119.5°$
	$B_0 = 10318.10$		$r_{SO} = 1.432$
	$C_0 = 8799.96$		

be used in subsequent calculations rather than the mass numbers which have been adequate for many earlier measurements. The variation of bond lengths round the pyridine ring is interesting in relation to reactivity of different positions. The T-shape of ClF_3 is distorted (two outer F atoms moved from the FClF line by $2\frac{1}{2}°$ each *towards* the central F atom) in a way which indicates the steric effect of two unshared pairs of electrons.

Stark effect

Microwave absorption requires an oscillating dipole and therefore for rotational changes requires that the molecule itself be a permanent dipole.

References p. 216–217

Under an applied static electric field it might be expected that rotational levels would be perturbed by interaction between the field and the dipole. Changes are indeed observed in the spectrum when the system is subjected to an electric field and this is known as the *Stark effect*.

The so-called first-order Stark effect is observed in symmetric top molecules. It can be shown that (3.120) must now be written:

$$F(J, K, M) = BJ(J+1) + (A-B)K^2 - \mu^\circ \frac{E}{h} \cdot \frac{MK}{J(J+1)} \tag{3.129}$$

M is the quantum number for the component of $\boldsymbol{J}$ in a fixed direction in space here defined by the field $\boldsymbol{E}$ of strength E. The perturbation term clearly expresses the fact that the molecule "takes notice" of this direction to an extent which is proportional both to the field and the molecule's sensing element, namely the dipole moment. The levels (3.129) are $(2J+1)$ in number since $M = 0, \pm 1 \ldots \pm J$. A rotational transition $J \rightarrow J+1$ has $(2J+1)$ first-order Stark components when $\Delta M = 0$.

$$\nu_J = F(J+1, K, M) - F(J, K, M) = 2B(J+1) + \frac{2\mu^\circ E}{h} \frac{MK}{J(J+1)(J+2)} \tag{3.130}$$

There are also allowed transitions in which $\Delta M = \pm 1$ but under the conditions normal for the measurement of the Stark effect the $\Delta M = \pm 1$ transitions are much weaker than the $\Delta M = 0$ and are often not observed.

Where the first-order Stark effect is observed (principally in symmetric top molecules) it is clear from (3.130) that a linear plot of ν_J against E gives an estimate of μ°. This assumes that the line under observation has been correctly assigned. Conversely the movement of the Stark components of a line with variation in E may assist in the identification of the transition. As explained on p. 35, the Stark effect may also provide a means of modulating microwave power for detection. In Fig. 84, the first-order Stark effect in $CH_3C{\equiv}CD$ is shown under varying fields. The plot of ν against E gives μ°.

In linear molecules the angular momentum $\boldsymbol{J}$ is always perpendicular to the molecular axis. Thus $K = 0$. In this case the first-order perturbation in (3.129) is zero and there is no first-order Stark effect. However a second order effect arises from the interaction between the field and a dipole induced by the field. On this account the perturbation, though smaller, varies as the square of E. In the case of transitions originating from $J = 0$ it turns out that

$$\nu_0 = 2B + \frac{4\mu^2 E^2}{15Bh^2} \tag{3.131}$$

In general the perturbation is of the same form but contains a cumbersome function of J and M. M^2 is involved rather than M, so that a transition from the Jth level has $J+1$ rather than $2J+1$ Stark components. This fact assists in assignment of a line. The dipole moment is obtained from the second-order Stark effect by plotting ν against E^2 rather than E. Usually unless E is very large a first-order Stark effect, if it is present, outweighs the second-order effect. Fields of up to 1000 volt per cm are required for second order effects and up to 100 volt per cm for first order.

By one or other of these methods dipole moments can be obtained with considerable accuracy. They are often quoted to 0.001 D which is an order

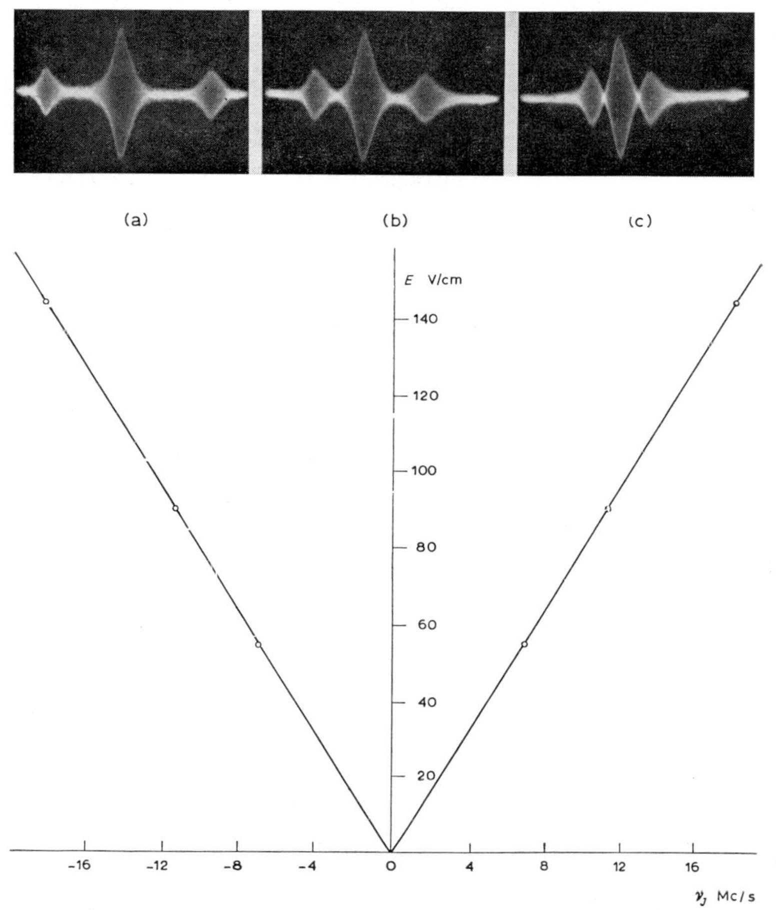

Fig. 84. First-order Stark effect in $CH_3C{\equiv}CD$. T = 20°C.
(a) $E = 144.94$ V/cm $\nu_J = \pm\ 18.23$ Mc/s (b) $E = 90.59$ V/cm $\nu_J = \pm\ 11.40$ Mc/s (c) $E = 54.35$ V/cm $\nu_J = \pm\ 6.84$ Mc/s
Slope = 7.90 V/cm/Mc/s

$$\nu = \frac{2\mu^\circ EKM}{hJ(J+1)(J+2)}; \qquad \mu^\circ = \frac{6}{1.0064 \times 7.90} \approx 0.75 \text{ Debye units.}$$

References p. 216–217

of magnitude better than is obtainable through the temperature dependence of polarization. Dipole moments have been measured in this way ranging from high values (in D), such as $^{39}K^{79}Br$ (10.41), $^{7}Li^{37}Br$ (6.2), CH_3CN (3.92), down to low ones, such as OF_2 (0.176), N_2O (0.166), SO_2F_2 (0.228), CO (0.1).

Centrifugal stretching

The accuracy of measurement of rotational spectra shows that equations (3.115), (3.116), (3.124), (3.125) are not adequate and that they should include a term in $(J + 1)^3$. Thus the dipole transition equations (3.115) and (3.124) give better experimental fit if they are amended to:

$$\nu_J(\omega_J) = 2B(J + 1) - 4D(J + 1)^3 \tag{3.132}$$

This implies that the levels themselves are better represented by:

$$F(J) = BJ(J + 1) - DJ^2(J + 1)^2 + \ldots \tag{3.133}$$

D is usually very much smaller than B. In CO, for example, $D_e = 0.189$ Mc/s by comparison with $B_e = 55{,}097$ Mc/s. The modification evidently takes account of a slight dependence of the moment of inertia upon J. In classical terms this means that I increases slightly as the molecule rotates more rapidly. The rotator is thus not strictly rigid and is subject to centrifugal stretching.

Interaction between vibration and rotation

The incidence of centrifugal stretching is, in a sense, an interaction with vibration. A change of I in going from one J-level to the next implies a change of internuclear distance which could then continue as a vibrational mode. This is shown up in the solution of the Schrödinger equation for a diatomic molecule acting under a Morse potential (p. 148) when rotation is considered in addition to vibration. The combined rovibrational levels turn out in the form

$$G(v, J) = \omega_e(v + \tfrac{1}{2}) - \omega_e x_e(v + \tfrac{1}{2})^2 + B_e J(J + 1) - D_e J^2(J + 1)^2 - \alpha_e(v + \tfrac{1}{2})J(J + 1) \tag{3.134}$$

in which ω_e and $\omega_e x_e$ have their former significance (p. 137); B_e is the rotational constant related to I_e, the value of I when $r = r_e$;

$$D_e = \frac{h^3}{128\pi^6 m^3 \omega_e^2 r_e^6} = \frac{4B_e^3}{\omega_e^2} \; : \; \alpha_e = 6\sqrt{\frac{\omega_e x_e B_e^3}{\omega_e^2}} - 6\frac{B_e^2}{\omega_e} \tag{3.135}$$

The fourth term in (3.134) is just the centrifugal stretching correction of (3.130) and the fifth term takes into account the differences in average moment of inertia in different vibrational states.

For CO the found values of B_e (above) and of $\omega_e = 2170.21$ cm^{-1} = 65.106×10^6 Mc/s (p. 144) give a value of $D_e = 0.165$ Mc/s (not to be confused with dissociation energy). The observed value is 0.189 Mc/s. The observed coefficient α_e is 524 Mc/s for CO. Calculation from (3.135) using $\omega_e x_e = 13.46$ cm^{-1} (p. 144) gives 477 Mc/s.

If the centrifugal stretching is neglected the rotational terms from (3.134) can be represented as:

$$F_v(J) = B_v J(J+1)$$

where

$$B_v = B_e - \alpha_e(v + \tfrac{1}{2}) \qquad (3.136)$$

The significance of B_e and B_0 $(= B_e - \frac{1}{2}\alpha_e)$ is thus explained. In the measurement of B in the ground vibrational state $B = B_0$. Spectra may sometimes be observed from pure rotational transitions within an excited vibrational state, giving B_v. Otherwise B values other than B_0 are obtained from rovibrational transitions (see below). In view of the size of α_e the difference between B_e and B_0 is evidently not negligible: about $\frac{1}{2}$ % for CO.

In polyatomic molecules the rotational constant is different for each vibrational level since there is an α term for each normal mode. Thus:

$$B_{v_1, v_2} \ldots = B_e - \alpha_1(v_1 + \tfrac{1}{2}) - \alpha_2(v_2 + \tfrac{1}{2}) - \ldots$$

and

$$B_{000} \ \ldots = B_e - \tfrac{1}{2}(\alpha_1 + \alpha_2 + \ldots) \qquad (3.137)$$

α is usually positive for stretching vibrations; for bending modes the average moment of inertia decreases with vibration with consequent increase in B, *i.e.* negative α. If the mode is doubly (species E) or triply degenerate (F) its contribution to B_v is $-\alpha(v + 1)$ or $-\alpha(v + \frac{3}{2})$.

Rovibrational transitions

When vibrational and rotational energies of a molecule are considered together the total energy is the sum of the $G(v)$ and $F(J)$ spectral terms, which are independent except is so far as $F(J)$ contains rotational constants which depend upon v, as (3.137). A vibrational change is in general accompanied by rotational changes so that

$$\omega = G(v+1) + F'(J') - G(v) - F''(J'')$$
$$= \omega_0 + F'(J') - F''(J'') \qquad (3.138)$$

$G(v+1) - G(v)$ represents the particular vibrational change ω_0 which may be called the *origin* of the band. The numerous lines which constitute the band arise from the numerous $F(J)$ terms. $F'(J')$ refers to the upper vibrational state and $F''(J'')$ to the lower, and they have the same forms as before (but subject to the interactions discussed in the previous section).

For diatomic molecules, and for linear molecules in those normal modes in which the dipole oscillates *parallel* to the molecular axis (the Σ modes), the selection rule $\Delta J = \pm 1$ operates. For pure rotation, of course, $\Delta J = -1$ could refer only to an emissive transition between the same two levels as are involved in the absorptive transition $\Delta J = +1$. In rovibrational transitions $\Delta J = +1$ means that $J' = J'' + 1$ (for either emission or absorption) and $\Delta J = -1$ means that $J' = J'' - 1$ (see Fig. 85). According to (3.106), (3.138) now becomes

$$\begin{aligned} \Delta J = +1: \quad \omega &= \omega_0 + B'(J''+1)(J''+2) - B''J''(J''+1) \\ &= \omega_0 + 2B' + (3B' - B'')J + (B' - B'')J^2 \qquad J = 0, 1, 2, \ldots \end{aligned} \tag{3.139}$$

$$\begin{aligned} \Delta J = -1: \quad \omega &= \omega_0 + B'(J-1)J - B''J(J+1) \\ &= \omega_0 - (B' + B'')J + (B' - B'')J^2 \qquad J = 1, 2, 3 \ldots \end{aligned} \tag{3.140}$$

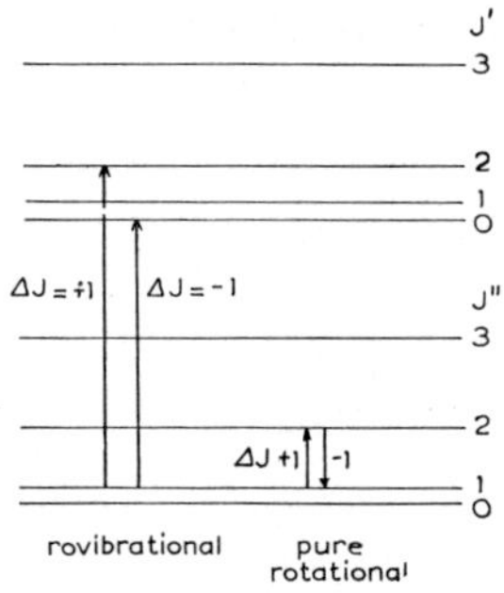

Fig. 85. $\Delta J = \pm 1$ in rotational and rovibrational transition.

where, after the first line, we have dropped the primes on J and consider J as referring to the initial state in absorption. Clearly the smallest J for $\Delta J = -1$ is $J = 1$. The B' and B'' are $B_{v'}$ and $B_{v''}$ in (3.137).

When the vibration of a linear molecule produces an oscillating dipole *perpendicular* to the internuclear axis the selection rule is $\Delta J = 0, \pm 1$ so that we may add:

$$\Delta J = 0: \quad \omega = \omega_0 + (B' - B'')J + (B' - B'')J^2 \qquad J = 0, 1, 2 \ldots \tag{3.141}$$

Rovibrational Raman transitions for diatomic molecules and parallel modes of linear molecules, have $\Delta J = 0, \pm 2$. For these transitions we have (3.141) and also:

$$\Delta J = +2: \quad \omega = \omega_0 + 6B' + (5B' - B'')J + (B' - B'')J^2 \qquad J = 0, 1, 2, 3 \tag{3.142}$$

$$\Delta J = -2: \quad \omega = \omega_0 + 2B' - (3B' + B'')J + (B' - B'')J^2 \qquad J = 2, 3, 4 \ldots \tag{3.143}$$

For perpendicular modes of linear molecules, $\Delta J = 0, \pm 1, \pm 2$ and the above five equations apply.

The rotational constants are not greatly different in the upper and lower vibrational states (they differ by the appropriate α in equation (3.137) and it is sufficient to illustrate the principal features of rovibrational bands by putting $B' = B'' = B$. It is customary to denote the lines for which $\Delta J = -2, -1, 0, +1, +2$ as O, P, Q, R and S branches respectively. To a first approximation we may therefore write:

$$\begin{aligned} O(J) &= \omega_0 + 2B - 5BJ & J &= 2, 3, 4 \ldots \\ P(J) &= \omega_0 \qquad - 2BJ & & 1, 2, 3 \ldots \\ Q(J) &= \omega_0 & & 0, 1, 2 \ldots \\ R(J) &= \omega_0 + 2B + 2BJ & & 0, 1, 2 \ldots \\ S(J) &= \omega_0 + 6B + 4BJ & & 0, 1, 2 \ldots \end{aligned} \tag{3.144}$$

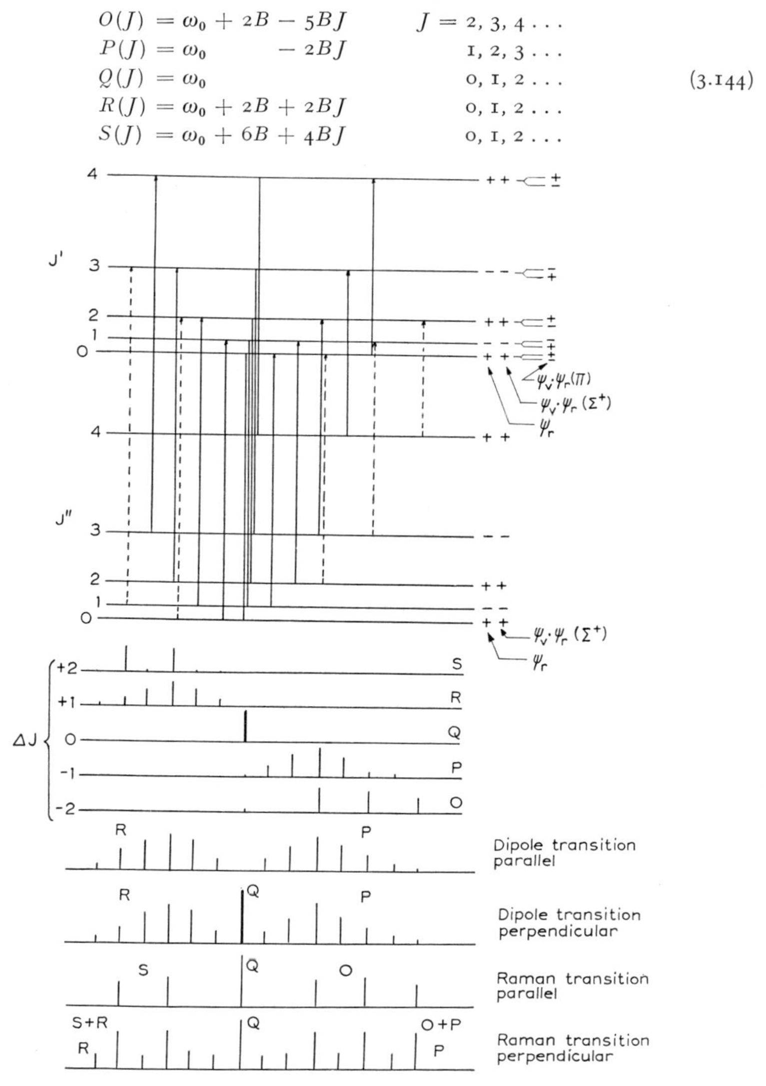

Fig. 86. Origin of rotational structure in vibration. Four complete bands are shown made up, in a manner depending upon the selection rules, from the individual branches shown above. The symmetries of ψ_r and $\psi_v\psi_r$ are indicated at the right of each level.

References p. 216–217

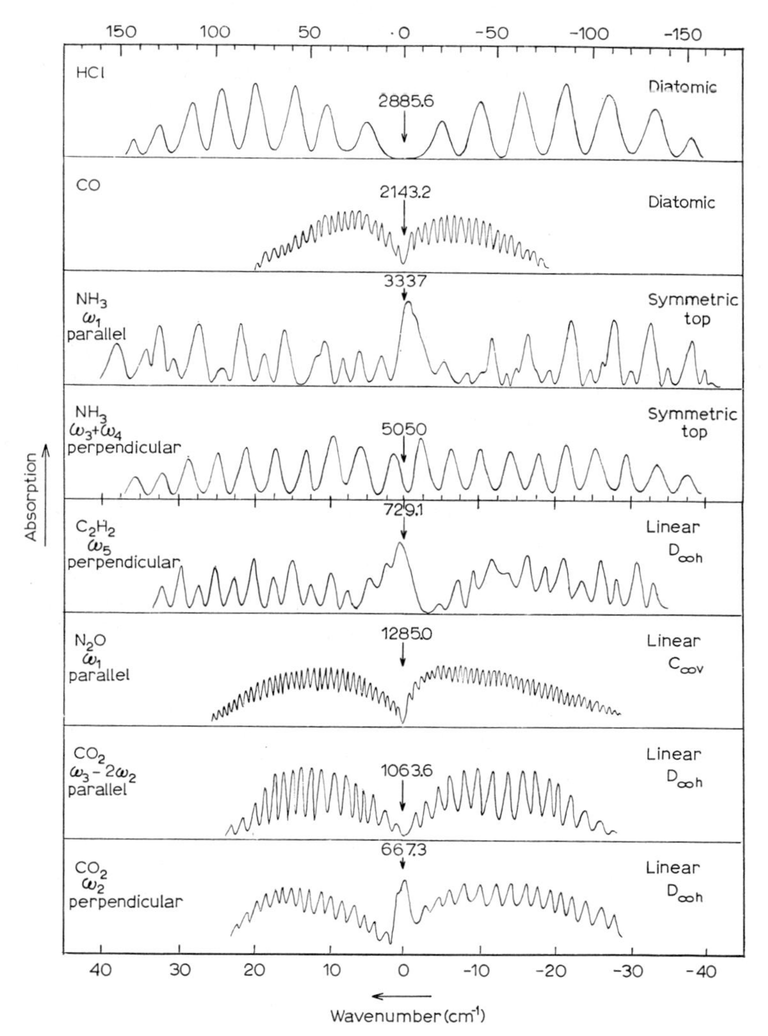

Fig. 87. Infra-red absorption bands in HCl, CO, NH_3, C_2H_2, N_2O and CO_2. The bands are brought to a common origin. The wavenumber scale of the bottom four bands is 4 times larger than that of the top four.

where $O(J)$ means the wave number of the line in the O branch arising from $J'' = J$ and so on. Fig. 86 illustrates this explanation of rotational structure.

In Fig. 87 the observed rotational structure of a number of infra-red absorption bands can be seen to conform to the above formulae. The absence of the Q branch ($\Delta J = 0$ not allowed) in the infra-red absorption of the diatomic molecules and the parallel Σ modes of the linear molecules is quite clear. The Π modes of the linear molecules have pronounced Q branches. Since $\Delta J = 0$ is allowed for the other classes of rotator the absence of a Q-branch in resolved rotational structure is a sure criterion of linearity. In a molecule known, and thus shown, to be linear the presence or absence of a Q-branch assists the characterization of vibrational states by identifying the Σ or Π modes. Our previous assignment of ω_1 and ω_2 for CO_2 is thus confirmed by the spectra in Fig. 87. The intensity alternation in the spectrum of C_2H_2 depends upon nuclear spin (p. 211). By the same cause the lines in CO_2 with odd J are completely absent, and hence the spacing of the rotational lines is about twice that for N_2O, although the moments of inertia are about the same. Rovibrational Raman spectra of C_2H_2 have been obtained under high resolution and provide examples of both types of band shown in Fig. 86.

A distinction between rotational transitions accompanying vibration and pure rotations is apparent from Fig. 87. The rotation of non-polar molecules such as CO_2 and C_2H_2 is observable in the rotational structure of infra-red bands even though there is no microwave absorption. This is also the case with spherical top molecules, whose pure rotation is not observed either in absorption or in the Raman effect. The tetrahedral group IV hydrides are the only spherical tops light enough for rotational structure to have been resolved.

Symmetric tops exhibit two sorts of band which are again determined by the direction of change of the dipole moment. When the associated vibrational change produces a transition moment *parallel* to the top axis, $\Delta K = 0$ and therefore the second term in equation (3.120) drops out when the difference in rotational constants in the upper and lower levels is neglected. With $\Delta J = 0, \pm 1$, P, Q and R structure given by (3.144) is to be expected. For vibrations whose transition moment is *perpendicular* to the top axis, $\Delta K = \pm 1$ and $\Delta J = 0, \pm 1$. Now, from (3.120), for $\Delta K = +1$ we find:

$$\begin{aligned} \Delta J &= -1, \quad \omega = \omega_0 - 2BJ + (A - B)(2K + 1) \\ \Delta J &= 0, \quad \omega = \omega_0 + (A - B)(2K + 1) \\ \Delta J &= +1, \quad \omega = \omega_0 + 2B(J + 1) + (A - B)(2K + 1) \end{aligned} \qquad (3.145)$$

This gives a set of *sub-bands* each with its own ${}^{R}P$, ${}^{R}Q$, ${}^{R}R$ structure but with different sub-origins depending upon the value of K. A similar set of sub-bands exists at lower frequencies for $\Delta K = -1$ each with its ${}^{P}P$, ${}^{P}Q$, ${}^{P}R$ structure (the superscript R or P refers to $\Delta K = +1$ or -1). Most usually, and especially when $A \gg B$, the prominent feature of each sub-band is the Q branch and the resultant spectrum is a set of lines, with no very clear band origin (4th spectrum in Fig. 87). The spacing between these lines is then $2(A - B)$ a quantity not obtained from microwave spectra of symmetric tops. The spacing between ${}^{R}Q(1) = \omega_0 + (A - B)$ and ${}^{P}Q(1) = \omega_0 - (A - B)$ is also $2(A - B)$, which accounts for the absence of any clear indication of the band origin.

Unresolved bands

Many molecules, of each rotational class, have moments of inertia which are too large to permit resolution of rovibrational structure. In this event the envelope of the band is recorded. Fig. 88 shows some typical contours. Each band of a spherical top molecule shows the same contour, a fact which serves to confirm the assigned symmetry. The intensity of the Q-branch in a symmetric top spectrum differs according to whether the band is a parallel or perpendicular one, and according to the ratio $A : B$.

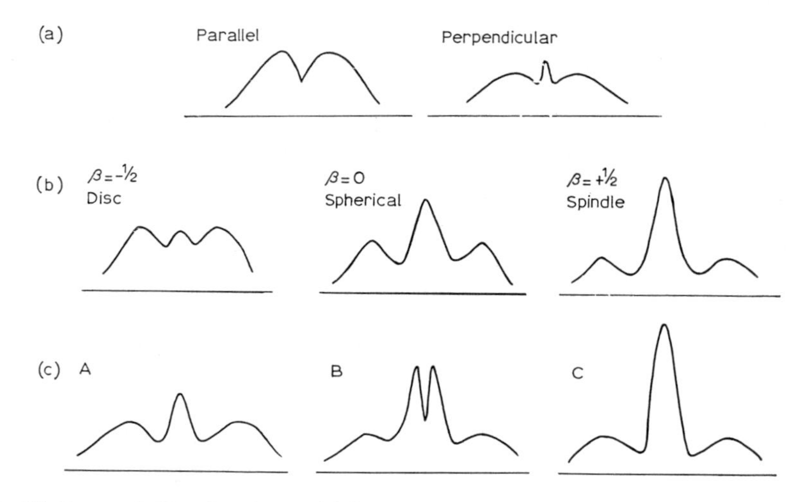

Fig. 88. Some rotational contours: (a) linear molecule; (b) symmetric tops, perpendicular mode, and spherical top, $\beta = (I^{b}/I^{a}) - 1$; (c) asymmetric top, $I^{a} : I^{b} : I^{c} = 0.69 : 0.75 : 1$.

In an asymmetric top molecule three types of band occur labelled A, where the vibration transition moment lies parallel to the I^{a}-axis, B when it lies parallel to the I^{b}-axis, and C when it is parallel to the I^{c}-axis. The shapes of the three types of band, and in particular the intensity of the Q-branch, depends upon the relative values of the rotational constants. The shapes of contours for symmetric tops are discussed by Gerhard and Dennison[23], and for asymmetric tops by Badger and Zumwalt[24]. A particular example of a B-type band is that shown in Fig. 85, where there are two Q maxima. This contour is observed in the spectrum of SF_4 and has been used[18] as evidence that that molecule has C_{2v} symmetry (asymmetric top) rather than C_{3v} symmetry (symmetric top).

Molecular dimensions from rotational structures

Rotational structure observed under high resolution reveals the need to take note of the different values of the rotational constants in the upper and lower vibrational states, and to allow for the possibility of centrifugal stretching. The analysis of the P and R branches now requires use of equations (3.139) and (3.140). If ω is plotted against J curvature is observed because of the J^2 term. The resultant curve can be fitted to the appropriate equation but a more accurate mode of plotting is to take differences. It can easily be shown from equations (3.139) and (3.140) that

$$R(J) - P(J) = 4B'(J + \tfrac{1}{2}) \tag{3.146}$$
$$R(J - 1) - P(J + 1) = 4B''(J + \tfrac{1}{2}) \tag{3.147}$$

where, as before, $R(J)$ means the wave number of the line in the R-branch originating the Jth level. These two differences plotted against J give straight lines whose slopes are respectively $4B'$ and $4B''$ and whose intercepts are $2B'$ and $2B''$. The sum

$$R(J - 1) + P(J) = 2\omega_0 + 2(B' - B'')J^2 \tag{3.148}$$

may be plotted against J^2 to give a straight line whose slope is $B' - B''$ and whose intercept gives ω_0, the origin of the band. Corresponding differences and sum equations can be used for perpendicular symmetric top bands involving K.

As an example of resolved structure in the infra-red, we give in Fig. 89 the structure of the band $(\omega_3 - \omega_2)$ for HCN[25]. (For this particular transition $R(O)$ should be absent, as is evidently the case.) The difference $R(J) - P(J)$ is plotted against J and the sum $R(J - 1) + P(J)$ against J^2. From the graphs $B' - B'' = -0.0099$, $B' = 1.4683$ and $B'' = 1.4782$ cm^{-1}. For this transition the lower level is that in which the bending mode is singly excited ($v_2 = 1$, $v_1 = 0$, $v_3 = 0$), *i.e.* $B'' = B_{010}$. The bending of the molecule may lead to two moments of inertia, and hence two values of B_{010}, depending whether the deformation is parallel to or perpendicular to the axis of rotation. The result is a doublet level ($v_2 = 1$) and the phenomenon is known as *l-doubling*. So far as the rotational spectrum is concerned the transition $\Delta J = \pm 1$ for $(\omega_3 - \omega_2)$ starts from one of the two levels and $\Delta J = 0$ from the other. The Q-branch is resolved in Fig. 89 and should fit an equation (3.141). It can be seen that $B' - B''_Q$ is negative but its value, -0.0174 cm^{-1}, is slightly different from $B' \quad B''_{PR}$. The difference $B''_{PR} - B'_Q = 0.0174 -$

References p. 216–217

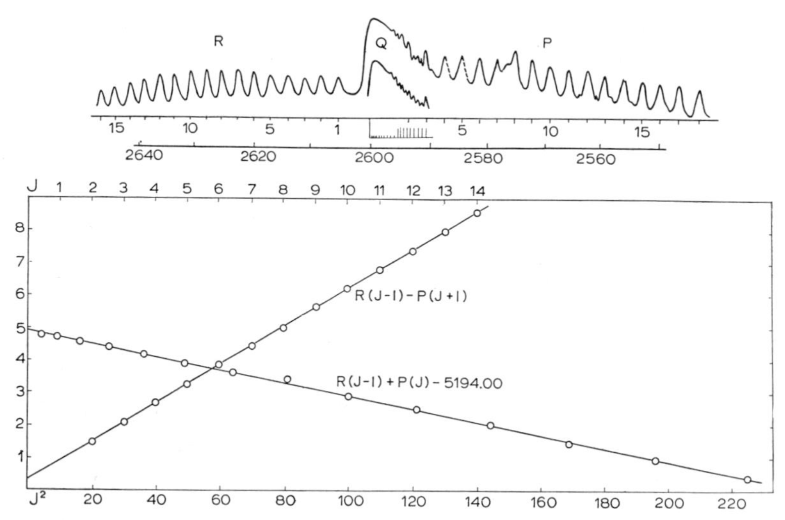

Fig. 89. Rovibrational bands (ω_3—ω_2) for HCN with (a) a difference plot $\frac{1}{10}$ $[R(J—1)—P(J+1)]^3$ against J and (b) a sum plot $[R(J—1)+P(J)—5194.00]$ against J^2.

0.0099 = 0.0075 agrees closely with the value 0.00743 obtained from microwave measurements[26]. $B'_{PR} = 1.4782$ cm^{-1} $= B^{PR}_{010}$. Other rotational bands, *e.g.* $3\omega_2$, for which the lower level is the ground state, also give $B'' = 1.4782$ cm^{-1} $= B_{000}$. Thus for one direction of the degenerate mode ω_2 there is no change in moment of inertia, compared with the ground state. The value of B_{000} in frequency units is 44,305 Mc/s which compares favourably with the microwave value given in Table 15. In order to determine the two internuclear distances in HCN it is necessary, as before, to measure B values similarly for DCN.

Rotational constants determined from rovibrational bands generally agree with microwave and rotational Raman values. Thus: for N_2O, B_{000} = 0.4203 cm^{-1} = 12,600 Mc/s and for FCN, B_{000} = 0.3515 cm^{-1} = 10538 Mc/s, results that are in agreement with the figures in Table 15. Table 18 gives some more examples of the agreement between pure rotational (microwave and Raman), rovibrational and rovibronic values of internuclear distances. Interesting discrepancies, which are as yet unexplained, appear in the spectra of the similar molecules $CH_2{=}C{=}O$, $CH_2{=}N{=}N$, $CD_2{=}C{=}CH_2$. In $CH_2{=}C{=}O$ the B_0 obtained for three parallel bands[27] is close to 0.3370 cm^{-1} in very good agreement with the microwave value 0.3371 cm^{-1}. In two other parallel bands which are assigned to modes in which the stretching of the $C{=}C{=}O$ chain is the principal motion, B_0 = 0.3403 and 0.3402 cm^{-1}. In $CH_2{=}N{=}N$ the parallel band connected with CH_2 deformation gives

$B_0 = 0.3698$ cm^{-1} but the two modes for C=N=N stretch yield $B_0 =$ 0.3724, 0.3726 cm^{-1}. And for CD_2=C=CH_2, but evidently not allene itself nor fully deuterated allene, two parallel bands give $B_0 = 0.2615$ and 0.2617 but from the parallel band involving the C=C=C antisymmetric stretch $B_0 = 0.2655$ cm^{-1}. The pure rotational Raman spectrum gives $B_0 =$ 0.2619 cm^{-1}. The value of $r_{CC'} = 1.309$ Å in allene, is considerably shorter than 1.339 Å found in ethylene and is therefore of considerable chemical interest. From a spectroscopic point of view, the interest lies in the high estimates of B_0 obtained from bands due to some X=Y=Z stretching vibrations. The discrepancy evidently does not extend to similar vibrations in FCN and N_2O.

Where rotational structure cannot be resolved it is still possible to obtain a rough measure of moment of inertia from the separation of the maxima in the P and R envelopes. For example the PR separation of rotational contours in the spectrum of NO_2F gives A = 11,300, B = 10,500 and C = 5,150 Mc/s. These are about 15 % too low by comparison with Table 17 but they would give an indication in the absence of more exact values.

TABLE 18

COMPARISON OF VALUES OF ROTATIONAL CONSTANTS AND DIMENSIONS OBTAINED BY VARIOUS SPECTROSCOPIC METHODS

	B (cm^{-1})				r (Å)			
	M	R	I	UV	M	R	I	UV
H_2		59.3392[a]	59.318*[b]	59.331[b]		0.75105		
		60.841	*60.809*	*60.848*		*0.74173*	*0.74166*	*0.74142*
HD		44.6678[a]		44.684[c]		0.74973		
		45.638		*45.655*		*0.74173*		*0.74136*
D_2		29.9105[a]		29.927[c]		0.74820		
		30.442		*30.459*		*0.74165*		*0.74164*
$^{16}O_2$	1.43773**[d]			1.437770[e]				1.210714
				1.445666	*1.20741*			*1.207404*
$^{12}C^{16}O$	1.922515[f]		1.922523[f]					
	1.922625[g]		1.92265[h]	1.9226[i]	1.13079		1.13075	1.1308
	1.931367		*1.93139*		*1.12823*		*1.12819*	*1.1282*
$^{127}I^{35}Cl$	0.1138940[j]			0.11389[j]				
	0.1141619				*2.32070*			*2.3207*
$^{14}N_2{}^{16}O$	0.419034[j]		0.41913[j]					
CH_3.C≡CH	0.28501[j]	0.29653[k]	0.2848[j]					

* from quadrupole rotation-vibration bands.
** from magnetic dipole rotational spectrum.
Numbers in Roman type give B_0 or r_0; those in italics give B_e or r_e. M, microwave absorption; R, rotational Raman; I, infra-red rovibrational spectra; UV, ultraviolet spectra, fine structure.

References p. 216–217

References:

(a) B. P. Stoicheff, *Can. J. Phys.*, 35 (1957) 730.
(b) G. Herzberg, *Can. J. Res.*, A28 (1950) 144.
(c) H. C. Urey and G. K. Teal, *Rev. Mod. Phys.*, 7 (1935) 34.
(d) M. Tinkham and M. W. P. Strandberg, *Phys. Rev.*, 97 (1955) 951.
(e) H. D. Babcock and L. Herzberg, *Astrophys. J.*, 108 (1948) 167.
(f) E. K. Plyler, L. R. Blair and W. S. Connor, *J. Opt. Soc. Am.*, 45 (1955) 102, — a comparison of microwave and infra-red measurements to determine the velocity of light.
(g) O. R. Gilliam, C. M. Johnson and W. Gordy, *Phys. Rev.*, 78 (1950) 140.
(h) G. Herzberg and K. N. Rao, *J. Chem. Phys.*, 7 (1949) 1099.
(i) K. N. Rao, *Astrophys. J.*, 110 (1949) 307.
(j) G. Herzberg, *Discussions Faraday Soc.*, 9 (1950) 208.
(k) B. P. Stoicheff, *Can. J. Phys.*, 33 (1955) 811.

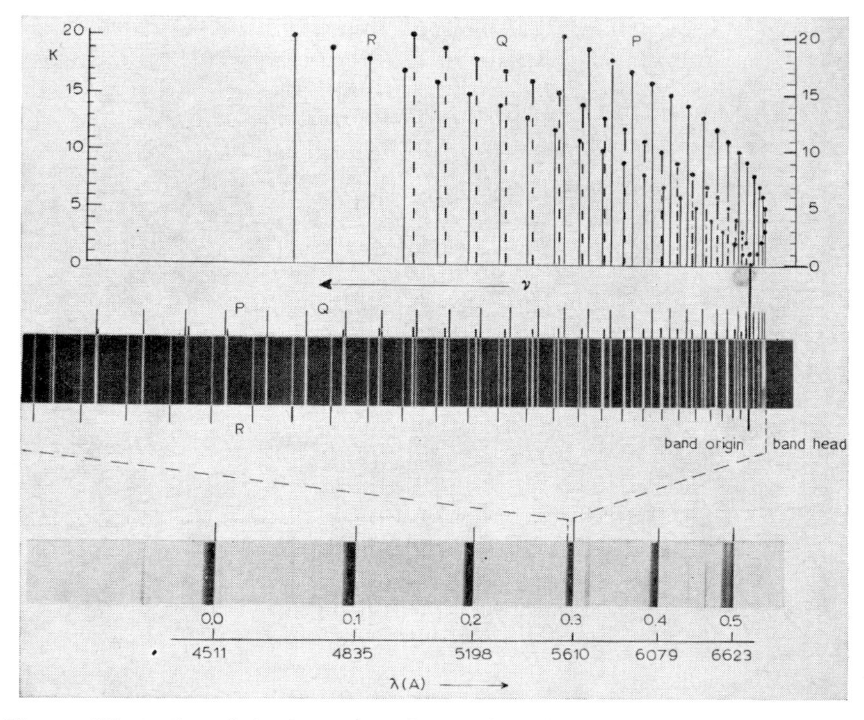

Fig. 90. Illustration of the formation of a band head in rovibronic spectra. Fortrat diagram and fine structure of an Angstrom band in carbon monoxide at 5610 A. The coarse vibrational structure of the band system is shown in the lower spectrum.

Rotation in electronic states

Some accurate values of the internuclear distances in diatomic molecules depend upon the high resolution analysis of rotational structure in electronic emission bands. The interpretation of these again makes use of equations (3.139) and (3.140) and (3.141), since the selection rules remain

$$\Delta J = 0, \pm 1$$

When, however, B' and B'' refer to different *electronic* states they may differ to a much greater extent than in purely vibrational transitions since a different electronic state implies a different strength of binding. The effect of large differences in rotational constants is to make the J^2 term outweigh the J term very rapidly. If, for example, B' is greater than B'', the P-branch (equation 3.140) starts off with its first few members going to lower frequency. But at a value of $J^2 = (3B' - B'')/(B' - B'')$ the branch turns back to higher frequencies and later numbers continue to diverge to higher frequencies. This gives rise to a so-called *band head* at a frequency $\omega_h = \omega_0 - (B' + B'')^2/4(B' - B'')$ which is much more prominent in the spectrum than the band origin. This behaviour of the P-branch and the parallel behaviour of the Q and R-branches is shown in Fig. 90. The band is shown *degrading to the violet* (*i.e.* to short wavelength) and this is an indication that B' is greater than B''. In such a case the moment of inertia in the upper state is less than in the lower state, or the internuclear distance in the upper state is smaller than in the lower state. It can easily be seen from equation (3.139) that a band *degrades to the red* (*i.e.* the R-branch forms a band head on the short wavelength side of the origin) if B'' is greater than B'. In this case the internuclear distance is greater in the upper state than in the lower.

Recognition of this phenomenon in rovibronic spectra is a necessary preliminary to assignment of rotational lines. When they have been correctly assigned then the differences and sums can be plotted against J or J^2 to give moments of inertia (as on p. 208). It is by this means that the internuclear distances of such molecules as O_2 ($r_e = 1.20739_8$ Å), Cl_2 ($r_e = 1.988$ Å), Br_2 ($r_e = 2.283_6$ Å) and I_2 ($r_e = 2.6666$ Å) have been established.

Nuclear Spin

We have already considered (p. 103) the basic phenomenon of nuclear spin and the effect of its interaction with an external magnetic field. Here we are concerned only with its effect on molecular rotational states. In this connection it appears in two ways: in determining the existence, or the probability of occurrence, of rotational states of symmetrical molecules and in the interaction between nuclear spin and rotational energy which leads to the hyperfine structure of rotational spectra.

References p. 216–217

Symmetry properties including nuclear spin

According to equations (3.104 p. 187) rotational eigenfunctions ψ_r either remain unchanged or change sign when all particles are reflected through the origin (= for θ write $\theta + \pi$). They are accordingly labelled + or — and it can be seen that the even levels (J = 0, 2, 4 . . .) are positive and the odd levels (J = 1, 3, 5 . . .) are negative. The vibrational eigenfunction ψ_v of molecules may be similarly analyzed. So far as linear molecules are concerned the ground vibrational state is symmetric and the excited Σ^+ states are also symmetric with respect to inversion. In consequence the products $\psi_v \cdot \psi_r$ for these vibrational states are also positive when J is even and negative when J is odd. For the degenerate Π vibrational state, however, each rotational level splits into two levels of slightly different energy for one of which $\psi_v \cdot \psi_r$ is positive and for the other it is negative, as shown in Fig. 86.

In symmetrical linear molecules (point group $D_{\infty h}$) it is found that the existence of some rotational states is limited, or altogether precluded, when nuclear spin is taken into account. Such a molecule contains at least one pair of *identical* nuclei and, as with electrons on p. 112, the eigenfunction describing the spin of two identical nuclei must allow for the indistinguishability. We may take as an example nuclei with spin quantum number $I = \frac{1}{2}$ (*e.g.* protons) so that with respect to a chosen direction, the angular momentum μ_2 may be either $\pm \mu_0$ (p. 92), which we may indicate as $\alpha(+)$ or $\beta(-)$. An eigenfunction which describes a possible spin orientation of two such nuclei as $\alpha_1\alpha_2$ or $\beta_1\beta_2$ is acceptable since exchange of the nuclei leads to an identical situation. On the other hand $\alpha_1\beta_2$ and $\alpha_2\beta_1$ are not acceptable solutions and must be replaced by linear combinations, $(\alpha_1\beta_2 + \alpha_2\beta_1)$ and $(\alpha_1\beta_2 - \alpha_2\beta_1)$. We now have four acceptable solutions of which three are symmetric with respect to inversion and one is antisymmetric.

$$\alpha_1\alpha_2, \quad \beta_1\beta_2, \quad (\alpha_1\beta_2 + \alpha_2\beta_1) \qquad \text{— symmetric}$$
$$(\alpha_1\beta_2 - \alpha_2\beta_1) \qquad \text{— antisymmetric}$$

For nuclei with $I = 1$ (*e.g.* deuterons) $\mu_z = +\mu_0$, 0, or $-\mu_0$ for which we write eigenfunctions α, 0, β. There are now six symmetric and three antisymmetric combinations.

$$\alpha_1\alpha_2, \quad 0_10_2, \quad \beta_1\beta_2, \quad (\alpha_10_2 + 0_1\alpha_2)(\alpha_1\beta_2 + \beta_1\alpha_2)(0_1\beta_2 + \beta_10_2) \qquad \text{— symmetric}$$
$$(\alpha_10_2 - 0_1\alpha_2), \quad (\alpha_1\beta_2 - \beta_1\alpha_2), \quad (0_1\beta_2 - \beta_10_2) \qquad \text{— antisymmetric}$$

For nuclei with $I = 0$ (*e.g.* ^{12}C, ^{16}O) the spin function can only be $^0\phi_1{}^0\phi_2$: that is, there is no antisymmetric spin function for nuclei with $I = 0$.

We now recall Pauli's principle that the total eigenfunction in odd nuclei (odd mass number) is always antisymmetric and in even nuclei (even mass number) is always symmetric. Accordingly for H · C≡C · H of H—H the total eigenfunction for the protons ($I = \frac{1}{2}$) must be antisymmetric and consequently only those spin functions which are symmetric will occur with antisymmetric $\psi_v \cdot \psi_r$ and vice versa. Thus in the ground vibrational state, and in all Σ^+ excited states,

antisymmetric nuclear spin function $\leftrightarrow$ even J
symmetric nuclear spin function $\leftrightarrow$ odd J

Therefore the odd J levels have a statistical weight of 3 compared with one for the even J levels. This is the explanation of ortho- and para-hydrogen. It is also the cause of the alternation in intensity of the lines in the rovibrational spectrum of C_2H_2 in Fig. 87, since the number of molecules in the odd J levels of the ground state is three times the number in the even J levels. The lines originating in odd J levels are therefore three times more intense.

For even nuclei Pauli's principle results in the associations (for the ground vibrational state):

symmetric nuclear spin function $\leftrightarrow$ even J
antisymmetric nuclear spin function $\leftrightarrow$ odd J

and thus for $D \cdot C{\equiv}C \cdot D$ and D_2 (where D has $I = 1$) the even and odd rotational levels have statistical weights of 6 and 3 respectively. In C_2D_2 the corresponding rotational spectrum shows a greater intensity in the lines originating from even J levels (in contrast to C_2H_2).

The even nucleus ^{16}O has $I = 0$ so that in $^{16}O{=}C{=}^{16}O$ the antisymmetric nuclear spin function does not occur. Consequently the odd rotational levels do not occur at all and the rotational spectrum of CO_2 (Fig. 87) has alternate lines missing. The spacing between the lines which are present is accordingly $4B$ rather than $2B$.

These observations are of great importance not only because they affect the estimate of the moment of inertia of CO_2 but because they are evidence for Pauli's principle and for the assignment of spin quantum numbers to nuclei.

Interaction with rotation

Nuclear spin may interact with molecular rotation by coupling between the magnetic moments of the two motions. There is, however, a larger interaction, and the one more frequently observed, which is electrical in nature. It depends upon the *nuclear quadrupole moment, Q*, of the nucleus. Nuclear theory cannot, as yet, predict values of Q and any theory of the nucleus must account for the experimental observation that only nuclei with $I \geqq 1$ have quadrupole moments. This excludes (see Table 8, p. 106) a number of common nuclei with zero spin (^{12}C, ^{16}O, ^{32}S) or with $I = \frac{1}{2}$ (^{1}H, ^{13}C, ^{15}N, ^{19}F, ^{29}Si, ^{31}P).

References p. 216–217

The quadrupole moment measures a deviation from spherical symmetry of the charge distribution on the nucleus itself, as is illustrated in Fig. 91(a). The different orientations of charge with respect to the spin axis account for positive and negative values of Q. From the chemical point of view the significance of the nuclear quadrupole moment lies in its use as a sensitive measuring device or probe embedded in the molecule. We will not discuss in detail the analysis of the hyperfine structure of microwave absorption spectra but note that such an analysis can lead to information about (a) the nuclear spins themselves and (b) the coupling between the nuclear spin and the molecular rotation. A number of nuclear spins were first determined in this way, *e.g.* ^{10}B ($I = 3$) ^{11}B ($\frac{3}{2}$), ^{35}Cl ($\frac{3}{2}$), ^{37}Cl ($\frac{3}{2}$).

The coupling between nuclear spin and molecular rotation depends upon the product eQq (*the quadrupole coupling constant*) in which e is the electronic charge, Q is the quadrupole moment and q is the gradient of the electronic field at the nucleus. In fact the energy of a rotational level is perturbed by an amount

$$W_Q = -eQq \cdot f(I, J, F) \tag{3.103}$$

where I is the nuclear spin quantum number, J the molecular rotational quantum number and F the quantum number for total angular momentum, $\boldsymbol{I} + \boldsymbol{J}$. The function $f(I, J, F)$ is a cumbersome algebraic combination of I, J and F: for all $J = 0$, $f = 0$. For $I = 1$ and $J = 1$, $f = 1/20$, $-\frac{1}{4}$ and $\frac{1}{2}$. Thus the $J = 0 \rightarrow 1$ transition of a molecule containing an atom with $I = 1$

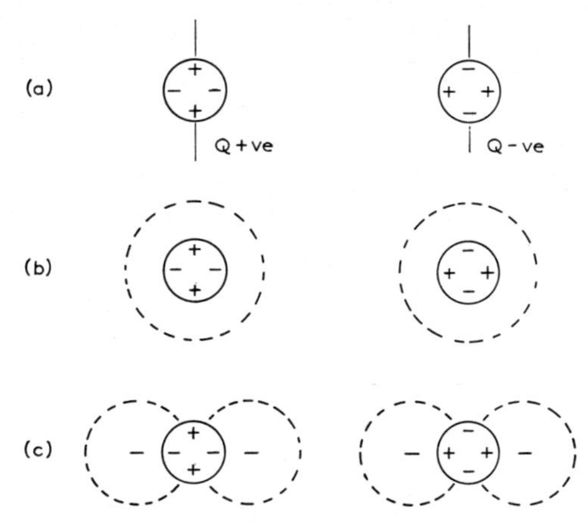

Fig. 91. (a) Nuclear quadrupole moments and their interaction with (b) *s* electrons, and (c) *p*-electrons. The sign here refers to charge, not to the sign of a wave function.

would be expected to have hyperfine structure comprising three lines separated by 0.45 eQq and 0.30 eQq. Higher transitions show more structure and so do nuclei of higher I. Nevertheless the method of assessing eQq is the same in principle.

Now if the electron distribution immediately surrounding the nucleus in question is spherically symmetrical, as is a closed shell or an s electron, the field gradient q at the nucleus is zero or, as illustrated in Fig. 91(b), the orientation of Q is immaterial in a spherical charge distribution. This can be seen more clearly if we write the potential V at the nucleus due to any electron at coordinates x, y, z with respect to the nucleus, or at distance r from it:

$$V = \frac{e}{r} = e(x^2 + y^2 + z^2)^{-\frac{1}{2}}$$

Then it is readily shown that

$$\frac{\partial^2 V}{\partial z^2} = e\,\frac{3z^2 - r^2}{r^5} = e\,\frac{3\cos^2\theta - 1}{r^3} \tag{3.104}$$

if θ is the angle between the z-axis and the radium vector r. q is then defined as the average value of $\partial^2 V/\partial z^2$ over the space distribution of the electron. We leave the details of the averaging[34] but it can be seen that a spherical charge distribution leads to $q = 0$. It is also very small for electrons well removed from the nucleus because of the $1/r^3$ term. Consequently only p electrons located in the nucleus in question make a significant contribution to q, and their interaction with Q is shown diagrammatically in Fig. 91(c).

In the rotational spectrum of gaseous NaCl and KCl eQq for ^{35}Cl ($I = \frac{3}{2}$) has very small values, < 1 and < 0.04 Mc/s as compared with -53 Mc/s in, say, HCl. It is clear that NaCl and KCl can, even in the gaseous state, be reliably regarded as ion pairs Na^+Cl^- since the Cl nucleus is evidently surrounded by a spherically symmetrical charge cloud. On the other hand, from radio-frequency atomic-beam spectroscopy, atomic chlorine, which is one p-electron short of the filled shell, has $eQq = 109.7$ Mc/s. Approaching this value is the coupling constant for ^{35}Cl in BrCl for which $eQq = -103.6$ Mc/s. The sign of eQq can be determined in molecular hyperfine structure but not in atomic beam experiments: the negative sign in BrCl indicates a defect of one p-electron from a closed shell configuration. In FCl, eQq for ^{35}Cl is as much as -146 Mc/s. It is evident that in these molecules (HCl, BrCl, FCl) the chlorine nucleus is embedded in an unsymmetrical charge distribution, such as must arise where there is a covalent bond. To regard this merely as a p-electron defect is obviously an oversimplification and account must be taken of redistribution of the remaining electrons. This can be done[28, 33] in terms of covalent-ionic resonance and of hybridization but goes outside the scope of this book.

References p. 216–217

REFERENCES

[1] F. A. JENKINS, *Phys. Rev.*, 31 (1928) 539.
[2] R. RYDBERG, *Z. Physik*, 80 (1933) 514.
[3] O. KLEIN, *Z. Physik*, 76 (1932) 226.
[4] A. L. G. REES, *Proc. Phys. Soc. (London)*, 59 (1957) 998.
[5] P. M. MORSE, *Phys. Rev.*, 34 (1929) 51; 42 (1932) 143.
[6] R. RYDBERG, *Z. Physik*, 73 (1932) 376.
[7] E. R. LIPPINCOTT, *J. Chem. Phys.*, 21 (1953) 2070; 26 (1957) 1678.
[8] J. W. LINNETT, *Trans. Faraday Soc.*, 36 (1940) 1123; 38 (1942) 1.
[9] R. T. BIRGE and H. SPONER, *Phys. Rev.*, 28 (1926) 259.
[10] H. BEUTLER and H. O. JÜNGER, *Z. Physik*, 101 (1936) 304.
[11] E. T. WHITTAKER, *Analytical Dynamics of Particles and Rigid Bodies*, Dover Publs., 1944.
[12] L. A. WOODWARD, *Quart. Revs. (London)*, 10 (1956) 185.
[13] H. MARGENAU and G. M. MURPHY, *The Mathematics of Physics and Chemistry*, 2nd. ed., Van Nostrand, New York, 1956, Chap. 15.
[14] W. BURNSIDE, *The Theory of Groups*, Cambridge Univ. Press, London, 1927.
[15] J. GAUNT, *Trans. Faraday Soc.*, 49 (1953) 1122.
[16] L. A. WOODWARD and H. L. ROBERTS, *Trans. Faraday Soc.*, 52 (1956) 615.
[17] R. E. DODD, *Trans. Faraday Soc.*, 55 (1959) 1480.
[18] R. E. DODD, L. A. WOODWARD and H. L. ROBERTS, *Trans. Faraday Soc.*, 52 (1956) 1053.
[19] R. C. LORD, D. W. ROBINSON and W. C. SCHUMB, *J. Am. Chem. Soc.*, 78 (1956) 1327.
[20] D. C. MCKEAN, R. TAYLOR and L. A. WOODWARD, *Proc. Chem. Soc.*, (1959) 321.
[21] J. W. LINNETT, *Quart. Revs. (London)*, 1 (1947) 73.
[22] B. P. STOICHEFF, *Advances in Spectroscopy*, Vol. I, Interscience, New York, 1959.
[23] S. L. GERHARD and D. M. DENNISON, *Phys. Rev.* 43 (1933) 197.
[24] R. M. BADGER and L. R. ZUMWALT, *J. Chem. Phys.*, 6 (1938) 711.
[25] I. R. DAGG and H. W. THOMPSON, *Trans. Faraday Soc.*, 52 (1956) 455.
[26] R. G. SCHULMAN and C. H. TOWNES, *Phys. Rev.*, 77 (1950) 421.
[27] P. E. B. BUTLER, D. R. EATON and H. W. THOMPSON, *Spectrochim. Acta*, 13 (1959) 323.
[28] W. J. ORVILLE-THOMAS, *Quart. Revs. (London)*, 11 (1957) 162.

Essential reference books for most of the contents of this chapter are:

[29] G. HERZBERG, *Spectra of Diatomic Molecules*, Van Nostrand, New York, 1950.
[30] G. HERZBERG, *Infra-red and Raman Spectra*, Van Nostrand, New York, 1945.

As well as giving a very full account of the theory, these books also contain detailed descriptions of the spectra of many molecules and comprehensive lists of references to the spectra molecules.

Many Raman spectra, with some treatment of vibrational theory, are also given by:

[31] K. W. F. KOHLRAUSCH, *Ramanspektren*, Akademische Verlagsges. Geest & Portig Kg., Leipzig, 1943.

The use of group theory in the elucidation of vibrational states is covered by:

[32] E. B. WILSON, J. C. DECIUS and P. C. CROSS, *Molecular Vibrations*, McGraw-Hill, New York, 1955.

The two books:

[33] W. GORDY, W. V. SMITH and R. F. TRAMBARULO, *Microwave Spectroscopy*, Wiley, New York, 1953.

[34] C. H. TOWNES and A. L. SCHAWLOW, *Microwave Spectroscopy*, McGraw-Hill, New York, 1955, deal thoroughly with a subject which had already provided an enormous collection of data in a little under ten years. See also:

[35] D. J. E. INGRAM, *Spectroscopy at Radio and Microwave Frequencies*, Butterworth, London, 1955.

Recent text books, which treat spectroscopic methods as well as others, are:

[36] P. J. WHEATLEY, *Molecular Structure*, Oxford Univ. Press, London, 1959.

[37] J. C. D. BRAND and J. C. SPEAKMAN, *Molecular Structure*, Arnold, London, 1960.

A valuable compilation of molecular parameters, to which spectroscopic studies have made an important contribution, is:

[38] *Interatomic Distances*, Chem. Soc. (London), Spec. Publ. No. 11 (1958).

For bond energies and dissociation energies, see:

[39] T. L. COTTRELL, *Strengths of Chemical Bonds*, 2nd ed., Butterworth, London, 1958.

[40] A. G. GAYDON, *Dissociation Energies*, Chapman and Hall, London, 1947.

[41] W. JEVONS, *Report on Band Spectra of Diatomic Molecules*, Phys. Soc. (London), 1932.

The structure of electronic spectra is the main topic of:

[42] R. C. JOHNSON, *Introduction to Molecular Spectra*, Methuen, London, 1949.

The Journals in which many papers are published dealing with the subjects of this chapter are:

The Journal of Chemical Physics,

Molecular Physics (commenced publication 1958),

Spectrochimica Acta (commenced publication 1941; Volumes 1 and 2 by Springer, 1941, 1944; Volume 3 by the Vatican Observatory, 1947-9; Volume 4 and subsequently by Pergamon Press),

Optics and Spectroscopy (*U.S.S.R*), (English Translation, commenced 1959, of *Optika i Spektroskopiya*).

Chapter 4

INTERACTIONS AND EMPIRICAL CORRELATIONS

Although it remains true that all observed spectroscopic frequencies represent transitions between characteristic states it is not true to say that a detailed interpretation of these states can always be given. The last two chapters have been concerned with those situations in which the states can be fairly completely characterized. We must now turn to spectroscopic observations in systems which do not permit that sort of detailed analysis but are rather interpreted by correlation and analogy. Spectroscopic features are found to be common to molecules which have some chemical features in common, be it a double bond, a methyl group or a particular configuration of fluorine atoms. It is then reasonable to anticipate that the correlation can be extended and to suspect the presence of the chemical feature whenever the spectroscopic feature appears. Guidance must be sought from more detailed interpretations of simpler systems to define the scope and limitations of the correlation: to that extent the correlation is not wholly empirical.

We also include in this chapter some discussion of the many cases in which well-recognised states are perturbed by local interactions within the molecule, or between molecules, or in a crystal field. Again the interactions, observed as frequency shifts, may be susceptible to theoretical analysis or they may be the subject of further empirical correlations.

Undoubtedly much of the use of spectroscopy in chemical problems falls within the scope of this chapter. To judge from the volume of literature available for illustration the empirical nature of the correlations hardly diminishes their usefulness and certainly provides the material for further theoretical speculation. Indeed the use of spectroscopy is now so commonplace that spectra are no longer regularly published in support of arguments for chemical constitution and it is not unusual to find the bald statement that "the infra-red, Raman and n.m.r. spectra were consistent with the proposed structure".

Chromophores

Very early in the development of organic chemistry it was recognised that colour could be imparted to hydrocarbons by substitution of appropriate chemical groups, notably the azo-, quinonoid and nitroso-groups. The name *chromophore* was given to such groups by Witt in 1876. Since the range of observation has extended on either side of the visible region many more groups have been found to give rise to characteristic absorptions. Infra-red, Raman and nuclear magnetic resonance spectra of molecules can be so characteristic as to be used as "finger-prints" for the certain identification of individual molecules. Electronic spectra are somewhat less selective, but often useful.

Ultraviolet and Visible Absorption Spectra

Although the study of band spectra of gaseous diatomic molecules has in many cases enabled a fairly detailed characterization of electronic states the process is much more difficult in polyatomic molecules. Some progress has been made in predicting electronic levels in hydrocarbons containing double bonds and aromatic rings. Some partial term schemes may be drawn up from the observed transitions and interpreted in terms of possible electron migrations. Ionisation potentials of some molecules have been measured (W. C. Price, R. Bralsford, P. V. Harris, R. G. Ridler, page 54 of Ref. 99) and in some cases bands have been found to conform to a Rydberg series as in atomic spectra. It is, of course, the extreme richness of vibrational and rotational structure, generally unresolved even in the vapour spectrum of larger molecules, which leads to broad bands in the ultraviolet. The absorption maximum may not accurately define the origin of the band and bands with different origins may overlap considerably. Spectra may be subject to frequency changes to some extent when they are obtained from liquids, and even more so when they have to be obtained from solutions (see p. 280). Interaction between electronic and vibrational states can often be sufficient to lead to chemical dissociation of the absorbing molecule (see p. 321): thus at 4000, 3000 and 2000 Å respectively the quantum corresponds to 71.5, 95.5 and 143 kcal/mole.

Much of the absorption spectroscopy practised by chemists only goes as far into the ultraviolet as the commercially available instruments will take it: that is, to about 2000 Å. Within that limit there is still scope for the empirical correlations which have been much used by organic chemists for elucidation of molecular constitution.

Single bonds and saturated molecules

In Chapter 2 we discussed the molecular orbitals available to a single electron in the hydrogen molecule-ion, H_2^+. The lowest two (Fig. 50) are the bonding orbital $\sigma_g 1s$ and the antibonding $\sigma_u{}^* 1s$. Although they are stable with respect to complete ionization and dissociation ($2H^+ + 2e^-$) it can be seen (Fig. 92) that $\sigma_g 1s$ is stable and $\sigma_u{}^* 1s$ is unstable with respect to disso-

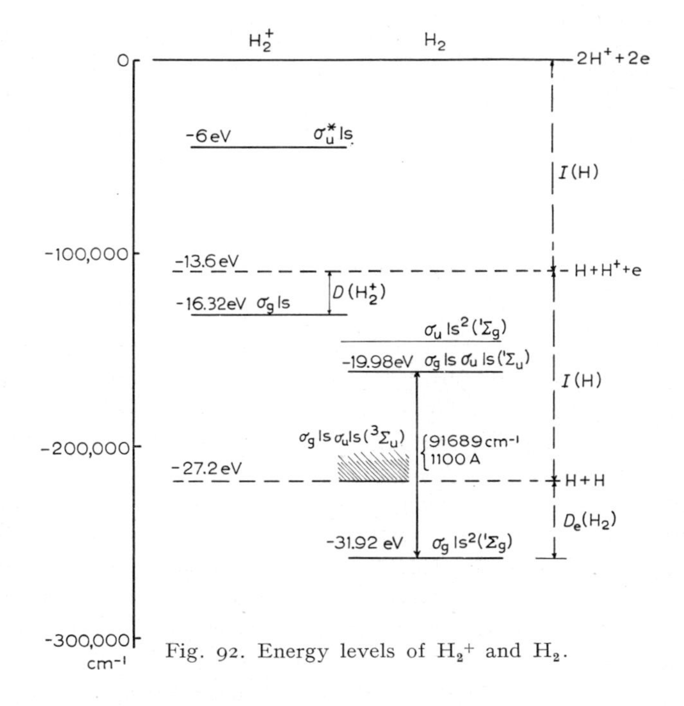

Fig. 92. Energy levels of H_2^+ and H_2.

ciation into H^+ and H. In the neutral molecule, H_2, the two electrons are subject (p. 112) to the principle of indistinguishability and to the restrictions of the Pauli principle. The ground state, $\sigma_g 1s^2$ ($^1\Sigma_g$), lies at 31.92 eV below the completely ionized and dissociated state ($2H^+ + 2e^-$) or 4.72 eV below the dissociated state, H + H. The configuration $\sigma_g 1s^1 \cdot \sigma_u 1s^1$ is represented by two states. Of these, the triplet state, $^3\Sigma_u$, has no minimum in its potential energy curve (Fig. 51) and is unstable. The singlet state, $^1\Sigma_u$, is potentially unstable to the extent of 7.2 eV with respect to H + H but has a potential energy curve with a minimum. It is thus stable with respect to a dissociation limit, say H(1s) + H(2p), involving an excited but not ionized hydrogen atom. There are further excited states of the hydrogen molecule of which the weakly stable state $\sigma_u 1s^2$ ($^1\Sigma_g$) is one.

A strict selection rule for electronic transitions, which we have seen to apply in the helium atom spectrum, is a prohibition against singlet-triplet transitions. This is simply because the interaction between the electromagnetic field and electron spin is small. In homonuclear diatomic molecules there is also a prohibition against g—g or u—u transitions. Consequently the only one of the upper states of H_2 shown in Fig. 92 which can combine with the ground state $^1\Sigma_g$ is the state $\sigma_g 1s \cdot \sigma_u 1s(^1\Sigma_u)$: the transition occurs at wavenumber 96,000 cm^{-1} or wavelength 1040 Å. In so far as the simple description of the states envisages a promotion of an electron from $\sigma_g 1s$ to $\sigma_u 1s$ this transition may be counted as a one-electron transition. In Mulliken's nomenclature it is a $N \rightarrow V$ transition (from the normal state but within the same valence shell and involving no change in principal quantum number). Reference to equation (2.71) (p. 113) shows that the state $^1\Sigma_u$ is described wholly in terms of combinations of atomic orbitals which place both electrons near to the same nucleus, *i.e.* in terms of "ionic" contributions H^+H^- and H^-H^+. This means that the transition is one in which the overall distribution of *both* electrons moves away from the centre of the molecule equally towards both nuclei. The movement of electric charge is parallel to the molecular axis, and so is the transition moment.

Methane and ethane are saturated molecules with all electrons involved in single covalent bonds. Absorption transitions are accordingly of the same type as in hydrogen and the separation of the levels is of the same order. Thus the first absorption band of methane occurs at 1250 Å, or 80,000 cm^{-1}. For ethane it is at longer wavelength, 1350 Å or 74,000 cm^{-1}, which suggests that the electrons of the C—C bond are involved; but this band continues to move to longer wavelength in the higher hydrocarbons.

Non-bonding electrons

Water vapour absorbs at a wavelength of 1720 Å (58,300 cm^{-1}) and methanol at 1840 Å (54,500 cm^{-1}). These absorptions correspond to transitions between closer levels than the above $N \rightarrow V$ σ—σ transitions. They have been interpreted as arising from the excitation of the lone pair electrons on the oxygen. It is probably the $p\pi$ rather than the $p\sigma$ electrons which are excited to higher bonding or antibonding σ orbitals in the same valence shell. The transitions are labelled $N \rightarrow Q$ and they are generally less intense than $N \rightarrow V$.

Clearly $N \rightarrow Q$ transitions may be expected from any molecules containing non-bonding electrons. Ammonia absorbs at 1920 Å or 52,200 cm^{-1}, methylamine at 2130 Å or 47,000 cm^{-1}. The hydrogen halides HCl, HBr and HI absorb at 2280, 1820 and 2080 Å (44,000, 55,000 and 48,200 cm^{-1}) respectively. The methyl halides CH_3Cl, CH_3Br and CH_3I absorb at 1725, 2040 and 2577 Å (58,000, 49,200 and 38,900 cm^{-1}).

$N \rightarrow R$ transitions

The above molecules also provide examples of transitions at shorter wavelengths in which the frequencies of a succession of maxima conform to series, known as Rydberg series by analogy with the atomic spectral series, of the form:

References p. 287–289

$$\omega = \omega_0 - \frac{R}{(n + \alpha)^2} \tag{4.1}$$

ω_0 is the series limit or ionization potential of the molecule. The implication of such series is that the higher excited electronic states of the molecule (involving higher quantum numbers) are similar to atomic states, a situation which is enhanced by the presence of the more electronegative elements. The transitions are labelled $N \rightarrow R$.

Double bonds

The chromophoric activity of many substituents in hydrocarbons arises from the introduction of unsaturation. Double bonds have electrons in molecular π-orbitals. The vacant antibonding π^*-levels and the occupied bonding π-levels lie closer together than the bonding and antibonding σ-states. That is to say, the transitions which occur at longer wavelengths (lower frequency) when double-bonded chromophores are introduced into the molecule may be interpreted as $N \rightarrow V$ π—π^* transitions. Ethylene itself absorbs strongly at 1650 Å or 61,000 cm^{-1}.

Other double bonds which commonly occur are the $>$C=O, $>$C=NH and —N=N— bonds. Acetone vapour absorbs in a broad band beginning at 3100 Å and having a maximum at 2600 Å. A further and stronger absorption occurs at 1950 Å. In hexane solution these bands occur at 2800 Å (35,800 cm^{-1}) and 1980 Å (53,500 cm^{-1}) as in Fig. 93. In acetaldehyde both

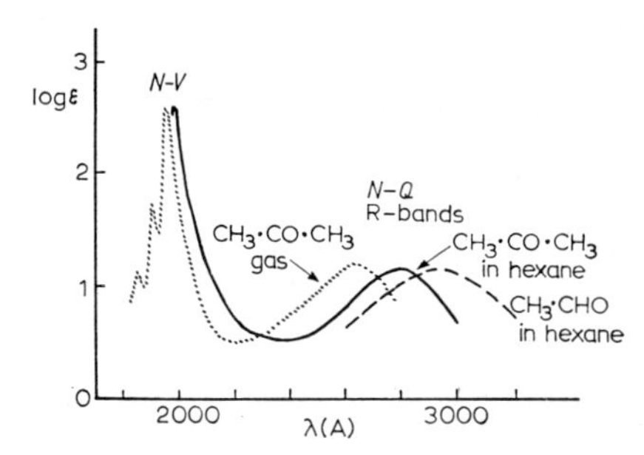

Fig. 93. Absorption of acetone (vapour and in hexane) and of acetaldehyde in hexane.

bands are at longer wavelength. Azomethane similarly has a weak absorption at about 3470 Å and a stronger one at shorter wavelength. A question of interpretation which arises here is whether the transitions are to be classified as $N \rightarrow Q$ (promotion of the electron from a non-bonding orbital on the oxygen) or $N \rightarrow V$ (promotion of a bonding π-electron). It is true that absorption of radiation in the first band leads to dissociation of these molecules and this might be thought to be evidence for an $N \rightarrow V$ transition to an antibonding orbital. However, the bond which is broken in acetone and acetaldehyde is the C—C bond adjacent to the carbonyl group. Calculations of the relative energies of the C=O π-orbitals and the unshared (non-bonding) orbitals of the O atom indicate that the π-orbitals are lower lying and that excitation from the π-orbitals to the antibonding π^*-orbital requires more energy than excitation from the non-bonding orbitals. Accordingly the longer wavelength band is assigned to the $N \rightarrow Q$ transitions and the shorter to the $N \rightarrow V$ transition. The photochemical decomposition, in either case, must evidently involve the transfer of energy within the molecule after absorption.

The azomethane absorption is similarly explained as an $N \rightarrow Q$ transition at longer wavelength and $N \rightarrow V$ transition at shorter wavelength.

Conjugation and isolation

In molecules which contain more than one chromophore the absorption spectrum may be the sum of the separate chromophores, or it may be the result of an interaction between the chromophores. Interaction occurs most markedly between double-bonded groups in *conjugation*. The simplest example of this is butadiene CH_2=CH—CH=CH_2 where there are two C=C double bonds separated by one single bond. In this case, in the planar configuration shown in Fig. 94, there is the possibility of mixing between four $p\pi$ orbitals where the separate ethylene double bond involves only two $p\pi$ orbitals. Four molecular π-orbitals may be formed from four π-orbitals and their possible shapes are also illustrated in Fig. 94. The two of the lowest energy will be occupied by two pairs of electrons, as shown.

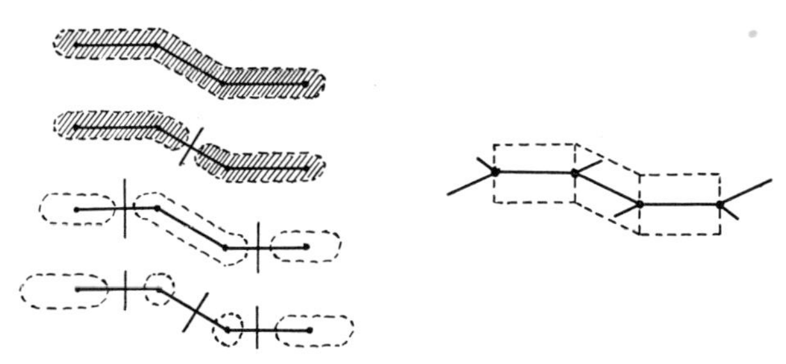

Fig. 94. Butadiene. $p\pi$ orbital mixing and possible molecular π-orbitals.

The effect of this conjugation on the $N \rightarrow V$ transitions of the molecule is considerable. The first (or longest wavelength) absorption maxima, which correspond to transition to the lowest lying excited states, shift to longer wavelengths as conjugation increases. Thus, ethylene (2000 Å; 50,000 cm^{-1}), butadiene (2100 Å; 47,800 cm^{-1}), hexatriene (2500 Å; 40,000 cm^{-1}), octa-

References p. 287–289

tetraene (3000 Å; 33,400 cm^{-1}). In the compounds C_6H_5—$(CH=CH)_n$—C_6H_5 the first absorption maximum[1,2] moves to longer wavelength as n increases, thus:

n	1	2	3	4	5	6	
λ	3190	3520	3770	4040	4240	4450	Å
ω	31,400	28,500	26,600	24,800	23,600	22,500	cm^{-1}

The longer chain compounds absorb at the blue end of the visible region and the compounds are therefore yellow or red. The yellow carotenes present in carrots and in butter are similar compounds containing eleven conjugated C=C double bonds (but no terminal phenyl groups) and lycopene, a darker red compound occurring in tomato and rose-hips, contains eleven conjugated C=C bonds with two additional double bonds not conjugated with the rest. Other series of compounds with conjugated double bonds show the same behaviour[3,4]. In Fig. 95 the wave-number of the first absorption band is plotted against the number (n) of open chain double bonds in the diphenyl polyenes and the polyenes.

A very simple account of this movement of the absorption band as the chain increases in length can be given in terms of the theoretical energy levels for an electron in a wire. We suppose the electron to be constrained to

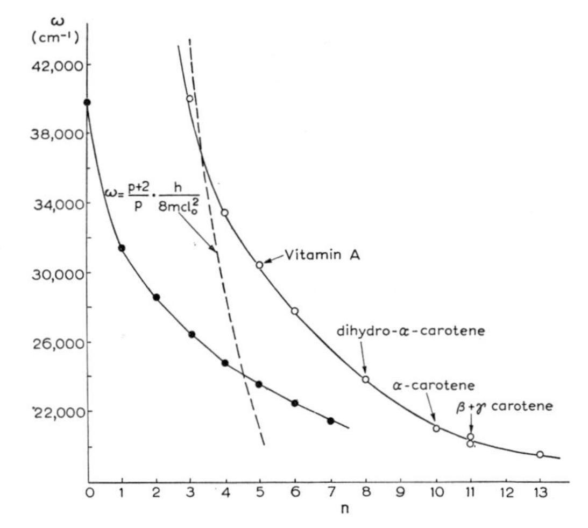

Fig. 95. The wavenumber of first absorption maximum in C_6H_5—$(CH=CH)_n$—C_6H_5 (filled circles) and H—$(CH=CH)_n$—H (open circles) plotted against n. The broken curve indicates the variation of ω with n predicted by simple electron-in-wire theory.

move in one dimension with zero potential inside the distance $x = 0$ to $x = l$ and with infinite potential outside those limits. The Schrödinger equation is

$$\frac{d^2\psi}{dx^2} + \frac{8\pi^2 mE}{h^2}\psi = 0 \tag{4.2}$$

to which the solutions are:

$$\psi_n(x) = A \sin\frac{n\pi}{l}; \quad E_n = n^2\frac{h^2}{8ml^2}; \quad n = 1, 2, 3\ldots \tag{4.3}$$

A is a constant, m is the mass of the electron, and n is a quantum number. The quantum restriction, of course, arises from the boundary conditions that $\psi = 0$ at $x = 0$ and at $x = l$.

We can imagine the π-electrons in a conjugated system as electrons constrained to move in one dimension within a length equal to that of the chain of carbon atoms and subject to a constant (zero) potential in that region. We may suppose the average C—C distance (double or single bond) to be l_0 and write the length of the chain as $l = pl_0$. Since the chain is imagined to be an alternating conjugated system there will be $\frac{1}{2}(p + 1)$ pairs of π-electrons. These pairs will, according to Pauli's principle, occupy the first $\frac{1}{2}(p + 1)$ levels and the transition to the first unoccupied level will accordingly be from $n = \frac{1}{2}(p + 1)$ to $n = \frac{1}{2}(p + 3)$ and hence:

$$\omega = \frac{\Delta E}{hc} = \frac{p + 2}{p^2}\,\frac{h}{8mcl_0^2} \tag{4.4}$$

If we take $l_0 = 1.40$ Å the values of ω obtained for ethylene, butadiene, hexatriene and octatetraene are 464,000, 86,000, 43,100 and 28,300 cm^{-1} as

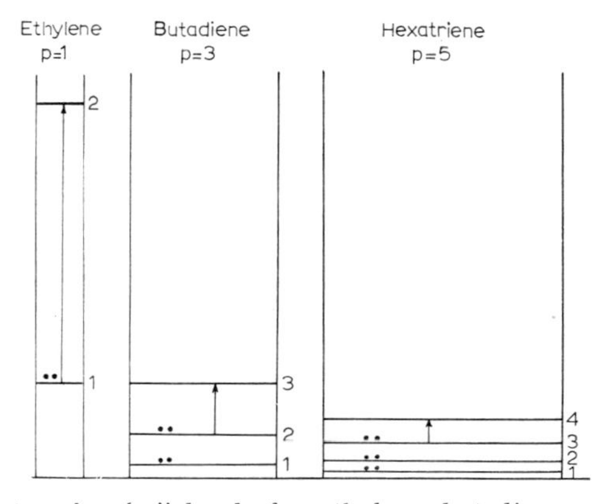

Fig. 96. "Electron-in-wire" levels for ethylene, butadiene and hexatriene.

shown in Fig. 96. These results (the broken line in Fig. 95) do not agree with the observed wavenumbers except in being of the correct order of magnitude. This is to be expected in view of the approximations used and the neglect of electron interaction or exchange. Nevertheless the trend of transition frequencies is clearly indicated and much better agreement is shown in refinements which take proper account of the variation in potential along the chain and of electron interaction. It is worth noting also the theory of Lewis and Calvin[5,6] which regards a polyene chain as a coupled set of n oscillators and predicts a relationship $\lambda^2 = k_1 n + k_2$ to which the data of Fig. 95 certainly conform reasonably closely.

The polyenes illustrate the way in which conjugation modifies the individual chromophores. Another classification of ultraviolet absorption spectra denotes the bands due to isolated chromophores as R-bands, generally weak in intensity. The more intense bands which arise from conjugation are known as K-bands. Conjugation can, of course, occur in other forms than in polyene chains as, for example, in the benzene ring, polynuclear aromatic systems,

the quinonoid system and other π-orbital systems in conjugation with these.

It is an important feature of ultraviolet absorption spectra that chromophores which are not conjugated but *isolated* do not give K-bands but a summation of R-bands. A $>CH_2$ group is usually sufficient to separate two chromophores, but $>O$ or $>S$ or $>NH$ are not. Hexacene (I) is a green compound and clearly absorbs well into the visible but the 6,15-dihydrohexacene (II) is colourless and its absorption spectrum is the sum of the spectra of anthracene and naphthalene.

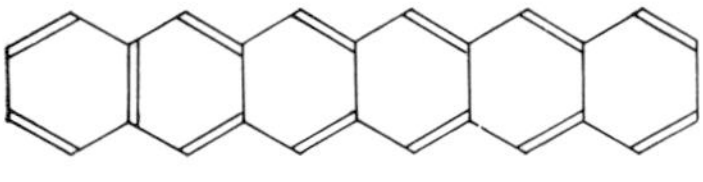

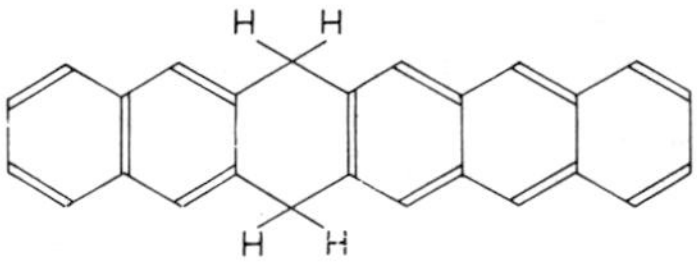

Aromatic hydrocarbons

Benzene is a cyclic conjugated polyene. It absorbs at 2600, 2000 and 1800 Å (Fig. 97) and in this case a considerable amount of theoretical work has gone into the correlation of transitions with electronic levels. The first treatment of the problem was due to Hückel[7] (see also Ref. 8,9) and consists essentially in setting up an approximate wave function to describe the behaviour of the six "aromatic" electrons. This is done by assuming the molecular

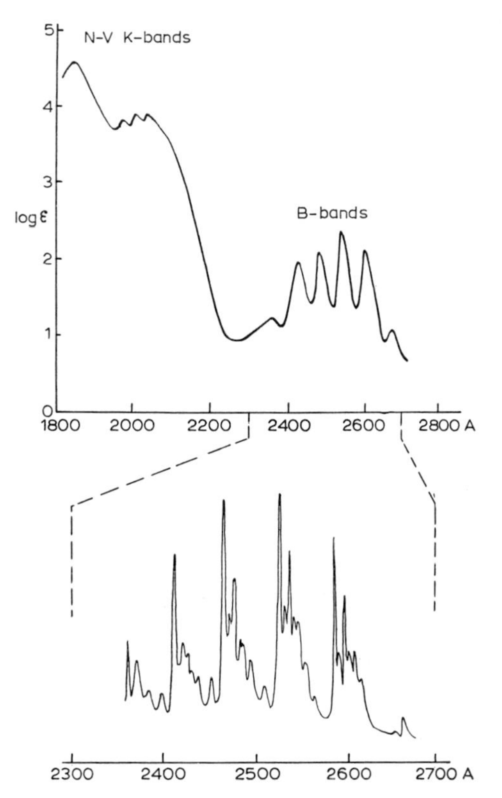

Fig. 97. Absorption spectrum of benzene, dissolved in ethanol (upper spectrum) and vapour (lower spectrum).

orbitals to be linear combinations of the *p*-orbitals of each carbon atom which lies perpendicular to the plane. Some apparently drastic assumptions are made to reach a "zeroth approximation" to the energies of the available orbitals. These are illustrated in Fig. 98 with accompanying indications of the shape of the electron distribution viewed from above the plane of the molecule.

To account for the observed transitions in benzene the effect of electron repulsion must be considered both for the ground state where the three pairs of electrons occupy the lowest three π-orbitals, and for the excited state, in which one electron is promoted to an antibonding orbital. In the latter case (Fig. 99) there are, in the first approximation, four equally probable distributions of the unpaired electrons between the four middle orbitals. Configuration interaction leads to four new states of which two are of equal energy. The electron distributions corresponding to each level have

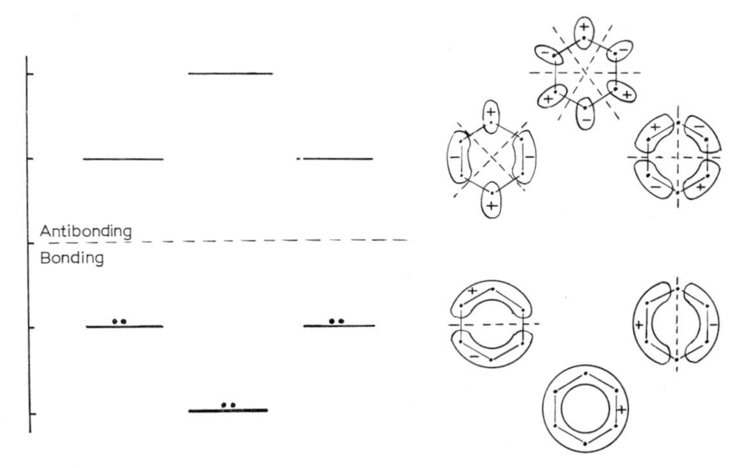

Fig. 98. Simple one-electron levels of π-electrons in benzene.

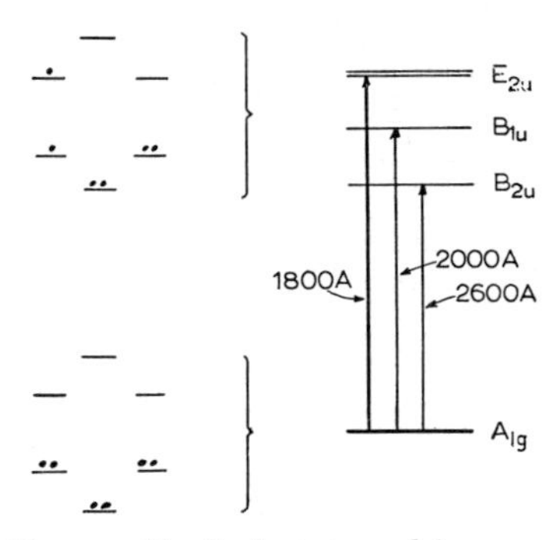

Fig. 99. Excited states of benzene.

symmetry properties relative to the symmetry of the molecule which enables them to be classified as belonging to the species B_{2u}, B_{1u} and E_{2u} of the point group D_{6h}. The ground state is totally symmetric (species A_{1g}). Now, if purely electronic transitions are considered, it can be shown from the symmetry of the states that transitions $A_{1g} \rightarrow B_{2u}$ and $A_{1g} \rightarrow B_{1u}$ are forbidden as dipole transitions but that $A_{1g} \rightarrow E_{2u}$ is allowed. (This is also true of transitions between vibrational states of those species.) It is possible however for interaction with vibration to lead to a partial breakdown in the selection rules permitting transitions to the B_{2u} and B_{1u} states. It is accordingly argued that the weak absorptions at 2600 Å (the band with vibrational structure, Fig. 97) and at 2000 Å are the transitions to B_{2u} and B_{1u} levels and that the strong absorption at 1800 Å is due to the fully allowed transition, $A_{1g} \rightarrow E_{2u}$. Each of these transitions would be classified as $N \rightarrow V$. That at

2600 Å, being particularly characteristic of benzenoid spectra, is sometimes referred to as a *B*-band.

As might be expected from the behaviour of polyenes, increasing the molecular size of condensed aromatic hydrocarbons shifts the absorption to longer wavelengths or lower frequencies. In phenanthrene the first absorption band is at about 3750 Å, naphthacene is red, pentacene is dark blue, hexacene is green and graphite is black.

Conjugated ketones — quinone

In Fig. 100 the u.v. absorption spectra of acrolein, $CH_2{=}CH \cdot CHO$, of diacetyl, $CH_3 \cdot CO \cdot CO \cdot CH_3$, and of *p*-benzoquinone are compared with acetone and acetaldehyde. It is evident that conjugation of the C=O group

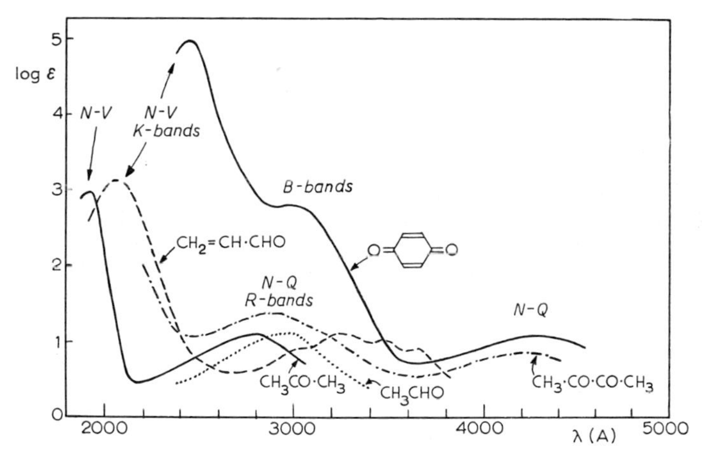

Fig. 100. Absorption spectra of carbonyl compounds; acetone, acetaldehyde, acrolein, diacetyl, quinone.

decreases the energy difference for the $N \rightarrow V$ transitions and intensifies them as it does in the olefines. Conjugation with C=C occurs in acrolein and the $N \rightarrow Q$ transition occurs at 3200 Å; the more intense $N \rightarrow V$ transition occurs at 2100 Å. Conjugation with another C=O as in diacetyl, $CH_3CO \cdot CO \cdot CH_3$, gives rise to an additional absorption at 4200 Å and the compound is yellow. The similar first absorption band at 4350 Å which makes *p*-benzoquinone golden yellow is probably also an $N \rightarrow Q$ transition, the upper state now being a π^*-orbital involving the whole molecule. Anthraquinone is similarly a yellow compound since it absorbs blue light in the visible region.

The carbonyl frequency is evidently suppressed in the carboxylic acids and also their esters and amides, for none of these absorb at wavelengths greater than 2300 Å.

Nitro and nitroso compounds

Nitromethane absorbs at about 2800 Å and more strongly at 2100 Å (Fig. 101). Nitrobenzene has a very intense band at 2520 Å, and weaker ab-

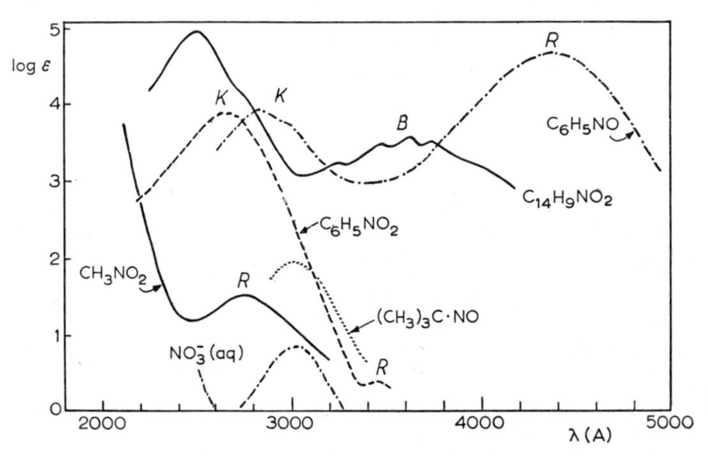

Fig. 101. Absorption spectra of nitro- and nitroso compounds; nitromethane, nitrobenzene, nitroanthracene, nitrate ion, nitroso-iso-butane and nitrosobenzene.

sorptions at longer wavelength which are responsible for the slight yellow colour. It is probable that in nitromethane the weak band at 2800 Å is again an $N \rightarrow Q$ transition and that at 2100 Å an $N \rightarrow V$. As with the carbonyl chromophore, the first band in nitromethane is assigned to $N \rightarrow Q$ promotion of an electron from a non-bonding orbital on the oxygen to a vacant π-orbital of the NO_2 group. Such is probably also the case in the nitrate ion. Possible shapes of the electron distribution described by the π-orbitals for NO_3^- and C—NO_2 are shown in Fig. 101. Four π-orbitals arise from four atomic p-orbitals perpendicular to the molecular plane in NO_3^- and there are three pairs of electrons left to fill them after σ-bonding orbitals and non-bonding orbitals have been filled. For nitrocompounds there are three π-orbitals and two pairs of electrons. In nitrobenzene, the π-orbitals of the NO_2 group undoubtedly mix with those of the benzene ring and the disposition of filled and vacant levels is altered. It seems probable that the intense band at 2620 Å is now an $N \rightarrow V$ (π—π^*) transition at lower frequency than it occurs in nitromethane and overlapping any $N \rightarrow Q$ transition. In the spectrum of 9-nitro-anthracene the interaction of the nitro group and anthracene is evident, both in the frequency shifts and in increased intensity.

Nitrosobenzene is green and nitrosobutane $(CH_3)_3C \cdot NO$ is blue. The latter is due to a weak band at 6650 Å additional to the N—Q transition (R-band) at 3000 Å. In nitrosobenzene there is a marked intensification and

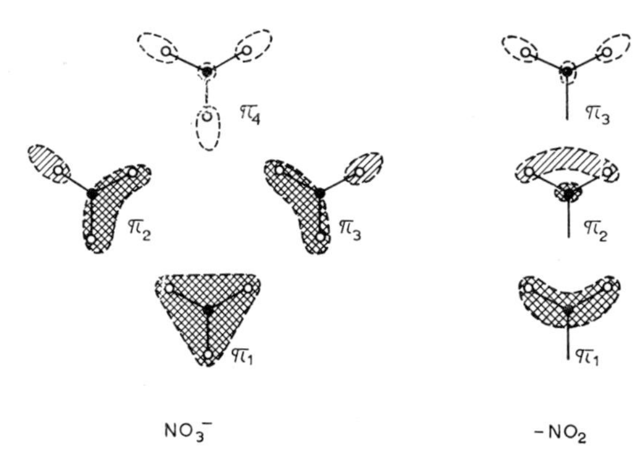

Fig. 102. π-orbitals of NO_3- and C—NO_2: symmetry of possible p-orbital combinations. The shaded areas indicate the orbitals which will be occupied by available electrons; and also the sign of the wavefunction in those cases.

the colour presumably arises from a broad strong band centred on 4400 Å and a further weak band in the visible.

Azo-compounds

The —N=N— chromophore is another group whose absorption is enhanced by conjugation. We have seen that, in isolation, the group gives two bands: one an $N \rightarrow Q$ promotion of nonbonding electrons and the other a stronger $N \rightarrow V$ absorption at shorter wavelengths. In conjugation, however, the $N \rightarrow V$ transition predominates. In azobenzene, in which the N=N is conjugated with two benzene rings and the π-orbitals extend over the whole molecule, the levels are brought closer together and the $N \rightarrow V$ transition occurs at 4450 Å. Azobenzene is therefore orange-red.

Dyes

The chromophores which we have discussed are the basis of many useful dyes. Generally absorption at longer wavelengths is obtained by conjugation of the doubly bonded groups and it is this rather than the groups themselves which is responsible for colour.

Further adjustment of levels is effected by the use of substituents, or modifying groups, called *auxochromes*. The utility of a dye is, of course, governed by chemical properties which determine its adherence to fabrics and its stability to chemical action and the action of light. The incorporation of chemical substituents designed to improve the utility of a dye may well affect the colour.

The following representatives of a great variety of dyestuffs contain examples of the chromophores mentioned above, mainly in conjugation. It

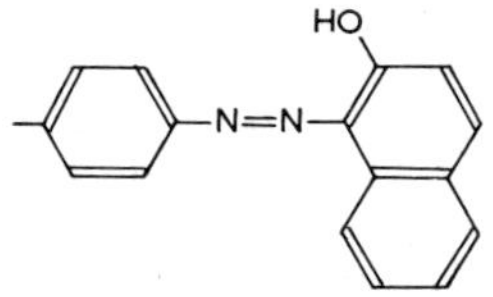

p-Nitraniline red (monoazo)

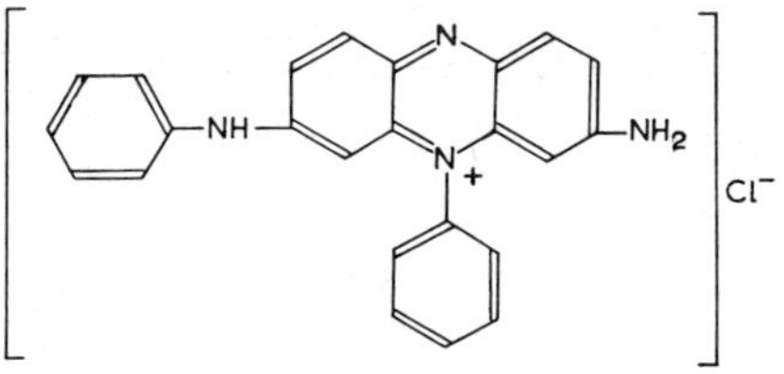

Mauveine (azine)

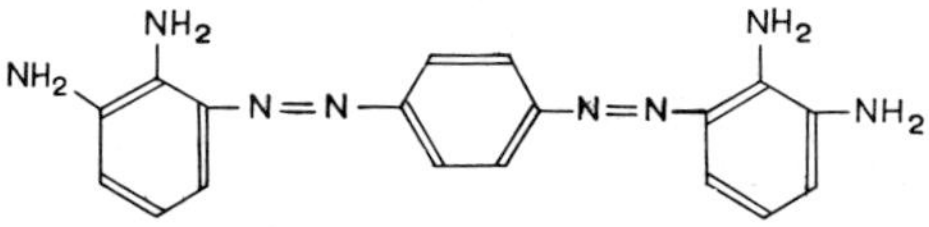

Bismarck brown (palyazo)

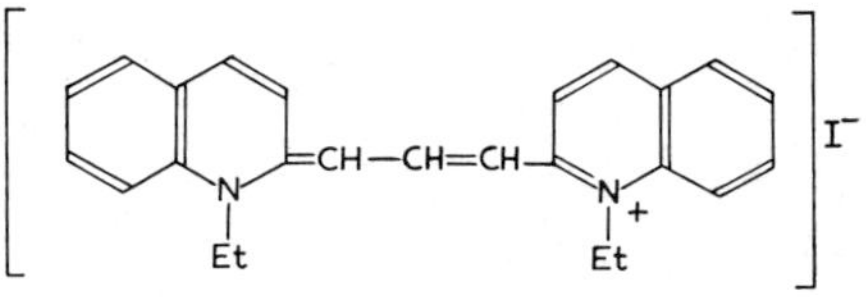

Sensitol red (cyanine)

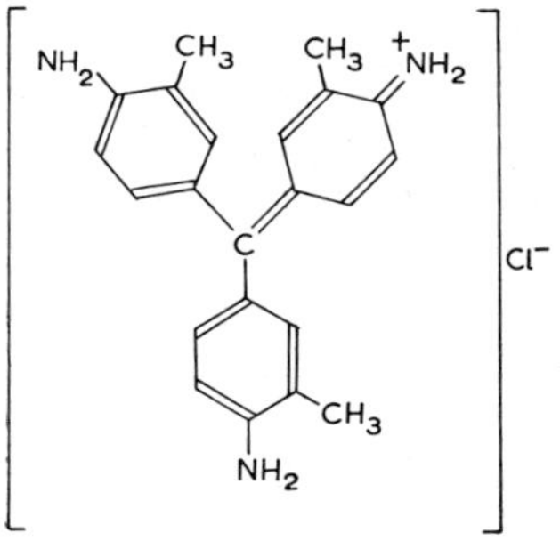

New fuchsine (triphenylmethane)

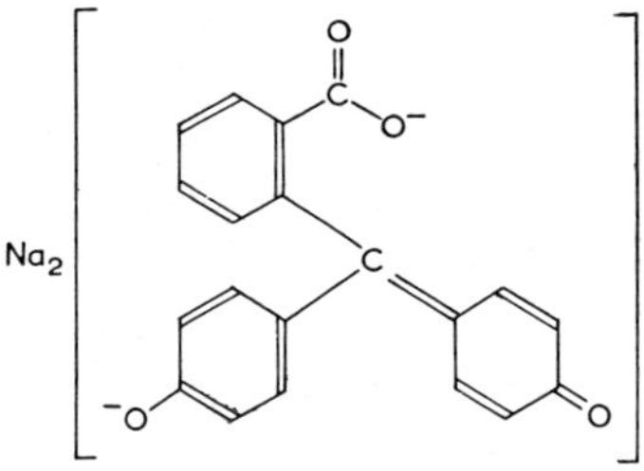

Phenolphthalein red (phthalein)

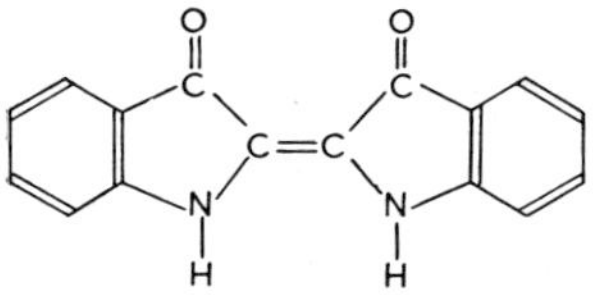

Indigo

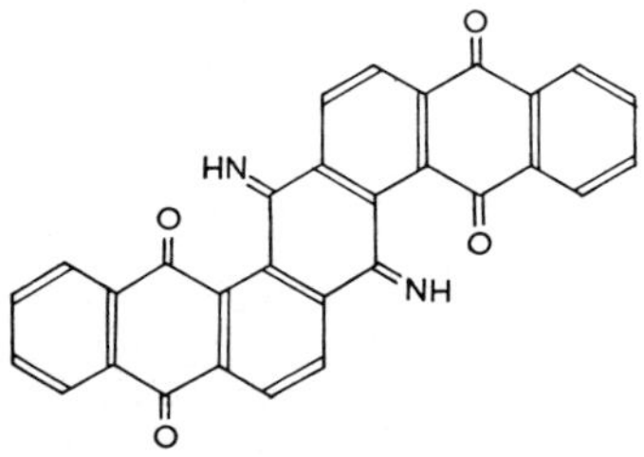

Indanthrone blue

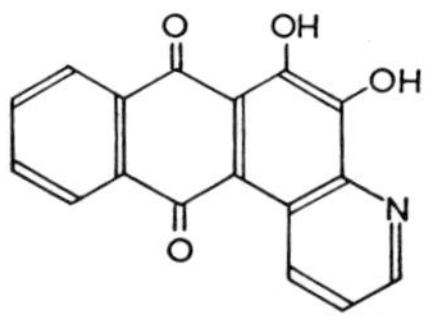

Alizarin blue

may be noted that the quinonoid structure =⟨ ⟩= can occur with NH as the attached basic group instead of O and in fact both occur in the blue vat dye indanthrone; but di-iminoquinone itself is colourless.

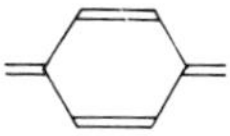

The poly-azo dye shown is one constituent of the mixture known as Bismarck brown. Similarly mauveine, the first aniline dyestuff to be prepared industrially by a method developed by Perkin in 1856, is a mixture containing phenylated safranines, including the one shown.

In each formula a particular disposition of double bonds and also of charge on the ionic species has the same significance as it does in say CH_2=CH—CH =CH_2 or ⟨ ⟩. As in those molecules so in the dyes there is a distribution of electronic charge over the entire length of the conjugated chain. Thus in the cyanine dye sensitol red the cationic charge is symmetrically distributed, no doubt mainly shared equally by the two nitrogen atoms. In the cation of new fuchsine the positive charge is evenly shared between the three *o*-toluidine groups.

Use and specificity of electronic spectra

It can be seen from the examples given that ultraviolet and visible absorption bands are too wide and too lacking in distinguishing feature to rival infra-red spectroscopy in specificity. Nevertheless they are still of value in elucidation of structure, especially if taken in conjunction with the chemical evidence.

The sort of problem in which a simple ultraviolet absorption spectrum may assist is in deciding between two possible isomeric arrangements of a known molecular formula. Thus acetonyl acetone, $CH_3 \cdot CO \cdot CH_2 \cdot CH_2 \cdot CO \cdot CH_3$, absorbs at about 2700 Å as does acetone, but at about twice the intensity. It may be concluded that in this molecule the two C=O groups are not conjugated but are isolated by two —CH_2 groups. The isomeric di-ethyl glyoxal (3,4-hexadione) absorbs at about 4000 Å, as does diacetyl. Such arguments are frequently used for eliminating some postulated molecular structures, but care should be taken with even this sort of diagnosis. Although the first absorption band of diphenylmethane at 2680 Å agrees with benzene, *p*-methyldiphenyl has its first absorption band at shorter wavelength, about 2550 Å, not longer. Separation of two carbonyls by one >CH_2 group does not achieve complete isolation because of the possibility of enolization.

A more sophisticated example[10–12] of the use of u.v. absorption spectra comes from alkaloid chemistry. The yellow alkaloid sempervirine forms a nitrate which shows differences in absorption spectrum between acid and alkaline solutions in ethanol (Fig. 103). The spectrum is typical of indoles and accordingly the formula (I) was suggested. Synthesis of this structure

(I) (II)

however gave a spectrum which differs from Fig. 103 and showed less change with pH. Reference to the infra-red spectrum then provided a further clue since there was no observable N—H frequency at 3460 cm^{-1} (see p. 239) whereas all indoles not substituted at the N-atom gave such a band. Accord-

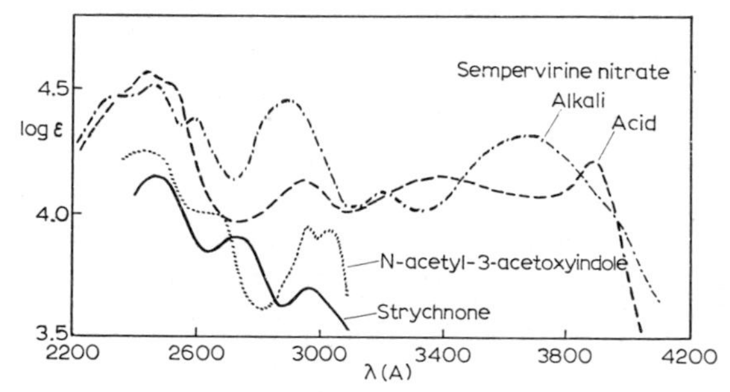

Fig. 103. Absorption spectra of sempervirine nitrate (in acid and in alkaline ethanol) of N-acetyl-3-acetoxy indole, and of strychnone.

ingly structure (II) was suggested for sempervirine. This structure was later synthesized and found to have the same absorption spectrum (and also infra-red spectrum) as the natural substance.

A related example[13, 14] is the final step in the elucidation of the structure of strychnine. We restrict attention to the question whether the structure is (III) or (IV). From strychnine there is obtained pseudostrychnine which, on

(III) (IV)

chemical evidence, contains a tertiary OH group on a carbon atom adjacent to the nitrogen atom, *i.e.* at the carbon atoms marked with * in (III) or (IV). Oxidation of pseudostrychnine by acid hydrogen peroxide yields the compound strychnone which has the u.v. spectrum shown in Fig. 103. It can be seen, by comparison with N-acetyl-3-acetoxyindole, to have a spectrum which is typical of N-acetyl indoles and it is entirely different from that of N-acetyl dihydroindoles or of strychnine itself. Strychnone is thus shown by its spectrum to contain the grouping (V) and not (VI). It therefore

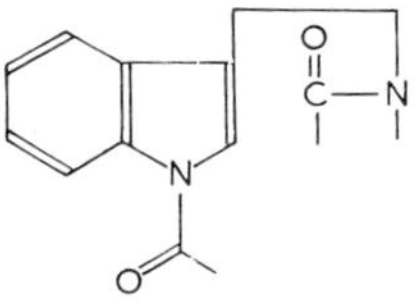

(V) 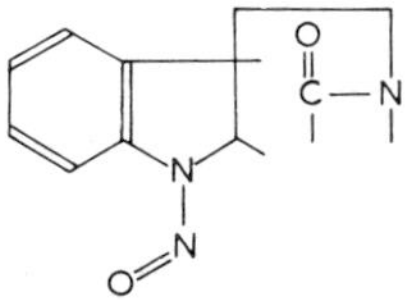 (VI)

appeared most likely that the correct structure of strychnine is (III) and not (IV). This brief account does not, of course, do any justice to the many years of chemical work which is disguised under the proposition that the structure is either (III) or (IV). But it is a fact that the spectroscopic evidence led to the suggestion of structure (III) as a small but important modification of the structure (IV) to which all chemical evidence seemed to point.

The effects of substituents upon the position of absorption bands has possible application in the elucidation of chemical structure and is discussed on p. 265. It also happens that steric effects can limit the conjugation of chromophores by limiting the degree of overlap of *p*-orbitals. For example the proximity of the benzene rings in *cis*-stilbene and in *cis*-azobenzene prevents the fully coplanar configuration which can occur in the *trans*-compounds. As a result of this restriction upon conjugation the *cis*-compounds absorb at wavelengths 200 Å lower than the *trans*.

Evidence that the spare *p*-orbital on the nitrogen in N-dimethylaniline is normally involved in the molecular π-orbitals comes first from the fact that the compound absorbs at 2950 Å but only at 2700 Å in acid solution, when the $[C_6H_5NH(CH_3)_2]^+$ cation is formed. The steric effect of two methyl groups when these are substituted in the positions ortho- to the $—N(CH_3)_2$ group is to rotate the $—N(CH_3)_2$ group so that the spare *p*-orbital can no longer mix with the benzene π-orbitals. Thus 2,6-dimethyl-N,N-dimethylaniline absorbs at 2600 Å. Many other examples of similar applications to the determination of structure are given by Gillam and Stern[102], and elsewhere[15, 16].

The chief reason for lack of specificity in ultraviolet absorption spectra is that the electronic transitions, which might be expected to be characteristic if they could be precisely determined, are in fact so heavily encumbered with accompanying vibrational transitions that the measurements of absorption maxima is possible only to within a few Angstrom units. This is not just a question of resolution. In absorption spectra of polyatomic molecules in the vapour phase rotational structure is to be expected as well as vibrational. In that case the transitions are sharp and resolution is the limiting factor. In liquids, however, despite the absence of rotational changes the effect of molecular interactions is to blur the vibronic levels and the bands have a natural width which no amount of resolution will overcome (see Fig. 97).

Collections [17–20] of ultraviolet and visible absorption spectra are to be found in International Critical Tables and Landolt-Bornstein. The American Petroleum Institute publishes a loose-leaf collection of u.v. spectra, mainly of hydrocarbons, and there is a tabulation of data on punched cards. In addition to these there is a valuable review[21] with tabulation of band data, but not illustrated; a number of books[22, 23, 102] on specific groups of compounds; and a valuable index[24] covering the literature for the period 1934 to 1954.

Infra-red Correlations

The number of vibrational modes in N-atomic molecules is $3N - 6$. For a molecule of any size this can be a large number. If the molecule is without any elements of symmetry or is of low symmetry all the normal modes may be expected to be active both in the infra-red and Raman spectra. If a number of combination bands (perhaps strengthened by Fermi resonance) are added to the fundamentals a fairly complex infra-red spectrum is to be expected. This is the case and it is found that the infra-red absorption spectrum, and in smaller degree the Raman spectrum, is a highly specific and characteristic "finger-print" of the absorbing substance. Quite apart from any detailed interpretation of the origin of the bands it is possible to identify a synthetic product with a natural material by comparing their infra-red spectra. There are many excellent examples of this procedure (a more sophisticated sort of mixed melting point) and we give two examples from carbohydrate chemistry.

Fig. 104 shows the identity of synthesized 1,β-D-ribofuranosyl-4(5)glyoxinylacetic acid hydrochloride with the natural product extracted from the urine of rats and mice after injection of histamine[25]. The samples used in each case were less than 2 mg incorporated in potassium bromide pressed discs from which they could be recovered. These spectra are plotted on an arbitrary frequency scale such as one gets with a prism instrument.

In Fig. 105 the natural product originally extracted from *Bacillus subtilis* has the spectrum (c), differing significantly from (b), the synthesized

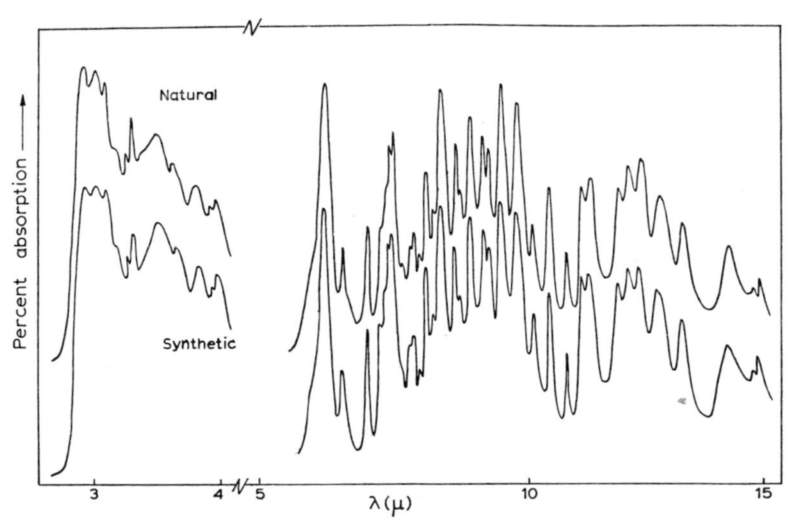

Fig. 104. Infra-red spectra of (a) natural and (b) synthesised 1,β-D-ribofuranosyl-4(5)-glyoxinylacetic acid HCl.

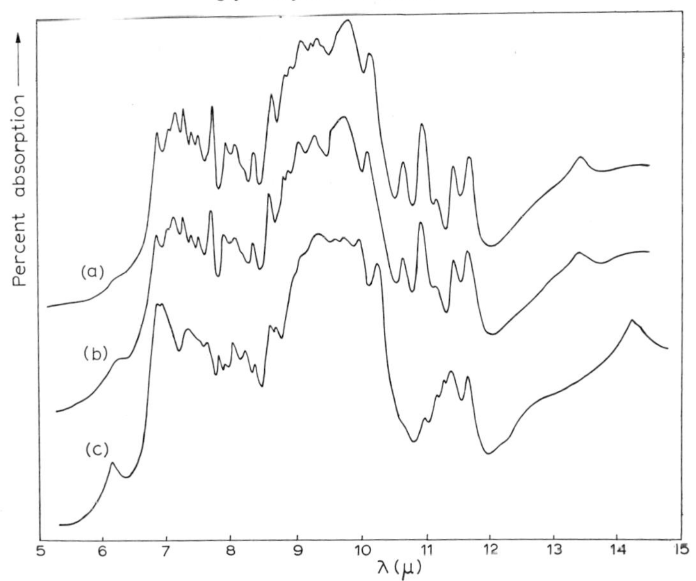

Fig. 105. Infra-red spectra of the anhydrous ribitol glucoside 4-O-p-D-glucopyranosyl ribitol; (a) natural, (b) synthesised. (c) is the ethanolate-hydrate of (a).

ribitol-4-D-glucoside[26]. In this case it proved necessary to recrystallize the natural material by the same procedure as had been used in preparing (b). The resultant spectrum (a) was then identical with (b).

References p. 287–289

Characteristic group frequencies

As a very rough approximation a bond, especially a terminal one, in a large organic molecule may be regarded as the bond in a diatomic molecule, R—X. The stretching frequency of the bond will be given by

$$\omega = \frac{1}{2\pi c}\sqrt{\frac{k}{m}} \quad \text{where } m = \frac{m_R m_X}{m_R + m_X} \tag{4.5}$$

If $m_R \gg m_X$ then $m \sim m_X$, and ω is independent both of m_R and of the nature of R if the force constant k is independent of R. These conditions are most nearly met and the approximation most nearly valid when X stands for a hydrogen atom. The force constant for C—H is almost the same whatever the nature of the rest of the radical R. There is a different force constant value for N—H but it is almost the same for all molecules containing that bond. Likewise for O—H. In accordance with this simple picture it is found that all molecules containing C—H, N—H or O—H bonds have absorption bands lying in the ranges 2850—3050 cm^{-1} for C—H, 3050—3500 cm^{-1} for N—H and 2860—3600 cm^{-1} for O—H. If the substances containing N—H and O—H are observed in sufficiently dilute solution to avoid intermolecular perturbations the ranges narrow to 3380—3500 cm^{-1} for N—H and 3600—3640 for O—H. These absorptions are therefore assigned to normal modes of the molecule which involve principally the stretching of the C—H, N—H or O—H bond.

These assignments are clearly confirmed by the replacement of hydrogen by deuterium. If the force constant remains unaffected by the isotopic substitution and if $m_R \gg m_H$, $m_{RD}/m_{RH} = m_D/m_H = \frac{1}{2}$ and hence the characteristic frequencies should be reduced by a factor $\sqrt{2}$. Such a reduction is found. A reduction by the full factor of $\sqrt{2}$ is not found when the X—H vibration is mixed with others and cannot be isolated.

Extension of this simple approximation to the case where X is a halogen atom, or where it is oxygen and the bond is a double one, is dubious. Still less are the assumptions valid when the bond in question is an internal one, such as $>C=C<$ or $\rightarrow C—C \leftarrow$ or when the vibration is a bending mode rather than stretching. Nevertheless the detailed examination of the spectra of thousands of organic compounds has made it possible to correlate certain frequency ranges with particular groups or bonds.

The correlations are, of course, quite empirical: their validity is statistical in character. The step into theory is taken when the frequencies are labelled not as characterizing a particular group but as characterizing a particular mode of vibration of that group. We have noted in the previous chapter that, in the classical sense, any vibration is a vibration of the whole molecule. A vibration of a particular part of the molecule only remains localized in that part if its normal frequency differs considerably from the frequencies of other modes of vibration. More precisely, any localized movement, say the stretching of particular bonds, will not in general be confined to one normal mode. The imagined movement may, by virtue of its symmetry, be assigned to a particular species or class of vibration, and all normal modes of vibration of the same symmetry class will partake of that movement. Only if the energy level of one member of a class is well removed from the levels of other members of the same class will its normal mode approximate to the stretching or bending of particular bonds.

In view of this general observation it is surprising that so many correlations have been found. It is true that the correlations are more satisfactory in the higher frequency region ($\omega > 1500$ cm^{-1}, $\lambda < 7\ \mu$) where the characteristic frequencies are relatively few in number. The correlations become less certain among organic molecules in the lower frequency region. Some group stretching frequencies that are very well established are given in Table 19.

TABLE 19

SOME REPRESENTATIVE GROUP STRETCHING FREQUENCIES

Group				
—O—H	stretch			3640—3600
$>$N—H	stretch			3500—3380
	(both these are lowered by hydrogen bonding)			
—C—H	stretch	in —C≡C—H		3305—3270
		attached to aromatic ring		3030
		attached to C=C		3090—3010
		in —CH_3	antisymmetric	2962
			symmetric	2872
		in $>CH_2$	antisymmetric	2926
			symmetric	2853
—C≡N	stretch			2260—2240
—C≡C—	stretch	in —C≡CH		2140—2100
		in —C≡C—		2260—2190
C=O	stretch			1900—1580
C=C	stretch			1680—1630
—N$<$(O, O)	stretch	antisymmetric		1560—1520
		symmetric		1360—1340
		in NO_3^-		1380—1350
—C$<$(O, O)	stretch	antisymmetric		1610—1550
		symmetric		1400—1300
$>$S=O	stretch			1060—1045
$>$S$<$(O, O)	stretch	antisymmetric		1335—1310
		symmetric		1160—1130
		in SO_4^{2-}		1130—1080
—C—F	stretch			1350—1200
$\geq$P=O	stretch			1300—1250
		in PO_4^{3-}		1100— 950
—Si—O	stretch			1100—1000
		in SiO_4^{3-}		1100— 900

(all wave numbers in cm^{-1})

Many more correlations (including bending and deformation modes) have been made with varying degrees of assurance and tables have been drawn up to assist identification. The tables given by Colthup[27,28] have been much quoted. Any who wish to make use of such correlations should be acquainted with the scope and limitations of the assignments: they should certainly consult either Jones and Sandorfy (Ref. 97, Chap. IV) or Bellamy[101]. With some minor changes in the frequency ranges the correlations, which have mainly come from infra-red data, are also good for Raman spectra. Relative intensities are different in the two methods so that bands prominent in one may be weak or even absent in the other but the frequencies, nevertheless, correspond.

Use of group frequency correlations

With the detailed assignments of some frequencies it is possible in favourable cases to make a fairly reliable diagnosis of molecular structure. An infra-red spectrum will indicate the presence or absence of certain groups and provide a guide to further chemical work. The number of assignments may be sufficient to suggest the entire structure, which could then be confirmed by synthesis of the postulated structure and comparison of spectra. One might, for example, isolate from caraway oil an odorous solid, of molecule formula $C_{10}H_{14}O$, and call it carvone. On heating with glacial phosphoric acid this compound isomerizes to an oil, of the same formula $C_{10}H_{14}O$, which can also be isolated from *Origanum hirsutum*: we may call it carvacrol. Chemical identification of these compounds in fact preceded any spectroscopic investigation but we may attempt an identification from the spectra shown in Fig. 106.

In carvone the first obvious features are the different C—H stretching frequencies near 3000 cm^{-1} and the carbonyl stretch at 1680 cm^{-1}. In carvacrol on the other hand there is the broad band at $\sim$ 3340 cm^{-1} attributable to O—H stretch (N—H would also appear here but the molecule contains no nitrogen) and no carbonyl band. The carbonyl band in carvone occurs at a frequency which implies some conjugation as in α, β-unsaturated ketones (1685—1665 cm^{-1}) or aryl ketones (1700—1680 cm^{-1}). The carvone C—H stretching frequencies occur as follows. One at about 3090 cm^{-1} suggests a terminal $>C=CH_2$ group: it is probably somewhat too high for aromatic C—H stretch ($\sim$ 3030 cm^{-1}) or for hydrogen attached to a substituted ethylenic link (3040—3010 cm^{-1}). There is, in any case, such a band at 3030 cm^{-1}. Bands at 2970 and 2890 cm^{-1} indicate a $—CH_3$ group, and at 2920 and 2850 cm^{-1} $>CH_2$ groups. The suggestion of a terminal $>C=CH_2$ group prompts one to look for the out-of-plane CH deformation frequencies at about 900 cm^{-1} for the $=CH_2$ and at about 990 cm^{-1} for CH if the group

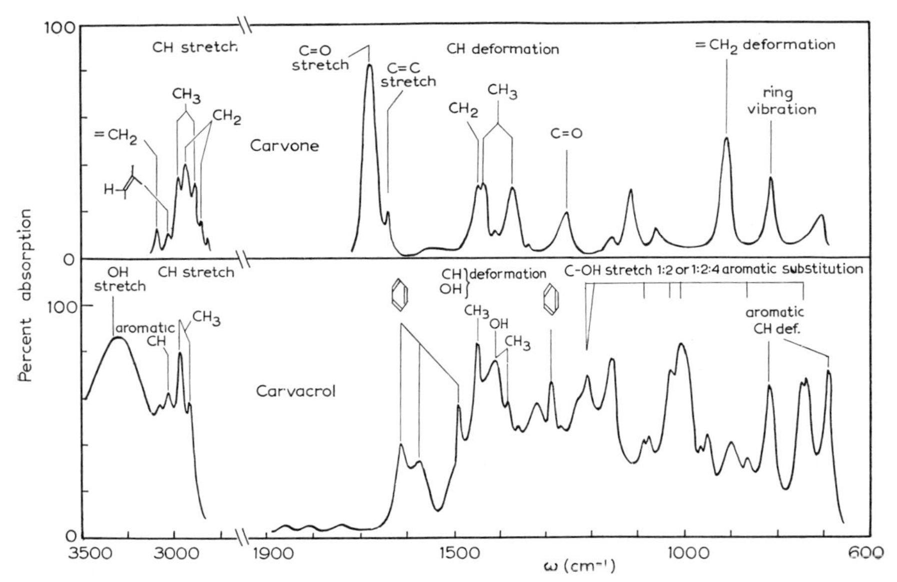

Fig. 106. Infra-red spectra of carvone and carvacrol.

is —$CN{=}CH_2$. The first is present at 905 cm^{-1}. The second is absent. The C=C stretching frequency is expected between 1680 and 1620 cm^{-1} or at lower frequency if conjugated. A weak band at 1640 cm^{-1} suggests that the C=C bond is not conjugated. C—H deformation frequencies of C—CH_3 and CH_2 are expected and appear at 1455 and 1440 cm^{-1} and another deformation frequency of the CH_3 group (the symmetrical one) appears at 1375 cm^{-1}. The two CH_3 deformations (antisymmetric and symmetric) are illustrated in Fig. 107 and also the antisymmetric and symmetric stretching modes. In practically all cases, and certainly among the lighter nuclei, the antisymmetric modes occur at higher frequency than the symmetric. The —NO_2 and —CO_2— stretching frequencies in Table 19 provide other examples of this, a fact which can, of course, only be established from the spectra of molecules simple enough to permit the full assignments described in Chapter 3.

Turning to carvacrol we find that the C—H frequencies are 3040 cm^{-1}, suggesting either $\rangle$C=CH— or an aromatic C—H, and the pair at 2955 and 2925 cm^{-1} which indicates —CH_3. The $\rangle CH_2$ frequencies have disappeared and might suggest that the isomerization has led to a conversion of the CH_2 group in carvone to aromatic or ethylenic C—H bonds in carvacrol. The CH_3 deformation frequencies are present but the OH deformation appears in the same region. The broad maximum at 1410 cm^{-1} and also the

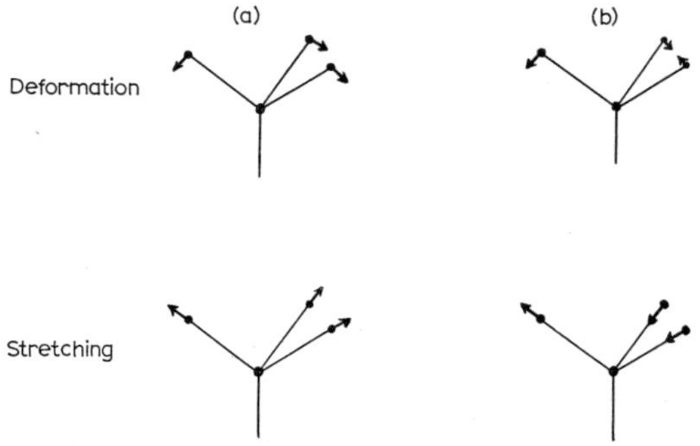

Fig. 107. Symmetric (a) and antisymmetric (b) deformation and stretching of CH_3 group.

underlying band at 1210 cm^{-1}, correspond to C—OH, the first to OH deformation and the second to C—O stretch: these frequencies suggest phenolic OH. If the indications of the CH and OH frequencies are that the molecule contains an aromatic ring then other characteristic frequencies are to be expected. Skeletal vibrations of the ring should occur between 1260 and 1350 cm^{-1} and indeed the bands which occur at 1615, 1575, and 1495 cm^{-1} compare well with many alkyl benzenes. Out-of-plane CH bending modes are also to be expected and the bands at 820, 740, and 690 cm^{-1} can be reconciled with monosubstituted or with 1 : 2 : 4- or 1 : 3 : 5-trisubstituted benzene.

These arguments go about as far as is safe with empirical correlations (and the advantage of knowing the answer.) The suggestions are that carvone contains —CH=C—C=O, $—CH_3$, $\rangle CH_2$, $\rangle C=CH_2$ and isomerizes to carvacrol which contains the benzene ring, phenolic —OH, $—CH_3$ but not $\rangle CH_2$, and may be 1 : 2 : 4 or 1 : 3 : 5 trisubstituted. It would therefore be reasonable to postulate that

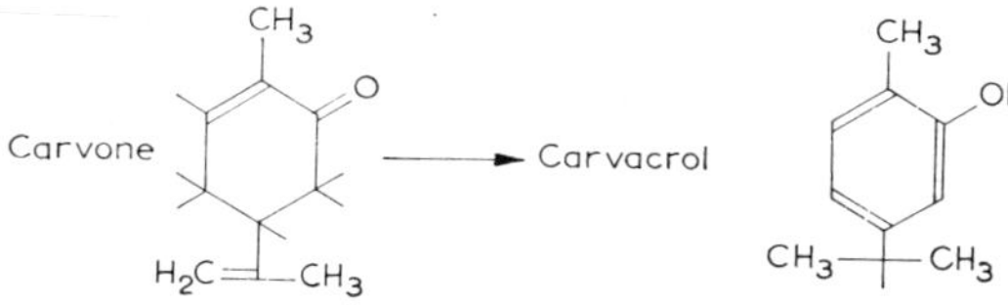

with some uncertainty about the position of substitution.

Knowledge of group frequencies is often used to establish small points related to structure. We have already noted the absence of N—H stretching frequencies in sempervirine. The presence of N—H stretching frequencies in the spectrum of $S_4N_4H_4$ confirms[29] the structure of that compound as $S_4(NH)_4$ or

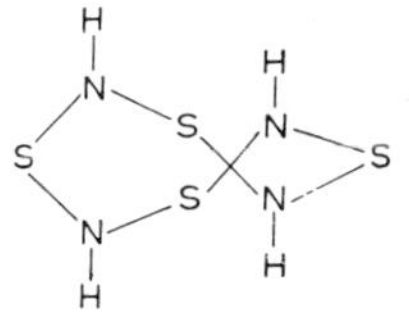

$$\begin{array}{c}\text{H–N}\langle\text{S}\rangle\text{N–H bridged via S–S to H–N}\langle\text{S}\rangle\text{N–H}\end{array}$$

The existence of a strong band at 1645 cm^{-1} in one form of polybutadiene singles it out as the 1 : 2 polymer, rather than the 1 : 4, since that frequency is characteristic of the $>CH{=}CH_2$ group:

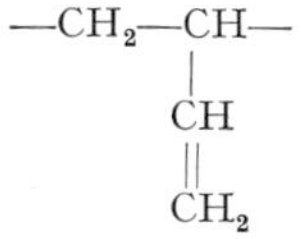

$$\begin{array}{c}-CH_2-CH-\\ |\\ CH\\ \|\\ CH_2\end{array} \qquad\qquad \begin{array}{c}-CH_2-CH\\ \|\\ CH-CH_2-\end{array}$$

1 : 2 polybutadiene 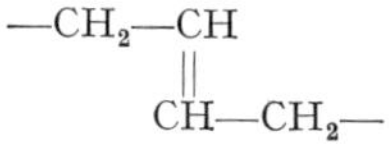 1 : 4 polybutadiene

In a comparison of the spectra of crystalline thiourea, $CS \cdot (NH_2)_2$, and the product, $(CSN_2H_3)_2TeCl_2$, of reaction of thiourea and tellurium tetrachloride, the complex had a band at 1635 cm^{-1}, not present in thiourea, and assigned to $>C{=}NH$ stretch. Both gave bands at 1618 cm^{-1}, assigned to $C{-}NH_2$; and thiourea gave a band at 1470 cm^{-1}, assigned to $N{-}C{=}S$ which was not present in the complex. These results were taken[30] to support the chemical evidence that the mode of attachment of the thiourea in these compounds is $Te{-}S{-}C\begin{smallmatrix}{\nearrow}NH\\ {\searrow}NH_2\end{smallmatrix}$.

These few examples must suffice to show the kind of use that can be made of infra-red spectra as an aid to the elucidation of chemical problems. The literature of preparative chemistry, both organic and inorganic, contains many more.

Specificity of infra-red spectra

In principle the existence of group frequency correlations limits the specificity of infra-red spectra in the sense that all molecules containing, say, a $C{\equiv}N$ group will show a band at about 2250 cm^{-1} and therefore that part of the spectrum is not uniquely characteristic of one molecule. This is partially true, so far as organic molecules are concerned, in the region above about 1500 cm^{-1} where the correlations are well established. But even in that region there are slight differences in position, shape and intensity of bands which are recognizably characteristic. In the region below 1500 cm^{-1} vibrations are more nearly whole-molecule vibrations and therefore more specific. The use of spectra, particularly in this region, as "finger-prints" is a demonstration of this specificity.

References p. 287–289

In so far as it is the whole shape of the spectrum which is important for identification the qualitative analysis of mixtures is not easy. It is often difficult to recognize the separate contributions of two overlapping spectra. The spectrum of a mixture is most useful when it is known what components are likely to be present. Then the spectrum may show whether the suspected components are present or absent. For such analysis, or for identification of pure compounds whose spectra are already known, there are various collections of spectra. The A.P.I. Index is a collection of infra-red, Raman and ultraviolet spectra, mainly of hydrocarbons, published at the Carnegie Institute of Technology, Pittsburgh, for the American Petroleum Institute, Research Project 44.

Three collections appear on cards: the D.M.S. collection (published jointly by Butterworth, London and Verlag Chemie, Weinheim), the Sadtler catalogue (published by Samuel P. Sadtler and Son, Inc., Philadelphia) and the Wyandotte-ASTM Index (published by the American Society for Testing Materials). These spectra are published on punched cards which can be sorted in various ways. In particular the DMS cards have holes corresponding to small frequency ranges (50 cm^{-1} up to 700 cm^{-1}; 25 cm^{-1} from 700 to 1000 cm^{-1}; 50 cm^{-1} from 1000 to 1800 cm^{-1}; 200 cm^{-1} from 1800 to 2800 cm^{-1}; 100 cm^{-1} from 2800 to 3700 cm^{-1}). The appropriate holes are punched for the frequency ranges in which strong bands appear. If now an unknown substance or mixture is found to have strong bands, in say, six of these ranges all spectra in the index having bands within these ranges can be readily found. If it is demanded that all six bands appear in the same substance (*i.e.* if the unknown is a single substance) the selection of cards can usually be narrowed down to a small number whose spectra may then be directly compared with that of the unknown. Similar provision for sorting is available in the other collections. The DMS cards can also be sorted in a number of other ways. The punched holes provide for finding all the substances with one or more of about 100 structural features, or for finding all the substances of the same carbon number.

References to a number of other sources of spectra are given by Jones and Sandorfy (Ref. 97, p. 328).

Raman Correlations

Much of what has been said of infra-red spectra in regard to correlations with particular groups is also true of Raman spectra, and need not be illustrated further. We may note that a type of vibration which is often not observed in the infra-red but is usually strong in the Raman is the totally symmetric stretching mode (see p. 173) and the following examples make particular use of such vibrations. In this section we are concerned only with empirical correlations for identification of some chemical species. Further examples are given later of the use of Raman spectroscopy both qualitatively and quantitatively (p. 300).

Aqueous solutions

For the present purpose we shall illustrate the usefulness of Raman spectroscopy in the field of aqueous solutions, where infra-red spectra are obtained only with difficulty.

A long standing anomaly in inorganic chemistry has been the existence of a compound, gallium dichloride, of empirical formula $GaCl_2$. Gallium is an element of group IIIB in which the expected valencies are three and one: an apparent valency of two is therefore anomalous. A Ga^{2+} ion would be expected to be paramagnetic: consequently, when solid $GaCl_2$ was found to be diamagnetic the structure $Cl_2Ga—GaCl_2$ was proposed. An alternative hypothesis is that the compound is gallium (I) tetrachlorogallata (III), $Ga^+[GaCl_4]^-$, containing univalent and tervalent gallium. The proposed anion is one which can be formed by solution of gallium trichloride in 6*M* hydrochloric acid and that solution is found to have a Raman spectrum containing the four lines expected of a tetrahedral species $[GaCl_4]^-$. They occur at: 114 cm^{-1} (strong, sharp and depolarized), 149 cm^{-1} (strong, sharp and depolarized), 346 cm^{-1} (very strong and strongly polarized) and 386 cm^{-1} (weak, diffuse and depolarized). Of these the very strong line at 346 cm^{-1} is identified as due to the symmetrical stretching mode and incidentally provides another example of a symmetrical mode having a lower frequency than the corresponding antisymmetric mode at 386 cm^{-1}. Gallium dichloride melts at 165° and its Raman spectrum has been observed in the liquid at 190°. Four lines were found at 115 cm^{-1} (strong, sharp and depolarized), 153 cm^{-1} (weak, diffuse and depolarised), 346 cm^{-1} (very strong, sharp and strongly polarized) and at 380 cm^{-1} (very weak, diffuse and depolarized.) Having regard to the different solvents and the accuracy of measurements ($\pm$ 3 cm^{-1}) the evidence is striking in favour of the formulation $Ga[GaCl_4]$ for the gallium dichloride[31,32]. The Raman evidence is against any formulation $Cl_2Ga—GaCl_2$, or ions derived from it. In this connection it is noteworthy that a single strong Raman line was obtained[33] as early as 1934 (six years after the discovery of the Raman effect) from solutions containing the mercurous ion as evidence for the formulation $[Hg—Hg]^{2+}$.

Perhaps the best known example of this kind of Raman investigation is the recognition of the linear triatomic ion, NO_2^+, in concentrated nitric acid, sulphuric, perchloric and selenic. The failure to recognize the species when the first spectra were obtained was principally because of the fortuitous proximity of Raman lines characteristic of HSO_4^- and NO_3^-. The Raman spectra of anhydrous nitric acid, Fig. 108(a), and of nitric-sulphuric mixtures, (b) and (c) were first measured in 1934—35. It was noted that a

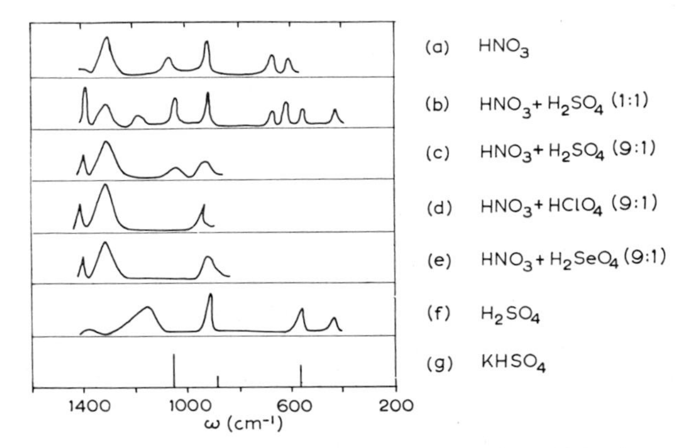

Fig. 108. Raman spectra of nitric acid, and mixtures with sulphuric, perchloric and selenic acids.

strong line appeared at 1400 cm^{-1} which was clearly not a component of sulphuric acid (f), but was weakly detectable in nitric acid. A likely explanation was that the species responsible for the 1400 cm^{-1} line was a product of self ionization of nitric acid (or an impurity) which was promoted by the addition of sulphuric acid. Inspection of Fig. 108(a), (b) and (c) shows that there is another broad line at 1050 cm^{-1} which appears to be associated with the line at 1400 cm^{-1}. This association persisted in solutions of N_2O_5 in nitric acid and also in nitric acid to which phosphoric oxide was added. The two lines were the only lines in the Raman spectrum of solid N_2O_5 and it was a reasonable conclusion that N_2O_5 was the species giving rise to both lines in all the solutions. The first observation which was not in accordance with this simple conclusion was that when N_2O_5 was dissolved in carbon tetrachloride or chloroform the Raman lines were found at 707, 860, 1240 and 1335 cm^{-1}, entirely different from the solid or solutions in nitric acid.

It has been suggested as early as 1903 by Euler that the active agent in aromatic nitration by nitric-sulphuric mixtures might be the nitronium ion, NO_2^+, and a good deal of kinetic evidence later accumulated to support the idea. If NO_2^+ is linear it is similar to CO_2 (with which it is isoelectronic) and we have already seen (p. 173) that the selection rules require two fundamental frequencies active in the infra-red and only one active in the Raman. It was therefore suggested[34] that the solution spectra could be accounted for by assigning the line at 1400 cm^{-1} to NO_2^+, and the line at 1050 cm^{-1} to either the nitrate ion or the bisulphate ion, both of which have Raman lines at that frequency. In agreement with this explanation the line at 1400 cm^{-1} is very strongly polarized, as is the corresponding Raman line at about 1390 cm^{-1} arising from CO_2. Finally the assignment was verified by the addition of perchloric acid (d) or selenic acid (e) to nitric acid, when the line 1400 cm^{-1} appears strongly without the appearance of the line at 1050 cm^{-1}. The relevant ionizations are as follows:

$$3HNO_3 \rightleftharpoons NO_2^+ + 2NO_3^- + H_3O^+$$
$$HNO_3 + 2H_2SO_4 \rightleftharpoons NO_2^+ + 2HSO_4^- + 2H_3O^+$$
$$N_2O_5 \rightleftharpoons NO_2^+ + NO_3^- \quad \text{in nitric acid}$$
$$4HNO_3 + P_2O_5 \rightleftharpoons NO_2^+ + 3NO_3^- + 2H_2PO_4^-$$
$$HNO_3 + 2HClO_4 \rightleftharpoons NO_2^+ + 2ClO_4^- + H_3O^+$$
$$HNO_3 + 2H_2SeO_4 \rightleftharpoons NO_2^+ + 2HSeO_4^- + 2H_3O^+$$

Non-aqueous solutions

Another example[35] of what is essentially empirical correlation is an investigation of complex formation between antimony pentachloride $SbCl_5$, and either phosphoryl chloride, $POCl_3$, or trimethylphosphine oxide, $PO(CH_3)_3$. The latter is expected to be a stronger donor molecule than the oxychloride. The Raman spectra shown in Fig. 109 were taken in 1,2-dichloroethane solution and the Raman lines due to the solvent are not in-

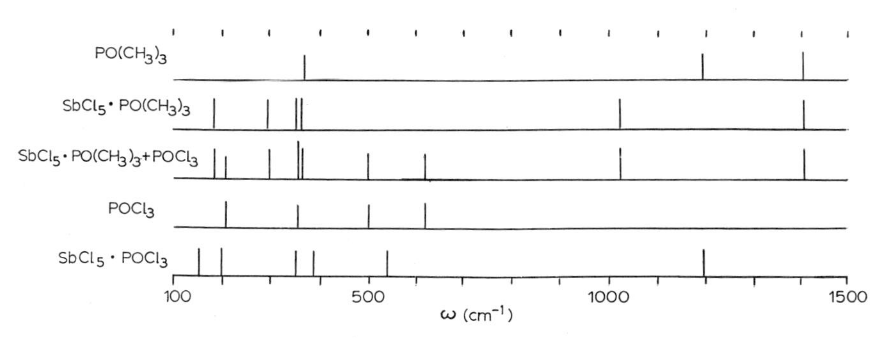

Fig. 109. Raman spectra of $POCl_3$, $PO(CH_3)_3$ and mixtures with $SbCl_5$.

cluded. It is evident that the spectrum of the equimolar mixture of $SbCl_5$ and $POCl_3$ does not contain the lines characteristic of $POCl_3$ and that complex formation may therefore be assumed. Similarly for the equimolar mixture of $SbCl_5$ and $PO(CH_3)_3$. On the other hand the spectrum of the solution containing equal amounts of the three compounds is the sum of $SbCl_5 \cdot PO(CH_3)_3$ and $POCl_3$: it is not the sum of $SbCl_5 \cdot POCl_3$ and $PO(CH_3)_3$.

References p. 287–289

It is therefore concluded that, as expected, $PO(CH_3)_3$ is a stronger donor than $POCl_3$ with respect to complex formation with $SbCl_5$.

Microwave Correlations

It is perhaps only in microwave absorption spectroscopy that single sharp transitions can be measured with such accuracy that each frequency is uniquely characteristic of the absorbing substance. The technique is, of course, restricted to polar molecules which are sufficiently volatile to give a vapour pressure of about 0.01 mm Hg. It is further desirable that the molecule should not have too many vibrational modes since interaction with the rotation leads to splitting of rotational lines and a weakening of each component. Hence only a few hundred molecules qualify for detection in this way. However, within that number positive identification of any one molecule should be possible by observation of one or two lines to a readily obtainable accuracy of 0.1 Mc/s. With that resolution, over an operating range of 20,000 to 30,000 Mc/s, there is room for 100,000 different characteristic frequencies.

A table of known microwave lines arranged in order of frequency is published by the National Bureau of Standards and a table of over 150 molecules, with molecular parameters but not actual observed lines, is given by Townes and Schawlow[36].

Nuclear Magnetic Resonance

According to the discussion of nuclear magnetic resonance in Chapter 2 a nucleus of spin I can adopt $2I + 1$ orientations with respect to a magnetic field H_0 and the energies of the different states relative to the energy in the absence of the field are given by

$$U_M = \frac{M}{I} \mu_0 H_0 \tag{4.6}$$

where μ_0 is the magnetic moment (*i.e.* the maximum observable value of the magnetic moment) of the nucleus and M may take the values $-I, -I+1 \ldots +I$. Equation (4.6) defines a set of $2I + 1$ energy levels. Nuclear magnetic resonance is observed when a quantum of radiation is absorbed to effect a transition between these levels. The transition is subject to the selection rule

$$\Delta M = \pm 1 \tag{4.7}$$

and hence there is expected one resonance frequency

$$\nu = \frac{\mu_0 H_0}{hI} \tag{4.8}$$

for any one set of levels given by (4.6).

Zeeman levels for nuclei of spin $I = \frac{1}{2}$ and $I = 1$ are shown in Fig. 110. Resonance absorption corresponds to a change in orientation of the nuclear

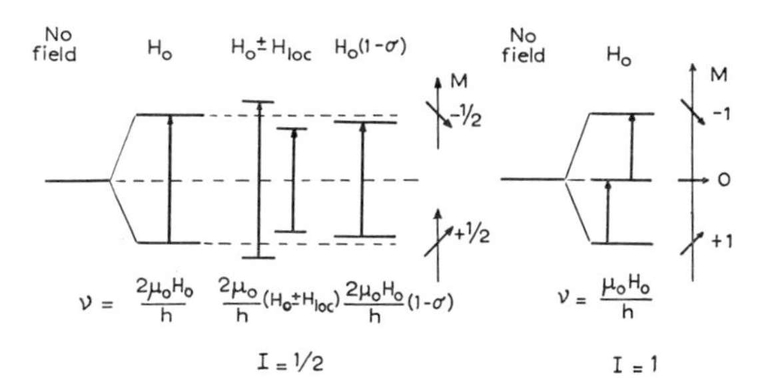

Fig. 110. Nuclear magnetic resonance for nuclei with spin $I = \frac{1}{2}$ and $I = 1$.

magnets; in the case when $I = \frac{1}{2}$ and there are only two orientations, the magnets "flip over" from an orientation with the field to one opposed to the field.

From Table 8 it can be seen that for a field of 10,000 gauss the expected resonant frequency for protons is 42.577 Mc/s, *i.e.* in the radio frequency range. For fluorine the frequency in the same field is 40.06 Mc/s; for ^{14}N, 3.07; for ^{29}Si, 8.56; for ^{31}P, 17.24 Mc/s. In practice resonance is observed by holding the frequency constant and varying H_0. Consequently resonant frequencies are often quoted in units of magnetic field, *i.e.* in gauss.

If this were all and the magnetic field experienced by the nucleus were indeed the same as the external applied field, H_0, nuclear magnetic resonance would have little chemical interest. In fact the effective magnetic field H at the nucleus differs from H_0 in three ways. The magnitude of the difference is in no case greater than about 0.5% of H_0 yet it is the accurate measurement of this difference which provides one of the newest sources of chemical information — mainly, but not all, structural in character.

Magnetic dipole interaction

If the sample is a solid all nuclei occupy practically fixed positions and those nuclei with magnetic moments contribute an addition to the external applied field which varies throughout the solid in a manner which is determined by the structure and symmetry of the solid. In principle any one nucleus experiences this magnetic field as the sum of contributions from all magnetic nuclei throughout the sample. In practice the effective local field of a nuclear magnet varies with distance from the nucleus as $1/r^3$ and hence

References p. 287–289

the major contribution to the effective local field experienced by a nucleus is that of its nearest neighbours. If a nucleus has one other like nucleus as its nearest neighbour the effective local field at either nucleus is

$$H_{\text{loc}} = \frac{3\mu_0}{4Ir^3}\,(3\cos^2\theta - 1) \tag{4.9}$$

where r is the internuclear distance and θ the angle between the internuclear line and the applied magnetic field. The field experienced by either nucleus is thus

$$H = H_0 \pm \frac{3\mu_0}{4Ir^3}\,(3\cos^2\theta - 1) \tag{4.10}$$

There is practically equal probability that the near neighbour will be in either of the possible nuclear orientations and hence for nuclei of spin $\frac{1}{2}$ the $+$ and $-$ signs are equally represented. For nuclei of spin 1 the situation $M = 0$ must also be included.

Suppose the crystal contains water of crystallization. The two protons provide the pair of nuclei of spin $I = \frac{1}{2}$. We should, by (4.8), expect two resonance lines at frequencies

$$\nu = \frac{2\mu_0 H_0}{h} \pm \frac{3\mu_0{}^2}{hr^3}\,(3\cos^2\theta - 1) \tag{4.11}$$

or separated by a frequency interval

$$\Delta\nu = \frac{6\mu_0{}^2}{hr^3}\,(3\cos^2\theta - 1) \tag{4.12}$$

In practice the frequency is kept constant and the external field H_0 varied. The values of H_0 at which resonance occurs are those which, corrected according to equation (4.10), give a value of H corresponding to the chosen frequency, *viz.* $h\nu/2\mu_0$. The separation between the two resonances is clearly

$$\Delta H = \frac{3\mu_0}{r^3}\,(3\cos^2\theta - 1) \tag{4.13}$$

and is independent of the values of H_0 and ν which determine the region in which the resonance is observed.

In Fig. 111 the two proton resonances[37] (see also Ref. 107, p. 178) are shown for a single crystal of the monohydrate of barium chlorate, $Ba(ClO_3)_2 \cdot H_2O$.

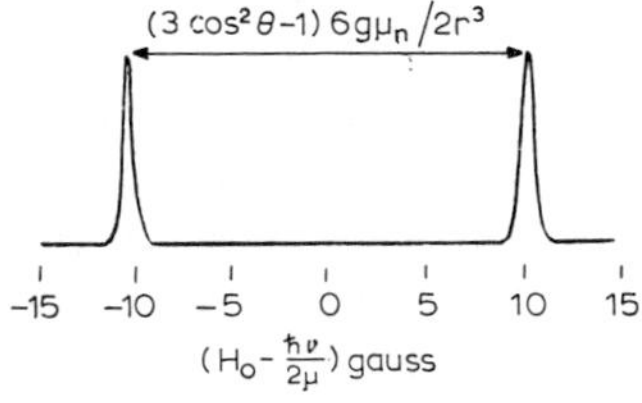

Fig. 111. Proton magnetic resonance in a single crystal of $Ba(ClO_3)_2 \cdot H_2O$.

The observed lines have finite breadth because of interaction with more distant nuclei. Their separation of 21 gauss is only 0.2 % of the usual applied field. In an earlier example[38] of this magnetic dipole interaction the lines for gypsum, $CaSO_4 \cdot 2H_2O$ are considerably broader because the density of water molecules in that crystal is considerably greater than in the barium chlorate and consequently the interaction with nuclei other than nearest neighbours is more prominent. When the single crystal is turned to different orientations in the field H_0 the separation of the lines varies in accordance with (4.13) and reaches a maximum of $6\mu_0/r^3$ when the interproton vector is parallel to the field ($\theta = 0$). This serves both to locate the interproton vector in relation to the crystallographic axes (or in relation to the faces of the crystal under observation) and to determine the interproton distance. For example, the separation observed for gypsum at $\theta = 90°$ is 106 gauss and hence $r = (3 \times 2.793 + 5.050 + 10^{-24}/10.6)^{\frac{1}{3}} = 1.58$ Å. The value is somewhat larger than in an isolated water molecule on account of the hydrogen-bonded attachment of the water to the sulphate ions.

When the sample is not a single crystal but a powder it may be assumed that all orientations are equally represented. The resultant spectrum is a comparatively broad band, extending over some 60 gauss, in which the doublet structure corresponding to a pair of protons can still be seen. When the protons occur in groups of three, as in methyl groups or in ammines, the magnetic dipole interactions are somewhat more complex. Each proton is affected by the other two and hence H_{loc} may be the sum of two contributions in either the positive or the negative sense again leading to $H = H_0 \pm H_{loc}$. On the other hand the two protons may be so orientated that their contributions to the local field at the third nucleus cancel each other; and $H = H_0$. Something of the resultant triplet structure remains even when the broadening effect of more remote nuclei and of random orientation in polycrystalline samples is taken into account. In tetrahedral arrangements of four nuclei the calculated shape of the line has become a broad and featureless hump. The theoretical line shapes for polycrystalline samples of crystals of these three types are shown in Fig. 112. By comparison of observed and calculated line shapes it has been shown that the monohydrates of nitric and perchloric acids[39] contain the oxonium ion H_3O^+ (Fig. 112(b)) or that infusible white precipitate[40], obtained by adding ammonia to a solution of mercuric chloride, is NH_2HgCl, conforming to Fig. 112(a), and not either $NHg_2Cl \cdot NH_4Cl$ (c) or $xHgO \cdot (1 - x)HgCl_2 \cdot 2NH_3$ (b).

In cases where the shape of the line is not informative or its theoretical shape is too difficult to calculate, much information can still be gained from the mean-square width (the so-called *second moment*) of the line. A recent review of the n.m.r. study of crystals is given by Andrew (Ref. 107, p. 177;

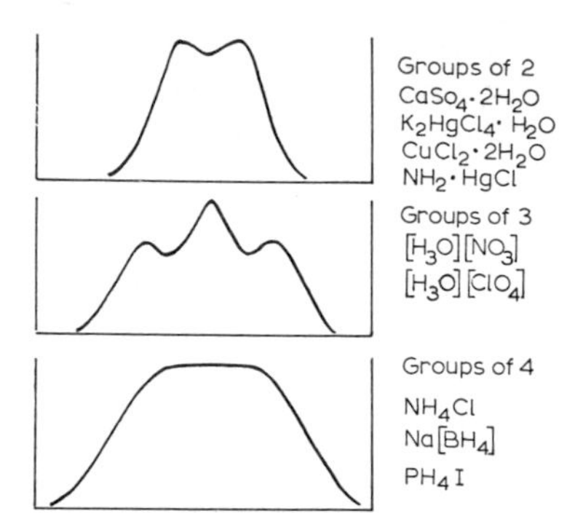

Fig. 112. Theoretical n.m.r. line shapes for polycrystalline samples of materials containing like nuclei of spin $\frac{1}{2}$ in groups of (a) two, (b) three (equilaterial triangles) and (c) four (tetrahedral).

Ref. 103). Particular attention has been given to proton resonance in the study of crystals because X-ray diffraction does not locate hydrogen atoms very readily. However the method is also applied to other magnetic nuclei.

N.m.r. in liquids and gases

The nuclear resonance lines obtained from liquids are considerably narrower than those from solids (say 10 milligauss rather than 10 gauss). The reason lies in the random orientation of molecules in the liquid state but that is not all for there is also random orientation in polycrystalline solids. The difference is that in the powdered crystals any one crystal has a fixed orientation and the average is taken over many crystals. In the liquid, however, each molecule is in motion and changes its orientation at a frequency comparable to rotation frequencies of free molecules, *i.e.* of the order of 10,000 Mc/s. Comparison with resonance frequencies of, say, 40 Mc/s shows that each molecule adopts a few hundred different orientations with respect to the field during the process of a nuclear transition. The time average of H_{loc} over that period for a single molecule (rather than the space sum over many crystals) is zero. Hence the broadening and splitting arising from direct magnetic dipole interaction in solids is not observed in liquids or gases.

On the other hand two other smaller perturbations become evident when line broadening due to direct spin-spin interaction is absent. They are effects due to magnetic shielding by electrons and to electron-coupled nuclear spin interactions: their time-averages are not zero.

Chemical shift

A perturbation of Zeeman levels arises from the influence of the external magnetic field upon the motion of the electrons in the molecule. As a result the nuclei are partially screened from the field H_0 by an amount σH_0 where σ is the *screening constant*. The field experienced by the nucleus is thus $H = H_0(1 - \sigma)$ — see Fig. 110. Chemically equivalent nuclei, such as three protons of a (freely rotating) methyl group, have the same screening constant but nuclei of the same element in different positions in the molecule are in different electronic environments and have different screening constants. The two protons in water are equivalent and there is one proton resonance frequency. In ethanol there are three equivalent protons in the CH_3 group, two in CH_2, and one in OH: σ increases in the order $\sigma_{OH} < \sigma_{CH_2} < \sigma_{CH_3}$. The value of H_0 at which resonance occurs for a fixed frequency ν is

$$H_0 = \frac{h\nu}{2\mu_0} \cdot \frac{1}{1 - \sigma} \tag{4.14}$$

and thus proton resonances for OH, CH_2 and CH_3 appear in that order as H_0 increases. The lines are identified by the relative intensities of the signals, 1 : 2 : 3. This and other spectra are illustrated in Fig. 113. In toluene there are only two sets of equivalent protons in ratio 5 : 3 and in acetaldehyde there are also two sets in the ratio 3 : 1.

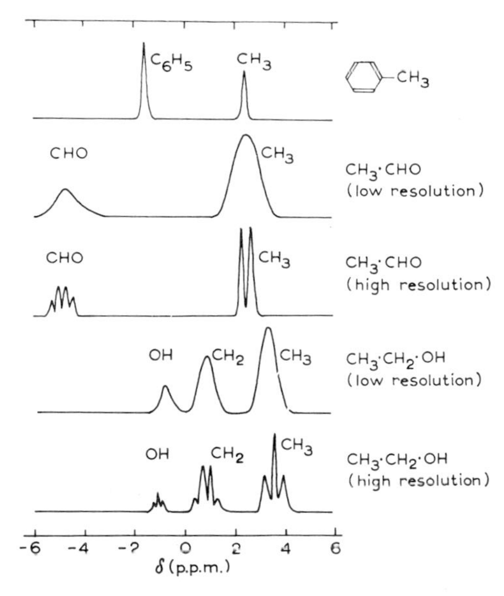

Fig. 113. Proton magnetic resonance in toluene, acetaldehyde and ethanol.

It is customary to measure σ relative to σ_r, the screening constant for a nucleus in a reference compound. If the corresponding resonance field strengths are H_0 and H_r we may define a ratio

$$\delta = \frac{H_0 - H_r}{H_r} = \sigma - \sigma_r \tag{4.15}$$

where δ is called the *chemical shift*. A positive chemical shift relative to the reference substance implies higher electronic screening and a resonance at a higher magnetic field. Relative to water the chemical shifts for the protons in ethanol are: -0.5 (OH), 1.0 (CH_2) and 3.5 (CH_3) $\times 10^{-6}$. These are values which require a relative accuracy of 1 part in 10^8 in the measurement of magnetic field (and also in the stability of the radio-frequency generator). The reference substance does not need to be present in the sample: it is often mounted in a capillary tube alongside the sample. However, in comparison of results based on various reference substances, allowance has to be made for the bulk diamagnetic susceptibility of the solvent and the reference. It may be noted that the chemical shift is a property primarily of the electrons and not the nuclei. It should therefore be the same for different isotopes of the same elements in the same position in the molecule, even though the resonance frequencies for the isotopes are different.

The compound diketene provides an example of the use of these ideas in a simple structural problem. At various times structures have been assigned as follows:

(a) $CH_3 \cdot CO \cdot CH{=}C{=}O$

```
(b) O=C—CH2                  (c) O=C—CH
      |  |                         |  ||
    H2C—C=O                      H2C—C—OH

(d) CH3—C=CH                 (e) CH2=C—CH2
        |  |                          |  |
        O—C=O                         O—C=O
```

The n.m.r. spectrum of the solid[41,42] favours (e), although it would also agree with (b) if this were not excluded on chemical and infra-red evidence. In the spectrum of the liquid proton resonances would be expected with relative intensities as follows: (a) two, 3 : 1; (b) one, 4; (c) three, 2 : 1 : 1; (d) two, 3 : 1; (e) two, 2 : 2. Thus it should be poassible to distinguish between all the structures except (a) and (d). In fact two resonances are found, at $\delta = -0.07$ and -0.16×10^{-6} referred to water, and of equal intensity. The structure (e) is thus unequivocally confirmed as the principal component in liquid diketene at room temperature. At higher temperatures the spectrum becomes more complex, indicating the formation of new species.

Correlations

Quantitative theories have been developed for chemical shifts and for spin coupling. In particular the spin-spin interactions are subject to considerations of symmetry: group theory enables considerable simplification in the quantum mechanical problem. The interested reader is referred to Pople, Schneider and Bernstein (Ref. 105, Chaps. 6—8; also Ref. 104).

The considerable collection of data which has been built up can be used as the basis for empirical correlations, without much regard for theoretical explanation. Table 20 gives a representative selection[43]* of chemical shifts for protons in various situations and it is clear that even with this limited selection a good deal of structural information can be got. Distinction between alkyl and aryl NH_2 and OH is very clear, and between protons attached to various carbon atoms.

TABLE 20

CHEMICAL SHIFTS FOR PROTONS IN VARIOUS POSITIONS

		δ relative to water	δ relative to benzene
$—CH_3$	alkyl	3.6 to 4.2 × $^{-6}$	5.4 to 6.0 × 10^{-6}
$—NH_2$	alkyl	3.1 to 4.2	4.9 to 6.0
$\rangle CH_2$	cyclic	3.2 to 3.8	5.0 to 5.6
$—CH_3$	aryl	3.0 to 3.7	4.8 to 5.5
$—CH_3$	attached to N	2.1 to 2.9	3.9 to 4.7
$—CH_3$	attached to O	~1.6	~3.4
$\rangle CH_2$	attached to C and X	0.9 to 3.3	2.7 to 5.1
$\rangle CH_2$	attached to O	1.3 to 1.7	3.1 to 3.5
$—NH_2$	aryl	1.2 to 1.9	3.0 to 3.7
—OH	alcoholic	−0.7 to 0.6	1.1 to 2.4
—CH	attached to C= and C—	−1.7 to −0.5	0.1 to 1.3
—CH	aromatic	−2.8 to −1.2	−1.0 to 0.6
—OH	phenolic	−2.7 to −2.1	−0.9 to −0.3
—CHO		−5.0 to −4.4	−3.2 to −2.6
$\rangle NH$	attached to O	~−5.1	~−3.3
—COOH		−7.2 to −5.6	−5.4 to −3.8

A good example[44] of the use of such correlations comes from an investigation of a rearrangement reaction of 5-nitrobornene. Whereas the action of cold dilute sulphuric acid on 5-nitrobornane leads to a ketone, with the evolution of nitrous oxide (Nef reaction),

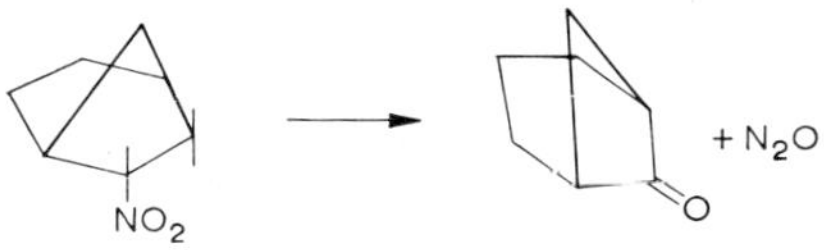

5-nitrobornene undergoes a rearrangement

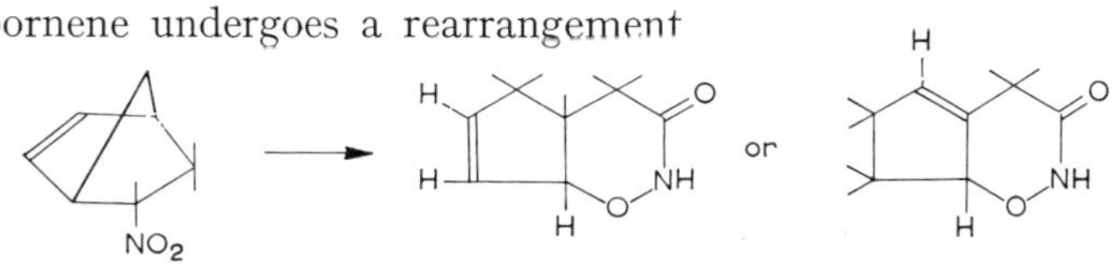

* In this paper, however, the chemical shift δ_{MSG} is related to that generally used and quoted here by $\delta_{MSG} = -\delta/10$.

Evidence that the product is the first alternative rather than the second comes from the proton resonance spectrum, Fig. 114. The resonance (relative to benzene) at $\delta = -3.3 \times 10^{-6}$ is due to $>$NH present in both. The com-

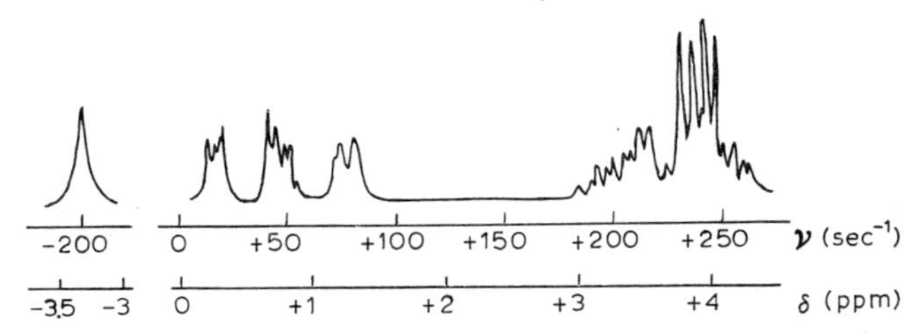

Fig. 114. Proton magnetic resonance spectrum of product of rearrangement of 5-nitrobornene. Reference, benzene resonance at 60 Mc/s.

plex region at $+ 4 \times 10^{-6}$ is due to $>CH_2$ protons or a ring junction $\geq$CH, and these are also present in both structures. There are, however, three proton resonances in the region appropriate to =C—H, or to a ring junction $\geq$C—H attached to O, in agreement with the first formula but not the second. The spectrum was measured in deuterochloroform to avoid interfering proton resonances. Additional evidence comes from the relative numbers of protons obtained from the integrated intensities of the various peaks. If the intensity of the NH resonance is taken as the unit intensity the three peaks at 0.3, 0.75 and 1.3×10^{-6} each have unit intensity and the group at 4.0×10^{-6} has overall intensity of 5 units. The second formula would require 6 units (= 6 protons) in the region assigned to $>CH_2$, without the ring junction $\geq$CH, and 7 units if that is included.

Characteristic chemical shifts have been measured and listed (Ref. 104, Chap. 2; Ref. 105, Chap. 12) for nuclear resonances of ^{13}C, ^{14}N, ^{17}O, ^{19}F, ^{29}Si, ^{31}P, ^{119}Sn, in a variety of compounds.

Electron-coupled nuclear spin interactions

Most nuclear magnetic resonance spectrometers are capable of sufficient resolution to give some (if not all) of the fine structure of resonance lines. Accordingly most structural arguments based on chemical shifts are augmented by fine structure data and we must therefore discuss the theoretical basis of the interpretation.

Although *direct* spin-spin interaction is averaged out by the tumbling motion of the molecules in a liquid, the much smaller spin-spin interactions, which are transmitted via the weak coupling between the nuclear magnets and electron spin, are not removed by random reorientation of the molecule. The observed interactions are not between equivalent nuclei (as are the direct dipole interactions) but are largely between adjacent groups. The high resolution spectra of Fig. 113 illustrates the phenomenon. In acetaldehyde the proton resonance assigned to the CH_3 group is split into two by the interaction with the single —CHO proton. Its resonance however is split into four by the adjacent group of three CH_3 protons. In ethanol the OH proton is only affected by the neighbouring CH_2 and not by the CH_3: similarly the CH_3 protons are only affected by the CH_2 group. Hence both the CH_3 and the OH signals are similarly split into triplets. The reason for the triplet character of the CH_3 signal is *not* the presence of three hydrogen atoms in the group.

We may examine some simple cases of interaction as follows: in each case the nucleus whose signal is being observed is affected by the spins of one or more neighbours.

(a) nucleus sees one neighbour of spin $I = \frac{1}{2}$, which may be in one of two possible orientations, $+\frac{1}{2}$ and $-\frac{1}{2}$: resonance line is a doublet. Examples: effect of CHO proton on CH_3 signal in $CH_3 \cdot CHO$; of H on D signal in HD; of H on F in CHF_3; of F on H in CH_3F; of P on F in PF_3; of one non-equivalent F on the other two F in ClF_3.

(b) nucleus sees one neighbour of spin $I = 1$, which may be in one of three possible orientations $+1$, 0 and -1; resonance line is a triplet. Examples: effect of D on CH_3 in $CH_3 \cdot CDO$; of D on H in HD; of ^{14}N on H in NH_3.

(c) nucleus sees two equivalent neighbours of spin $I = \frac{1}{2}$, for which the possible combination of orientations fall into three groups:

$$\begin{array}{ccc} +\frac{1}{2}\ +\frac{1}{2} & +\frac{1}{2}\ -\frac{1}{2} & -\frac{1}{2}\ -\frac{1}{2} \\ & -\frac{1}{2}\ +\frac{1}{2} & \end{array}$$

The middle two affect the nucleus similarly: resonance line is a triplet, intensity ratio 1 : 2 : 1. Examples: effect of CH_2 protons on either CH_3 or OH protons in $CH_3 \cdot CH_2 \cdot OH$; of pair of H nuclei on single proton in $CHCl_2 \cdot CH_2Cl$; of two equivalent F on the other F in ClF_3.

(d) nucleus sees three equivalent neighbours of spin $I = \frac{1}{2}$, for which the possible combinations of orientations fall into four groups

$$\begin{array}{cccc} & +\frac{1}{2}\ +\frac{1}{2}\ -\frac{1}{2} & +\frac{1}{2}\ -\frac{1}{2}\ -\frac{1}{2} & \\ +\frac{1}{2}\ +\frac{1}{2}\ +\frac{1}{2} & +\frac{1}{2}\ -\frac{1}{2}\ +\frac{1}{2} & -\frac{1}{2}\ +\frac{1}{2}\ -\frac{1}{2} & -\frac{1}{2}\ -\frac{1}{2}\ -\frac{1}{2} \\ & -\frac{1}{2}\ +\frac{1}{2}\ +\frac{1}{2} & -\frac{1}{2}\ -\frac{1}{2}\ +\frac{1}{2} & \end{array}$$

Resonance line is a quarter, intensity ratio 1 : 3 : 3 : 1. Examples: effect of CH_3 protons on CHO in $CH_3 \cdot CHO$; of F on P in PF_3.

(e) nucleus sees four equivalent neighbours of spin $I = \frac{1}{2}$ for which the possible combinations fall in five groups. Resonance line is a quintet intensity ratio 1 : 4 : 6 : 4 : 1. Example: effect of H on ^{14}N in $^{14}NH_4^+$ in solution.

The separation of components in this fine structure is the *coupling con-*

References p. 287–289

stant J_{YX} between the nucleus Y and neighbours X. It is usually expressed in frequency units and measures the extent to which the electrons are liable to transmit the influence of X to Y, and vice versa. For HD J_{HD} is 280 sec^{-1}; for CH_4, $J_{HH'}$ is 12.4 sec^{-1}; and for $>C=C<^{H}_{H}$, *cis* $>CH=CH<$ and *trans* $>CH=CH<$, $J_{HH'}$ is respectively 1—2, 8—11 and 17—18 sec^{-1}. Coupling with nuclei other than protons can be considerably bigger; $J_{CH'}$ in C_6H_6 is 159 sec^{-1}, J_{NH} in $^{14}NH_3$ is 46 sec^{-1}, J_{PF} in PF_3 is 1,400 sec^{-1}. The values of coupling constants clearly provide information about electron distributions in molecules but the interpretation is not within the scope of the present book.

The above simple rules for deducing the fine-structure components of resonance lines breaks down when the magnitude of the splitting is of the same order as the chemical shift measured in frequency units, *viz.* $(H_0 - H_r)2\mu_0/h$. Such a situation may occur where there are two resonances from nonequivalent nuclei of the same element. Since, however, the chemical shift in frequency units depends upon H_0, whereas J is independent of H_0, it is often possible to adjust H_0 (and consequently ν) in order to separate the two effects.

The fluorine spectrum of ClF_3, shown in Fig. 115, is a case in point[45]. The spectrum obtained at 10 Mc/s ($H_0 \sim 2500$ gauss) is a complex one

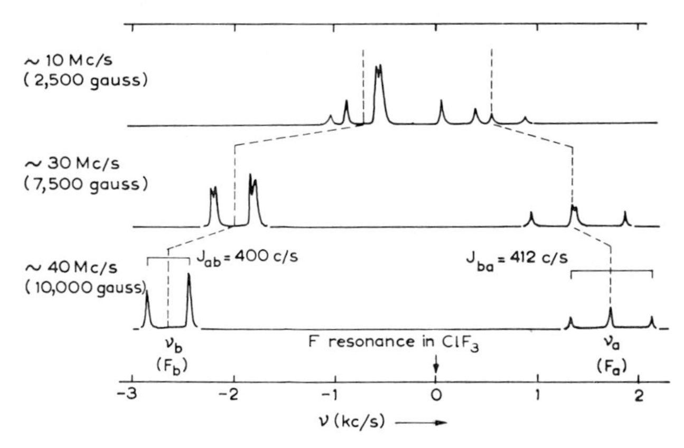

Fig. 115. Fluorine resonance in ClF_3 at different field strengths.

containing 8 maxima which can be seen to arise from mixing of chemical shifts and spin-spin interaction, since on increasing the field the two effects can be sufficiently separated to permit a simple analysis. At 10,000 gauss the difference in chemical shift has been raised to 4367 sec^{-1} ($= \nu_a - \nu_b : \delta_a - \delta_b = 109 \times 10^{-6}$). The doublet splitting of ν_b corresponds to the F_b resonance affected by the single F_a ($I = \frac{1}{2}$) whereas the triplet splitting of ν_a corresponds to the resonance of F_a affected by the two nuclei F_a ($I = \frac{1}{2}$). These results are in good agreement with the symmetry for ClF_3 but do not, of

course, indicate the value of the $F_a\hat{Cl}F_b$ angle.

Another example of fluorine resonance which confirms a structure of interest is found[46] in SF_4. On the evidence of infra-red and Raman data this molecule is assigned a structure (a) in the point group C_{2v} (see p. 167). From the point of view of n.m.r. spectra the molecule contains two pairs of equivalent fluorine nuclei. This is entirely confirmed by the spectrum in Fig. 116 which has two resonance lines of equal intensity ($\delta_a - \delta_b = 48 \times 10^{-6}$) both of which are triplets (intensity 1 : 2 : 1) with the same splitting. This spectrum could not arise from the tetrahedral structure (b) or the C_{3v} structure (c).

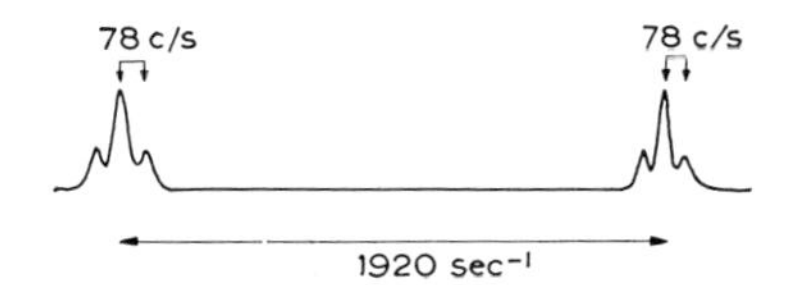

Fig. 116. Fluorine resonance in SF_4 at −90° C.

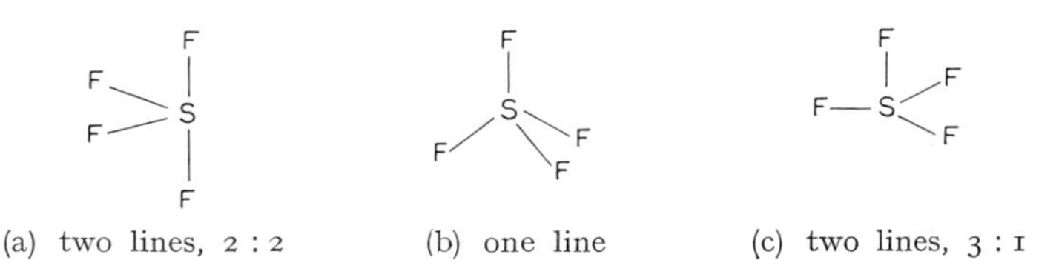

(a) two lines, 2 : 2 (b) one line (c) two lines, 3 : 1

As an example of the way in which multiplet splittings can overlap Fig. 117 shows part of the proton resonance spectrum[47] of propylene at

40 Mc/s. The whole extent of the figure is 72 c/s. The chemical shifts identify these resonance lines as arising from the Zeeman splitting of the nuclear levels of the vinyl hydrogen atoms. The group on the left of the figure has been analyzed in terms of the resonance of H_a split by H_c ($J_{ac} = 18$ c/s), a

further doublet splitting of each by H_b ($J_{ab} = 8.7$ c/s), and a quartet splitting of each of the resultant four components by the protons of the CH_3 group ($J_{ad} = 6.6$ c/s)

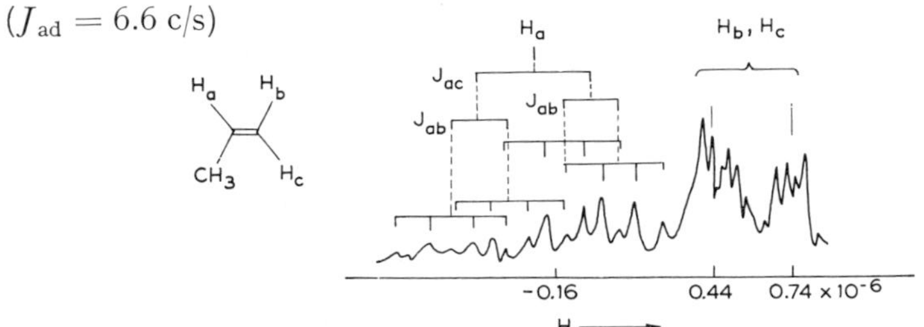

Fig. 117. Proton resonance in propylene.

Diborane, B_2H_6, is a molecule which presented problems of chemical bonding when it was thought to have a structure similar to ethane. The hypothesis of structure involving hydrogen-bridge bonds,

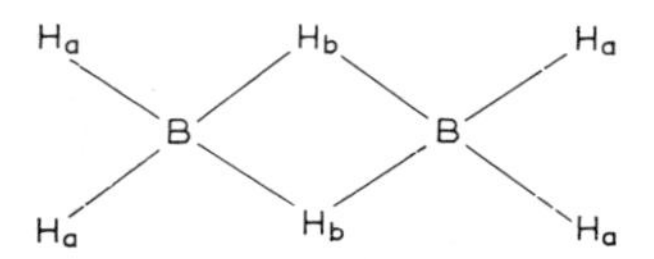

first put forward in 1921, revived in 1940—43 and now generally accepted, receives striking confirmation from the structure of the proton resonance spectrum[48]. From Fig. 118 it is immediately evident that all hydrogen atoms

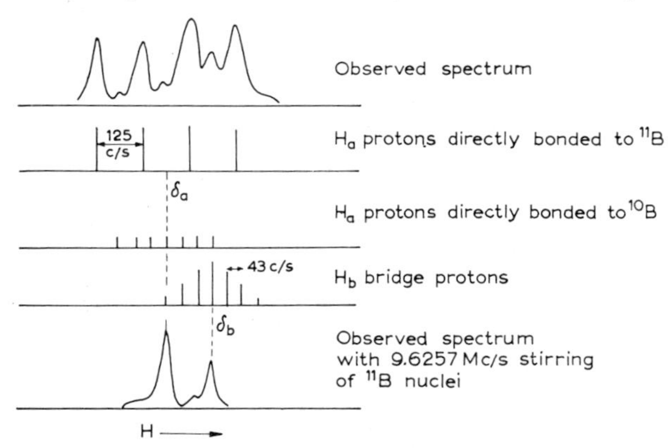

Fig. 118. Proton resonance in diborane at 30 Mc/s.

are not equivalent since the spectrum is not symmetrical. If the structure is as depicted the spectrum can be largely reconstructed from the three contributions indicated in Fig. 118. In the first place the protons H_a will be affected by their directly attached boron atoms. Those attached to the more abundant ^{11}B atoms (81 %, spin $I = \frac{3}{2}$) will see with equal weight each of the four orientations $-\frac{3}{2}$, $-\frac{1}{2}$, $+\frac{1}{2}$, $+\frac{3}{2}$. Those attached to the less abundant ^{10}B atoms (19 %, $I = 3$) will see with equal weight each of seven orientations, -3, -2, -1, 0, $+1$, $+2$, $+3$. The chemical shift δ_A is the same for each case. The bridge protons H_b (different chemical shift δ_b) will see an equivalent pair of (mainly) ^{11}B nuclei resulting in a septet in the intensity ration 1 : 2 : 3 : 4 : 3 : 2 : 1 (which it is left to the reader to verify).

The lower spectrum in Fig. 118 is an example of a technique sometimes employed in n.m.r. spectroscopy, in which the orientation of the ^{11}B nuclei is destroyed by irradiation of the sample at the characteristic frequency for ^{11}B (9.6257 Mc/s) while the proton resonance is still being observed at 30 Mc/s. This double irradiation method effectively "stirs" the ^{11}B nuclear magnets and the spectrum reduces to two lines at δ_a and δ_b in the intensity ratio 4 : 2. The structure due to ^{10}B remains and is just discernible.

N.m.r. fingerprints

It should be evident from the above discussion that n.m.r. spectra are uniquely characteristic of the molecules from which they are derived. Elements can be identified by their resonant frequencies if they have magnetic nuclei, but carbon, oxygen and sulphur are important elements which do not have magnetic nuclei in appreciable abundance. In general, however, the resonance spectrum for each magnetic nucleus in a substance is a reliable and useful fingerprint.

There are three qualifications which ought to be added to that statement. The first is evident from the example of ClF_3. The appearance of a spectrum which involves both chemical shifts and multiplet splitting can depend markedly upon the magnetic field (and, consequently, the resonant frequency). Comparison of n.m.r. spectra for the purpose of identification should therefore be done at the same field strength. The second qualification relates to the condition of the sample. If the sample is a solid it has either to be liquefied or brought into solution in order that sharp spectra may be obtained from it. The effect of solvents on nuclear resonance spectra has been described as being, in many cases, very dramatic. Comparison of spectra should therefore be done with the same solvent. Thirdly, there are some cases in which the spectrum varies with temperature. Thus the spectrum of SF_4 given in Fig. 116 could only be observed at $-90°C$. At temperatures above that the multiplet structure was destroyed and then the two resonance lines merged as a result of rapid exchange of fluorine atoms. It is important therefore that liquid samples should be compared at the same temperature. Any publication of standard n.m.r. spectra will have to state the precise conditions under which the spectra have been obtained.

References p. 287–289

Intramolecular Perturbations

Spectra at all frequencies are subject to perturbations which we may divide into two classes. The spectrum of a molecule or ion may be affected by various changes in environment including changes of solvent or changes in concentration: these are intermolecular perturbations and are discussed later in this chapter. On the other hand there are differences in spectra which arise from substitutions within the molecule. So long as the differences are small and the affected transitions are recognisably of the same origin we may regard this phenomenon as a perturbation and look for some correlation between the substituent and the frequency shift. The dividing line between a shift of frequency having the same assignment and a new frequency with a different assignment is not always clear.

Electronic spectra

The combination of chromophores in conjugation within one molecule is in a sense an intramolecular perturbation. There are, however, smaller differences which are caused by substitution of groups not recognisably chromophores in their own right, and referred to as *auxochromes*. The study of auxochromes is, of course, of considerably practical importance in the dyestuff industry. Certain groups were recognized as being able to intensify the colour of a chromophore, an effect which might arise from an increase in intensity (*hyperchrome*) or from a shift of frequency further into the visible, *i.e.* towards the red, to longer wavelength or to lower frequency (*bathochrome*). The opposite effect of lightening colour might arise from decrease in intensity (*hypochrome*) or a shift away from the red (*hypsochrome*). Bathochromic groups in dyestuffs are —NH_2, —OH, —SO_3H and —COOH: they evidently have the effect of raising the energy levels of the chromophore, the bonding orbitals more than the antibonding, and thus decreasing the energy of the transition.

The effect of various substituents upon the benzene N—V transitions has been studied fairly extensively[49], (see also Ref. 97, p. 629), as relating to a system most likely to be interpretable. So also has the effect of alkyl substitution upon $N \rightarrow Q$ as well as $N \rightarrow V$ transitions. In benzene the effect of NH_2, OH and even more of NO_2 is to increase the extent of the π-orbital system, and this evidently has the effect of shifting the absorption to longer wavelength. It may be regarded, alternatively, as a movement of negative charge into the benzene ring, which thus lowers the energy required to remove an electron to higher orbitals. In any case there is an observed correlation between bathochromic shift and the ortho-para directing nature of these substituents. The methyl group also has the effect of moving some electrons into the benzene ring. It increases the wavelength of the *N*—*V* transition in benzene and lowers the ionisation potential from 9.24 eV in benzene to 8.92 eV in toluene. The effect of progressive alkyl substitution on many other molecules, including water, ammonia, and hydrogen chloride, is to lower the ionization potential and to increase the wavelength of the first absorption band.

We will not discuss further the interpretation of auxochromes in electronic spectra. It should, however, be noted that any proper interpretation must take account of the effect of the auxochrome on both levels. While the bathochrome shift caused by NH_2 on the benzene ring is evidently correlated with extension of the π-orbitals, there are basic molecular systems in which such changes as increasing the size of the conjugated system, alkylation and replacement of a double bond by a triple bond lead to hypsochromic shifts, *i.e.* shifts to shorter wavelength[50]. Among these are molecules such as benzo- and dibenzofulvene relative to fulvene; *p*-naphto-(yellow) and *p*-anthraquinodimethane (colourless) relative to *p*-benzoquinodimethane (orange); 5, 6-benzoazulene relative to azulene; and 2-, 4- and 6-methylazulenes relative to azulene.

Another intramolecular effect which should be noted is that of steric hindrance[51] (see also G. H. Beavan & E. A. Johnson in Ref. 99, p. 78). There are many examples; *e.g.* in dye molecules and in methyl biphenyls.

Vibrational spectra

It has already been noted that vibrational frequencies assigned to certain groups, particularly C—H and C=O are sensitive to the chemical nature of the molecules. These are again basically electrical effects but they operate upon the potential energy function: *i.e.* a change in the electron distribution which constitutes the chemical bond results in a change in force constant.

So far as C—H stretching vibrations are concerned, if the frequencies 2926 and 2853 cm^{-1} in the CH_2 groups in a paraffin chain are taken as the norm, then there is evidently some slight stiffening in terminal methyl links at 2962 and 2872 cm^{-1}. There is a noticeable increase in methylene frequency in the cycloparaffins in which there is ring strain. Thus in cyclopentane the frequencies are hardly changed but in cyclopropane the CH frequency has shifted to 3040 cm^{-1}. Attachment of the carbon to nitrogen appears preferentially to displace the methyl 2872 cm^{-1} to higher values, whereas attachment to oxygen raises the 2962 cm^{-1} band of methyl. Attachment to the halogens (but not fluorine) raises both methyl bands by about 86 cm^{-1}. Proximity to a double bond or to one aromatic ring also raises the C—H stretching frequency to the region of 3100 cm^{-1}.

Considerable attention is given by Jones and Sandorfy (Ref. 97, pp. 443—498) to the factors which influence C=O stretching frequencies. Among them can be discerned both the electrical and the mass effects of α-substituents, conjugation, ring strain and vibrational coupling. The electrical effects of substituents or of conjugation are those which tend to change the character of the bond. This can be seen in the series:

$CH_3 \cdot CO \cdot CH_3$	1715	$CH_3 \cdot CO \cdot OCH_3$	1735	$CH_3 \cdot O \cdot CO \cdot O \cdot CH_3$	1756
		$CH_3 \cdot CO \cdot CF_3$	1769	$CF_3 \cdot CO \cdot CF_3$	1801
		$CH_3 \cdot CO \cdot Cl$	1810	$CH_3 \cdot CO \cdot F$	1866
$Cl \cdot CO \cdot Cl$	1827	$Cl \cdot CO \cdot F$	1876	$F \cdot CO \cdot F$	1924

where it is suggested that the electron-attracting substituents tend to polarise the C=O bond in the sense

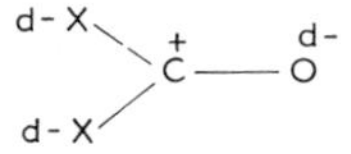

d-X, d-X, C+, O d-

and, on that account, to increase the frequency. On the other hand substitution of the NH_2 group lowers the frequency to 1690 cm^{-1} (in acetamide) which is taken to imply a redistribution of charge approaching

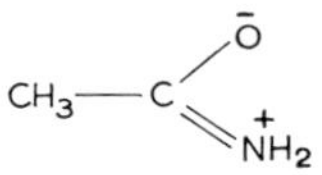

CH_3—C, $\bar{O}$, $\overset{+}{N}H_2$

This is certainly in accordance with the shift to lower frequency which is also observed with conjugation of the C=O group as in α, β-unsaturated ketones (1677 cm^{-1}), acetophenone (1690 cm^{-1}) and 1,4-benzoquinone (1667 cm^{-1}). It is plausible to consider conjugation as diminishing the double bond character of the C=O link and thus reducing its force constant. However, the explanation of frequency shifts in terms of the inductive (as shown for COX_2) and mesomeric effects (as shown for $CH_3 \cdot CO \cdot NH_2$) are fraught with ambiguity. Thus the explanation usually given for the short C—X distance in the halogen cyanides, XCN, is the participation of the structure X^+=C=N^- analogous to the shift in charge shown for acetamide. The bond lengths would suggest increasing participation of that structure on going from X=Cl to X=F and yet the force constant for C≡N stretch *increases* in the same direction. The CF distance in COF_2 is also shorter than normal (say in CH_3F) and the analogous explanation would be the participation of a structure $\overset{F}{+F}\!\!>\!C\text{—}O^-$. By analogy with XCN this would be expected to increase the CO stretching frequency but by analogy with acetamide would be expected to decrease it. Until these bond systems can be described in sufficiently good quantum mechanical approximation it is perhaps better to seek, and to make use, of further empirical correlations.

It is found that in a series of compounds containg the grouping R—CO—O—R the changes in R which lead to lowering in the C=O stretching frequency also bring about an increase in the frequency assigned to the C—O—C stretch in region 1270—1200 cm^{-1}. The extreme case of this correlation is the carboxylate anion $R\text{—}C\!\left\langle\begin{matrix}O\\ \overline{O}\end{matrix}\right.$ in which, in the symmetrical stretching frequency of a structure more properly regarded as $R\text{—}C\left\langle\begin{matrix}O\\O\end{matrix}\right\}^-$, the C=O frequency has been reduced and the C—O frequency raised to the common value of 1400 cm^{-1}.

An interesting series of correlations has been observed between the reactivity of substituted benzene derivatives and characteristic frequency shifts. Thus the rate of esterification of the substituted benzene acids

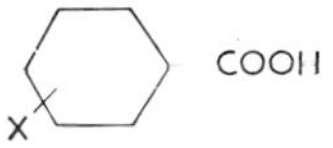

$X\text{-}C_6H_4\text{-}COOH$

can be shown to conform to a relationship[52–54]

$$\log k - \log k_0 = \sigma\rho \qquad (4.16)$$

where k is the rate constant ($= k_0$, for X = H), σ is a constant called the *Hammett σ-factor*, depending only upon the nature and position of X (whether *o*-, *m*-, or *p*-) and ρ is a constant, which is characteristic of the particular reaction. The σ-factor is related to the effect of X on the electron distribution in the molecule and is more-or-less transferable to other reactions whose rates might be expected to be similarly affected by the substituent X. For a number of derivatives (X=NH_2, OCH_3, CH_3, H, Br, Cl, NO_2) there is a significant increase[55] in the C=O stretching frequency with increase in σ as can be seen from Fig. 119. The increase of ω and of σ is in the direction of increasing electron-withdrawing powers of the substituent X, as judged by the direction of substitution in C_6H_5X. Similar correlations[56] (also Ref. 99, p. 165) have been studied in $X \cdot C_6H_4 \cdot Y$ when Y is CO · OEt, CHO, CO · CH_3, CO · C_6H_5, CN and in each the C=O or C≡N stretching frequencies increase with σ. The OH stretching frequency, when Y is OH or CO · OH, decreases with increase in σ, and the NH stretching frequency when Y is NH_2 or $NHCH_3$ increases. The NH frequency when Y is CO · NH · CH_3 is unaffected by changes in X.

References p. 287–289

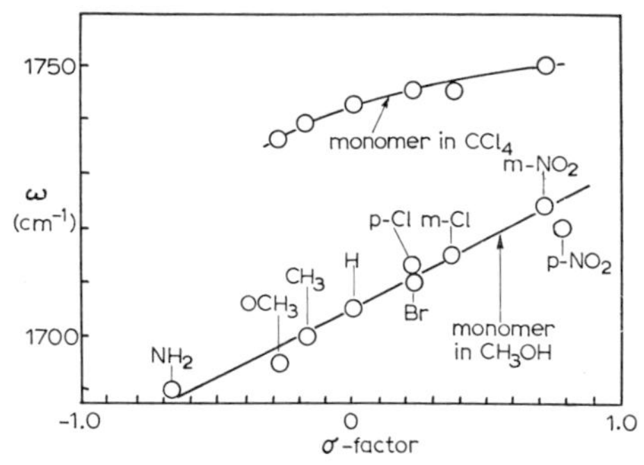

Fig. 119. Variation of C=O frequency with σ-factor for esterification in various substituted benzoic acids.

A significant dependence upon σ of the intensity of these various bands has also been noted. Since, however, intensity depends upon the rate of change of dipole moment with variation in the normal coordinate $(\partial\mu/\partial q)$ whereas frequency depends upon force constant there is no reason to expect that frequency and intensity will be similarly affected by X. There is no change in intensity of the NH band when Y is $CO \cdot NH \cdot CH_3$; the OH band when Y is OH increases in intensity and the NH band when Y is NH_2 diminishes. The CN band markedly diminishes in intensity on going from *p*-$CH_3 \cdot O \cdot C_6H_4 \cdot CN$ to C_6H_5CN (increase in σ) and the same is observed for the $—N^+{\equiv}N$ band in *p*-$CH_3O \cdot C_6H_4 \cdot N_2^+ \cdot BF_4^-$ and $C_6H_5 \cdot N_2^+ \cdot BF_4^-$.

Complexes and Ionic Association

Although the ultraviolet and visible absorption spectra of polyatomic inorganic ions have been known and used for a long time it is only fairly recently that much progress has been made in an understanding of the transitions which are being observed. The justification for including them in the present chapter is that the interpretation is essentially one of interaction between a set of attached groups, the so-called *ligands,* and the atomic levels of a central atom. The free ion of any transition element has partially filled degenerate levels which come under the action of an electric field when the ligands are brought up to it from various preferred directions. The result is a Stark splitting of the levels in a manner which depends upon the strength of the ligand field and the symmetry of the complex. The absorption spectrum of a complex ion in the visible region usually arises from transitions between these separated levels. As has often been the case, the theory which was primarily developed to account for spectra has been found helpful in explaining the chemistry as well. Here, however, attention is confined to the spectra and to an introduction to the underlying principles.

Crystal field theory

The basis of the so-called *crystal field theory* can best be seen in a discussion of an atomic configuration comprising filled shells plus one d-electron. From Chapter 2 we can immediately assign a spectral term to this electron state. The filled shells contribute nothing to $\boldsymbol{L}$ or to $\boldsymbol{S}$ but the d-electron has $l = 2$ and $s = \pm \frac{1}{2}$. Consequently $L = 2$, $J = \frac{5}{2}$ or $\frac{3}{2}$ and the state is ^2D, with doublet components $^2D_{\frac{3}{2}}$ and $^2D_{\frac{5}{2}}$. We have seen that hydrogen-like energy levels do not depend upon any quantum number but n. The alkali-metal spectra, however, show that the presence of electrons in lower filled shells affects the energy of upper levels so that they depend also upon the quantum number l. That is to say, the configurations $1s^2\,2s^2\,2p^6\,3s^1$ ($^2S_{\frac{1}{2}}$), $1s^2\,2s^2\,2p^6\,3p^1$ ($^2P_{\frac{1}{2}}$, $^2P_{\frac{3}{2}}$) and $1s^2\,2s^2\,2p^6\,3d^1$ ($^2D_{\frac{3}{2}}$, $^2D_{\frac{5}{2}}$) have different energies whereas the excited levels for one electron in the $3s$, $3p$ or $3d$ states in the hydrogen atom are of the same energy. Configuration interaction splits the hydrogen level degeneracy into S, P and D states but there remains the further degeneracy of the P and D states in which the quantum number m may take different values. We have seen (p. 92) that the space quantization implied by the quantum number m can be called into play by the application of a magnetic field (Zeeman effect): similarly it can be observed in the presence of an electric field (Stark effect).

It was convenient in Chapter 2 to discuss the effects in terms of angular momentum without reference to electron distribution. Now, however, it is useful to recall the form of the atomic orbitals, predicted by the Schrödinger equation for the hydrogen atom. We are interested only in the angular dependence of the wave functions. The s-orbitals are spherically symmetrical and the p- and d-orbitals depend upon angle as follows (see Fig. 120 for θ and ϕ):

$$
\begin{array}{llll}
 & & \psi & \psi\psi^* \\
p & l = 1,\ m = 0 & \frac{1}{2}\sqrt{\frac{3}{\pi}}\ \cos\theta & \frac{3}{4\pi}\ \cos^2\theta \\
 & l = 1,\ m = \pm 1 & \frac{1}{2}\sqrt{\frac{3}{2\pi}}\ \sin\theta e^{\pm i\phi 2} & \frac{3}{8\pi}\ \sin^2\theta \\
d & l = 2,\ m = 0 & \frac{1}{4}\sqrt{\frac{5}{\pi}}\ (3\cos^2\theta - 1) & \frac{5}{16\pi}\ (3\cos^2\theta - 1)^2 \\
 & l = 2,\ m = \pm 1 & \frac{1}{2}\sqrt{\frac{15}{2\pi}}\ \sin\theta\cdot\cos\theta\cdot e^{\pm i\phi} & \frac{15}{8\pi}\ \sin^2\theta\cdot\cos^2\theta \\
 & l = 2,\ m = \pm 2 & \frac{1}{4}\sqrt{\frac{15}{2\pi}}\ \sin^2\theta\cdot e^{\pm 2i\phi} & \frac{15}{32\pi}\ \sin^4\theta
\end{array}
\tag{4.17}
$$

The solutions, as obtained in terms of explicit dependence upon m, are such that the electron distribution given by $\psi\psi^*$ does not distinguish between

References p. 287–289

$m = +1$ and $m = -1$. The angular dependence of $\psi\psi^*$ is shown in Fig. 120(a). Each diagram is to be rotated about the z axis to complete the description,

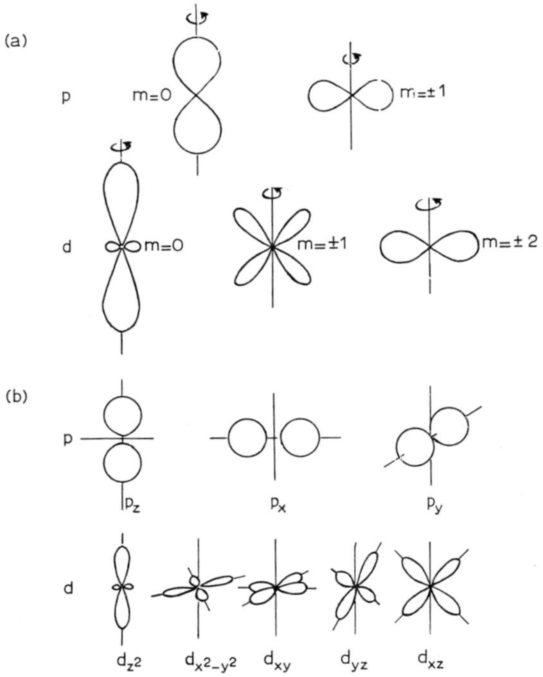

Fig. 120. p- and d-orbitals: (a) dependence upon m, (b) equivalent orbitals. The curves show angular dependence of $\psi\psi^*$.

An external electric field "sees" two p-electron distributions and three d-electron distributions, whereas a magnetic field sees respectively three and five orientations of angular momentum. Consequently under the action of an external electric field P states split into two and D states into three, the perturbation being generally proportional to the square of the field strength and the state with $|m| = 1$ lying above $m = 0$. The resultant splitting of spectral lines is the Stark effect.

In the degenerate-field free state of p and d electrons the choice of orbitals is not unique and an equally acceptable set of solutions may be obtained from linear combinations of the given set. From the p-orbitals it is possible to obtain a new set of orbitals which are identical except for their axes of symmetry. They are the equivalent orbitals, commonly described as p_x, p_y and p_z, shown in Fig. 120(b). From the d-orbitals it is not possible to form five independent equivalent orbitals, but it is possible to achieve four of the same shape ($d_{x^2-y^2}$, d_{xy}, d_{yz}, d_{xz}) which differ only in mutual orientation leaving a fifth (d_{z^2}) of different shape. These will be seen to be more satisfactory for the present purpose.

In an octahedral complex ion there is an electric field which can be simulated by placing six negative charges of equal magnitude at equal distance from the central atom, as in Fig. 121(a). The model may be refined by placing six dipoles in position to approximate more closely to the attachment

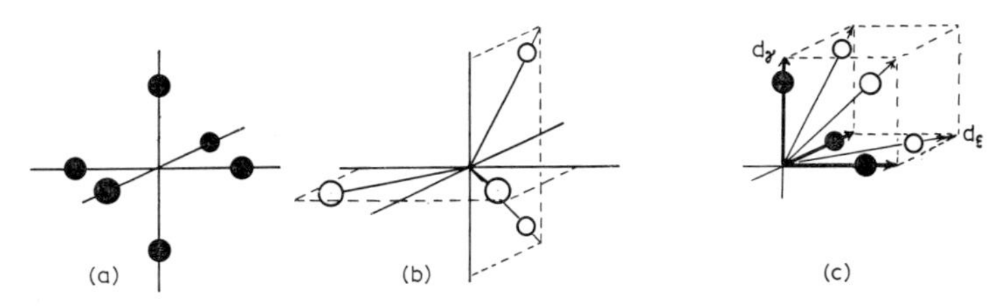

Fig. 121. Octahedral and tetrahedral crystal fields: (a) six negative charges octahedrally disposed, (b) four negative charges tetrahedrally disposed, (c) relation to d_ε and d_γ of equivalent octahedral (●) and tetrahedral (○) positions.

of ligands, such as H_2O or NH_3. The simplest case to consider is the configuration d^1 (^2D). The octahedral potential field is applied as a perturbation to all members of the degenerate set of orbitals comprising the D term, and, by standard quantum mechanical methods, the perturbation energy E is evaluated. It is thus shown that the five-fold degenerate D term is split into two new levels, one three-fold degenerate and one doubly degenerate, as shown in Fig. 122. The new levels are given labels appropriate to their symmetry

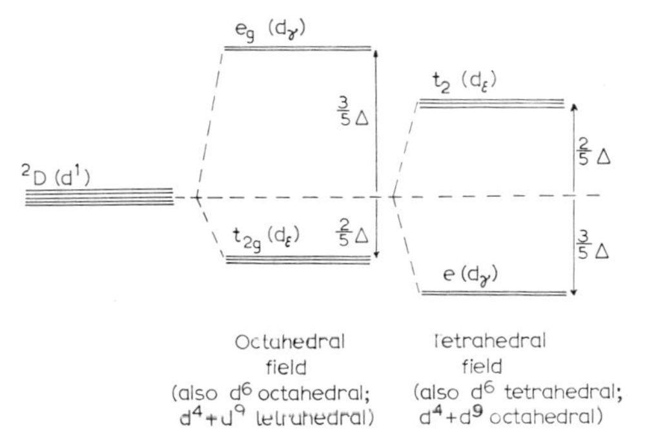

Fig. 122. Splitting of degenerate ^2D levels into e_g and t_{2g} levels under octahedral and tetrahedral fields.

in relation to the octahedral symmetry of the complex (point group O_h). In designating vibrational levels it is customary to label the three-fold degenerate species as F (or f) but in this context T (or t) is preferred, presumably to avoid confusion with the designation F for the seven-fold degenerate field-free atomic levels for which $l = 3$. The spacing Δ between the levels is in principle calculable from the model (it depends upon the magnitude of the six negative charges and inversely upon this distance from the centre) but is generally found from the spectra. The lower level t_{2g} lies $\frac{2}{5}\Delta$ below the field-free term and the upper level $\frac{3}{5}\Delta$ above.

The new d-orbitals which are associated with the new levels are found to be just those chosen as most nearly equivalent, thus:

triplet level (t_{2g}) d_{xy}, d_{xz}, d_{yz} denoted $d\varepsilon$
doublet level (e_9) d_{z^2}, $d_{x^2-y^2}$ denoted $d\gamma$

It can be seen from Fig. 120 and Fig. 121(c) that the d_γ orbitals lie along the axes upon which the ligands are assumed to be placed, whereas the d_ε orbitals avoid them. The qualitative argument follows that the d_ε state will lie lower in energy than the d_γ state. In the same way it can be seen that the opposite situation arises in a tetrehedal field produced by four negative charges placed as in Fig. 121(b). In this case the t_2 state (d_ε orbitals) lies above e (d_γ orbitals) as shown in Fig. 122.

A configuration closely related to d^1 is d^9. Whereas in d^1 there is one electron which is, in the octahedral case, accommodated in the d_ε orbitals in d^9 there are four pairs and one single electron. The field-free term is ^2D, as for d^1, but now the configuration of the ground state is $(d_\varepsilon)^6(d_\gamma)^3$ and of the upper state $(d_\varepsilon)^5(d_\gamma)^4$. An alternative description of d^9 is d^{10}-plus-a-positive-hole. The hole can be accommodated in a doubly-degenerate lower level or triply-degenerate upper level. Hence the order of levels for d^9 (octahedral) is as for d^1 (tetrahedral). The levels available for the configuration d^6 (d^5, d^1) are as for d^1 and for d^4 are as for d^9.

For other configurations the levels are more complicated just as they are in the field-free atom. For example, configuration d^2 in the field-free atom may give the terms ^1S, ^1D and ^1G (spins paired) and the terms ^3P and ^3F in which there are two unpaired electrons. In the octahedral field ^3F becomes three levels T_{1g} (ground state), T_{2g}, A_{2g}. ^3P gives a higher T_{1g} level. For further detail of crystal field theory see Refs. 57-59.

We are now in a position to understand some of the spectra of complex ions. We should expect relatively simple spectra from ions with d^1 configuration. The $[Ti(H_2O)_6]^{3+}$ ion is just such an ion. Its absorption spectrum shows one weak band which has been identified[60] as arising from the simple transition $t_{2g} \rightarrow e_{2g}$. If this is the case then $\omega = \Delta = 20{,}300$ cm^{-1} (4900 Å). There is reason to believe that Δ remains at about 2×10^4 cm^{-1} for all such 3+ ions in the first row of transition elements. For $[Mn(H_2O)_6]^{3+}$ (d^4) $\Delta =$ 21,000 cm^{-1} and for $[Cr(H_2O)_6]^{3+}$ (d^3) $\Delta = 17{,}400$ cm^{-1}. In 2+ ions Δ

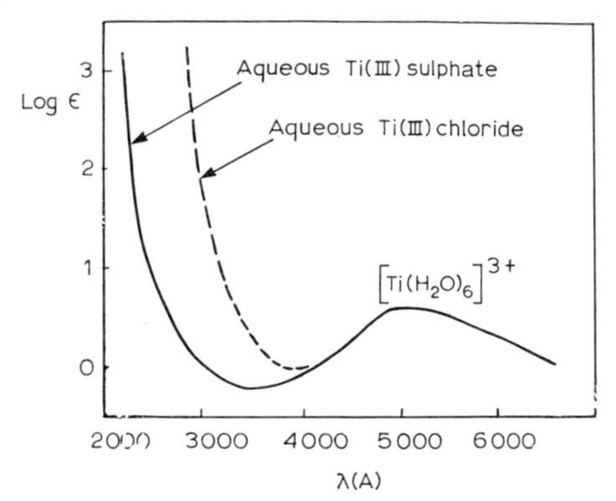

Fig. 123. Absorption spectrum of $Ti(H_2O)_6^{3+}$ in aqueous solution.

should be somewhat less, say about 10^4 cm^{-1} (10,000 Å): in $[Cr(H_2O)_6]^{2+}$ (d^4) $\Delta = 13{,}900$ cm^{-1} (7200 Å) and in $[Cu(H_2O)_6]^{2+}$ $\Delta = 12{,}600$ cm^{-1}.

As an illustration of a spectrum with more than one transition of this type the absorption of the hydrated nickel ion (which is responsible for the familiar green colour) is shown in Fig. 124. Spectra are given[61, 62] of the

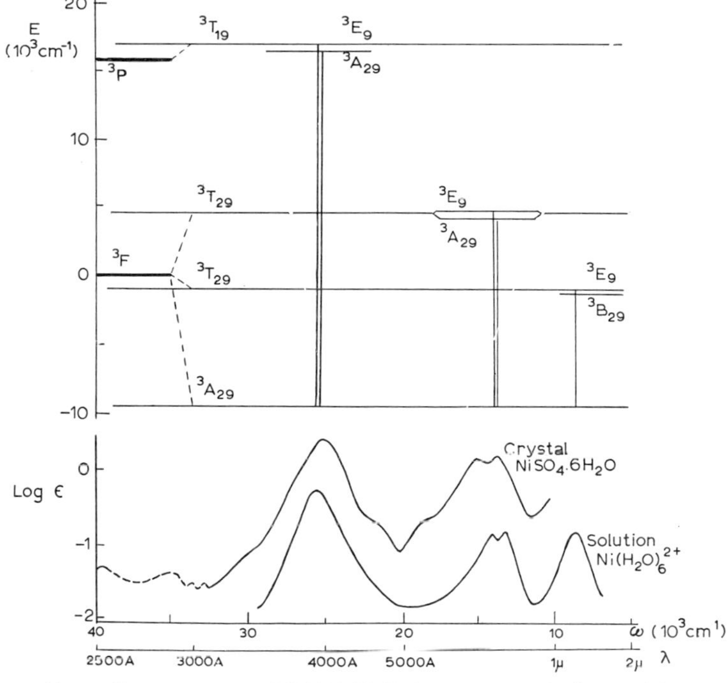

Fig. 124. Absorption spectrum of $Ni(H_2O)_6^{2+}$ in aqueous solution and in crystalline $NiSO_4 \cdot 6H_2O$.

References p. 287–289

aqueous solution and of the crystalline nickel sulphate hexahydrate, $NiSO_4 \cdot 6H_2O$. It is immediately apparent that the two spectra are very similar and identify the hydrated ion in solution with that known to exist in the crystalline hydrate, namely $[Ni(H_2O)_6]^{2+}$. The three main peaks are consistent with the crystal field prediction of four levels for the d^8 configuration and are so interpreted in the figure.

The nickel sulphate solution spectrum shows one band as a doublet (seen in the figure) which is interpreted in terms of coupling between spin and orbital motions of the electron. The crystal spectrum shows further features. Some of the smaller maxima and shoulders are assigned to transitions from the triplet state $^3A_{2g}$ to singlet levels derived from the atomic terms ^1D and ^1G. The other feature not shown in Fig. 124 is the fine structure of the bands at 13,500 and 25,300 cm^{-1}. Under good resolution and with the sample cooled to $-205°$ to improve the band structure doublets are observed about 300 cm^{-1} apart, which also shift considerably on warming the sample to room temperature. Evidently the distortion of the $[Ni(H_2O)_6]^{2+}$ ion in the crystal leads to a lower symmetry of field (tetragonal) and thus to further splitting of the levels. One of the particular contributions of the electrostatic theory has been to show that, even in solution, an octahedral or tetrahedral structure for a complex ion may not be the most stable one. The less regular structure which is adopted will have a larger number of excited state energy levels and consequently a more elaborate absorption spectrum: this is often the case.

Selection rules

Selection rules apply to these electronic transitions as they do in other atomic spectra. There is, however, considerable relaxation of the rules so that forbidden transitions are often observed, albeit weakly. The general selection rule which prohibits singlet-triplet transitions is fairly well obeyed. *Spin-forbidden* transitions do occur but they are very weak: an example occurs in the $[Ni(H_2O)_6]^{2+}$ spectrum above.

Where a complex has a centre of symmetry transitions within the same *d*-state cannot lead to changes in dipole moment since each of the orbitals concerned also has a centre of symmetry. This is a case of the so-called *Laporte selection rule.* Tetrahedral complexes are not centro-symmetric and *d—d* transitions are therefore allowed. In octahedral complexes relaxation of the Laporte rule evidently occurs by way of molecular vibrations which disturb the octahedral symmetry.

As a general rule therefore spin-forbidden transitions are very weak; Laporte-forbidden *d—d* transitions are somewhat stronger: Laporte-allowed *d—d* transitions are more intense but even these do not have extinction coefficients much above 50. We shall see that the fully allowed transitions which generally occur at shorter wavelengths are 1000 times more intense. Nevertheless most of the absorptions of transition metal complexes occurring in the visible region are *d—d* transitions and are responsible for a great many of the colours in inorganic chemistry.

Spectrochemical series

It had been known before the interpretation in terms of Stark splitting that ligands could be arranged in order of their effect upon the spectrum of a complex. Thus, in the *spectrochemical series*,

$$I^- < Br^- < SCN^- < Cl^- < NO_3 < F^- < OH^- < H_2O < NCS^- < NH_3 < CN^-$$

the ligands are arranged in such order that on moving from left to right the effect on the first (long wave) absorption maximum is to move it to shorter wavelength. (This is, of course, towards higher energy and implies an increase in Δ on going from left to right). Where a *d—d* transition for an aquo-ion lies in the near infra-red, as is the case with $[Cu(H_2O)_4]^{2+}$, replacement of the water by ammonia leads to a shift into the visible and an intensification of colour, first to blue as in $[Cu(NH_3)_4]^{2+}$. The CN^- group achieves large shifts and hence a variety of colours.

Ligand field theory

A clear distinction is not always made in current usage between the terms crystal field theory and *ligand field theory*. The former owes its name to its first application to a treatment of magnetic properties of crystals. While consideration of purely electrostatic effects has been strikingly successful in the interpretation described above, account must also taken of covalent bonding between the ligands and the central atom. Such bonding may arise from simple σ interaction of atom and ligand orbitals, or it may also involve Π interactions. The extent to which covalent bonding must be considered also depends upon the ligand and this is expressed in the so-called *nephelauxetic series*, which shows the extent of covalent bond formation decreasing in order

$$F^- > H_2O > NH_3 > SCN^- > Cl^- > CN^- > Br^-.$$

In tetrahedral ions the electrostatic perturbations lead to d_ε orbitals which, in the tetrahedral symmetry, can interact strongly with p orbitals. Spectral transitions are no longer purely *d—d* and covalent bonding is of greater importance. This is not to say that electrostatic theory is necessarily invalidated since a well-localized σ-pair of electrons may still be treated as a simple electrostatic perturbation. On the other hand in the spectra of many strongly bound complexes and especially the tetrahedral ones, stronger absorptions

References p. 287–289

of higher energy predominate which are usually classed as charge-transfer spectra.

Charge-transfer spectra[63]

There is a variety of systems which exhibit spectra generally regarded as charge-transfer spectra. The absorptions are characterized by high intensity and are evidently fully-allowed transitions. The common feature is that the upper state can be thought of as one in which an electron has been "transferred" from one atom to another rather than elevated to higher orbitals of the same sort as the ground state.

One of the earliest examples of recognition of electron transfer is in the spectra of alkali halides. Fig. 125 shows the absorption spectra[64, 65] of gaseous sodium halides and of crystalline potassium and rubidium halides. It was soon recognized that the separation of the peaks in the gaseous bromides and iodides (respectively ~3200 and 8000 cm^{-1}) corresponded to the se-

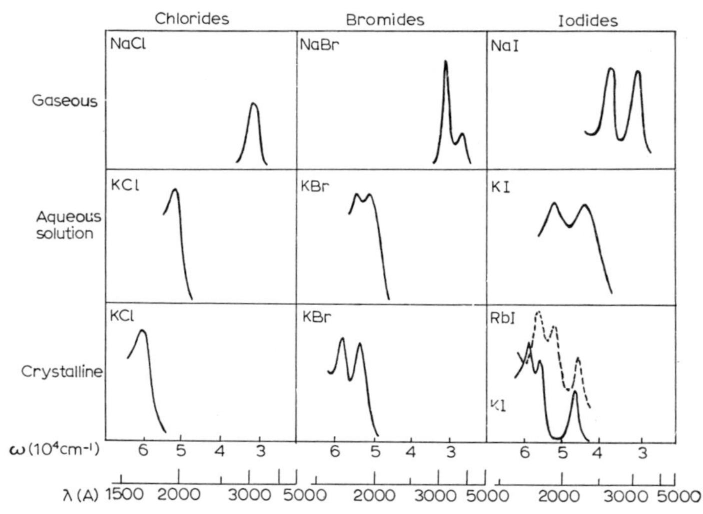

Fig. 125. Absorption spectra of gaseous sodium halides, of potassium halides in aqueous solution, and of crystalline potassium and rubidium halides.

parations 3700 and 7600 cm^{-1} known from the spectra of atomic bromine and iodine to be the separation of the $^2P_{\frac{1}{2}}$ and $^2P_{\frac{3}{2}}$ states. For chlorine the separation is much smaller. It appeared therefore that the upper state in these transitions could be fairly accurately represented by the molecule Na · X in contrast to the $Na^+ \cdot X^-$ of the ground state (Fig. 126). There are similar separations in the crystal spectra, even though they are a shorter wavelength, and in crystalline RbI there is a further separation corresponding to the first

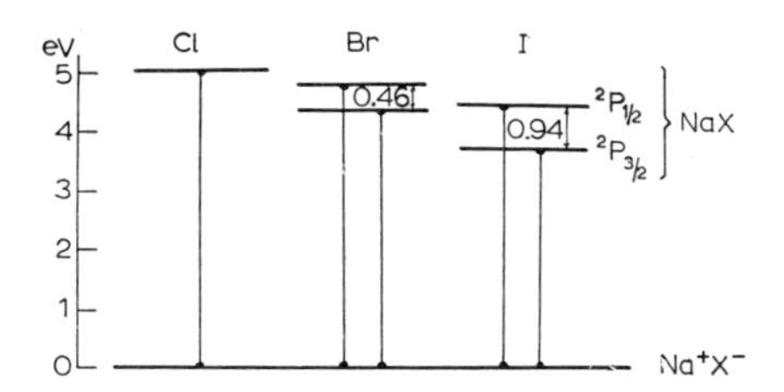

Fig. 126. Charge transfer transitions in gaseous sodium halides.

S—P difference in atomic rubidium. The reason for the doublets appearing at different wavelengths in gaseous and crystalline spectra is, among other things, due to the stabilization of the ionic state in the crystal field.

Charge-transfer spectra of ions in solution

In addition to the weak *d—d* absorption bands shown in or near the visible region by complex and hydrated ions of the transition elements, there are very strong absorption bands in the ultraviolet, which are shown by the majority of ions and not only by transition element complexes. All are classified as charge-transfer spectra, a term which thus covers electron transfer from ion to solvent, from ion to ion in an ion-pair, from ligand to central atom in a complex ion and from ligand Π-bond to central atom — a considerable range of distances over which transfer of the electron is taking place. In a sense all observable electronic dipole transitions are charge-transfer spectra and it is not easy to define a point between the alkali halide spectra, for which the term is entirely appropriate, and absorption by the sulphate ion, where the term seems no more or less applicable than in say, the spectrum of benzene.

In solutions of the halide ions (Fig. 125) the position of the maxima is intermediate between the gaseous and the crystalline halides. There is a doublet in the iodide spectrum which corresponds, as before, to the atomic $^2P_{\frac{1}{2}}$—$^2P_{\frac{3}{2}}$ separation and thus it appears that the upper state is again a neutral halogen atom. The electron in this instance is presumably transferred to the solvent sheath. If this is the case the position of the absorption peaks should be markedly dependent upon the solvent, as is the case. The first band of the iodide spectrum moves[66] about 100 Å to shorter wavelength in isopropyl alcohol and about 200 Å to longer wavelength in acetonitrile.

An analogous interpretation may be given to the u.v. spectra of OH^- (absorption begins in aqueous solution about 2150 Å, λ_{max} 1850 Å), SH^- (begins 2700 Å, λ_{max} 2300 Å), and perhaps also CN^- (begins 2200 Å, λ_{max} < 2000 Å). Other common anions require other explanations.

References p. 287–289

Ionic association — complex ions

There is no reason to suppose that such species as $PbCl^+$, $FeCl^{2+}$, or $Fe(CNS)^{2+}$ are anything other than complex ions. Although they have been described as ion pairs, that designation ought to be reserved for the species discussed in the following section. It is generally assumed that ions written as $FeCl^{2+}$ are aquocomplexes, say $[Fe(H_2O)_5Cl]^{2+}$ from which the halide ion has replaced a water molecule but it is conceivable that they should be formulated as $[Fe(H_2O)_6]^{3+}Cl^-$. Evidence on that question is not always clear and may come better from other types of experiment[67]. For the present we will regard them as complex ions.

The spectra of lead halides in dilute aqueous solution are markedly affected by increasing concentrations of halide ions, and species such as $PbCl^+$, $PbCl_2$, $PbCl_3^-$ and $PbCl_4^{2-}$ (or $[Pb(H_2O)_3Cl]^+$, $[Pb(H_2O)_2Cl_2]$, $[Pb(H_2O)Cl_3]^-$ and $[PbCl_4]^{2-}$) may all be expected. By observation of the changes in the absorption with changes in the halide concentration, the spectra of the ions PbX^+ (or $[Pb(H_2O)_3X]^+$) have been observed[68,69] (Fig. 127). Like the spectra of the free halide ions in aqueous solution (Fig. 125) they move to

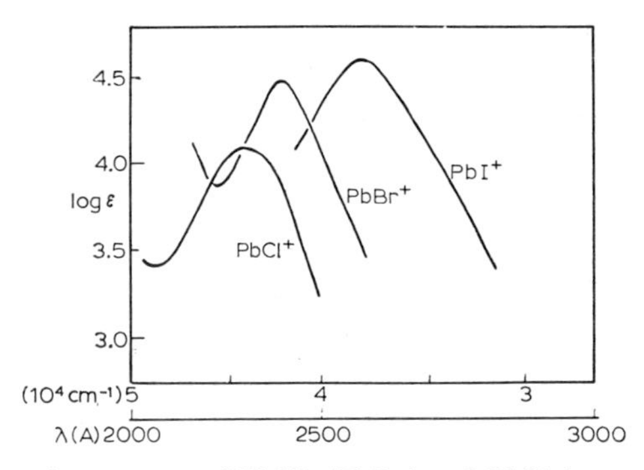

Fig. 127. Absorption spectra of $PbCl^+$, $PbBr^+$ and PbI^+ in aqueous solution.

longer wavelength on going from Cl to I, but they all occur at longer wavelength than the free halides and are therefore clearly associated with the lead complex. Similar behaviour can be found in many halogen complexes. Fig. 128 shows the absorption spectra[70] of the complex ions $[Co(NH_3)_5X]^{2+}$ where $X = NH_3$, F, Cl, Br or I: the charge transfer bands move to longer wavelength with change in X, so that in the iodo-complex the weaker *d—d*-transitions are almost obscured.

The decrease in wavelength with change in ligand is clearly correlated with the decreasing electron affinity of the ligand, or an increase in the reducing power of the X^- ion. Some of the complexes are intimately concerned

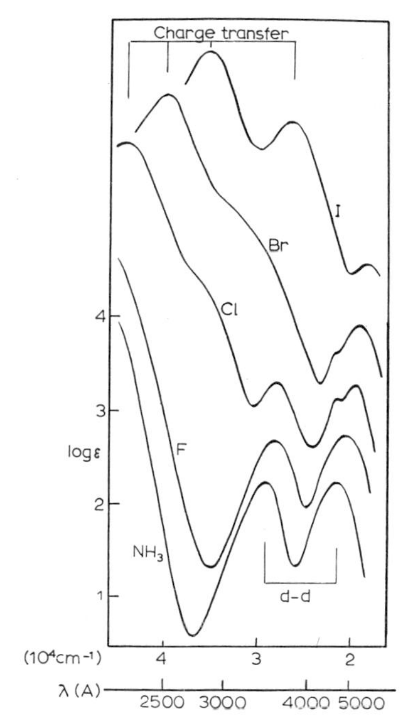

Fig. 128. Absoption spectra of $[Co(NH_3)_5X]^{n+}$ where X = NH_3 ($n = 3$) and X = F, Cl, Br, I ($n = 2$). Each spectrum is raised half a unit of log ε above the lower one.

with photochemical reactions in aqueous solution[71]: in the presence of readily polymerized monomers irradiation of FeX^{2+} complexes with ultraviolet light of wavelength corresponding to the charge transfer band leads to polymerization initiated by X atoms. This is direct evidence for the nature of the upper state in these transitions. In the absence of such reagents the upper state will generally revert to the initial state unless the ligand is so readily oxidisable and the reduced state of the metal so stable that a permanent oxidation–reduction reaction occurs. Under these conditions the charge-transfer band has usually moved into the visible and the reaction is sensitive to visible light. Cupric solutions in the presence of Cl^- and Br^- give charge-transfer bands at long wavelengths and extrapolation to the iodide suggests that the charge-transfer band for the iodide would be at very low energy (low frequency). In fact CuI_2 is unstable with respect to CuI and $\frac{1}{2}I_2$, but it is not clear whether this reaction occurs through thermal activation or is induced by a charge-transfer absorption. The ferric-iodide system is similar.

It may be noted also that as the number of ligands of the same kind increases round a central atom the charge-transfer absorption moves to longer wavelength. Thus as the concentration of chloride ions is increased the colour

References p. 287–289

of a copper solution, first blue on account of slight absorption in the red and the green from a *d—d* band of $[Cu(H_2O)_4]^{2+}$, gives way to green as replacement of water by chloride ions progressively moves a charge-transfer band into the visible from the ultraviolet side.

Ionic association — ion pairs

There are cases where spectra are observed, and classed as charge-transfer spectra, which clearly require the presence of two partners to an association but where there is evidence against covalent binding of a ligand to the central atom. Thus, when iodide is added to solutions containing the hexammino-cobalt ion $[Co(NH_3)_6]^{3+}$, an intense new band appears[72] which is ascribed to the ion-pair $[Co(NH_3)_6]^{2+}I^-$ and not to $[Co(NH_3)_5I]^+$.

There are many other systems in which ionic associations are thought to occur, the evidence coming from thermodynamic measurements, from conductivity, from nuclear magnetic resonance, from Raman spectra. They range from hydroxides such as Tl^+OH^-, $Ca^{2+}OH^-$, through halides and thio-cyanates such as $UO_2^{2+}Cl^-$ and $UO_2^{2+}CNS^-$ to complex cyanides such as $Ca^{2+}Fe(CN)_6^{3-}$. It is by no means certain in each case how the ions are associated. There is considerable evidence, for example, that thallous hydroxide is incompletely dissociated in aqueous solution. Mercuric chloride is so little dissociated that a volumetric procedure for analysing chloride can be based upon it. The Raman spectra of covalently bound $HgCl_2$ molecules can be readily observed, but in aqueous thallous hydroxide no Raman line could be detected[73]. A covalently bound TlOH molecule would almost certainly have exhibited a Raman line due to Tl—O stretching but theoretical considerations indicate that an ion pair Tl^+OH^- would go undetected. The evidence is therefore that in this case an ion pair exists, but it leaves undecided the question of hydration of the ions.

The question of ion-pair *versus* covalent association in thallous hydroxide depends upon negative evidence. In the case of dimethyl thallic hydroxide[74], where there is also conductimetric and kinetic evidence of association, there is again no new Raman line which can be assigned to a Tl—O covalent bond. On the other hand every Raman line of the free $(CH_3)_2Tl^+$ ion, as measured in solutions of the perchlorate, shows a shift of from 3 to 27 cm^{-1} to lower frequencies in strong aqueous solutions of the hydroxide. This is taken to be evidence of a positive kind that the associated species consists of electrostatically bound pairs.

Similar arguments are put forward for electrostatically bound pairs in certain ionic associations of the uranyl ion. Frequency shifts (to lower wavenumber) are observed[75] in the Raman spectra of aqueous solutions of uranyl chloride, sulphate, and nitrate relative to the Raman frequency of the uranyl ion in solutions of the perchlorate. The characteristic Raman lines of the nitrate ion are also affected by the presence of cations such as Li^+, Be^{2+}, Ca^{2+}, Cd^{2+}, Cu^{2+}, Ce^{3+} and Th^{4+}, in concentrated solution. The shifts are not all to lower wavenumber but it has been shown[76] that they are towards the frequencies found in the crystalline nitrates. There is evidence that the passage from unbound to bound nitrate is a discontinuous one since under certain conditions two lines are observed.

Anions

There remain among those spectra classified as charge-transfer spectra the strong ultraviolet or visible absorptions of anions such as sulphate, chromate, perchlorate, permanganate, nitrate, sulphite, thiosulphate, thiocyanate — some of which are shown in Fig. 129. They are undoubtedly intense bands, and are considered as arising from electron transfer from molecular Π-orbitals to a higher state more closely centred on the central atom. In the nitrate ion it is probable that the first absorption is a Π—Π^* transition ($N \rightarrow V$ in the nomenclature used on p. 221) and that the second is the charge-transfer band.

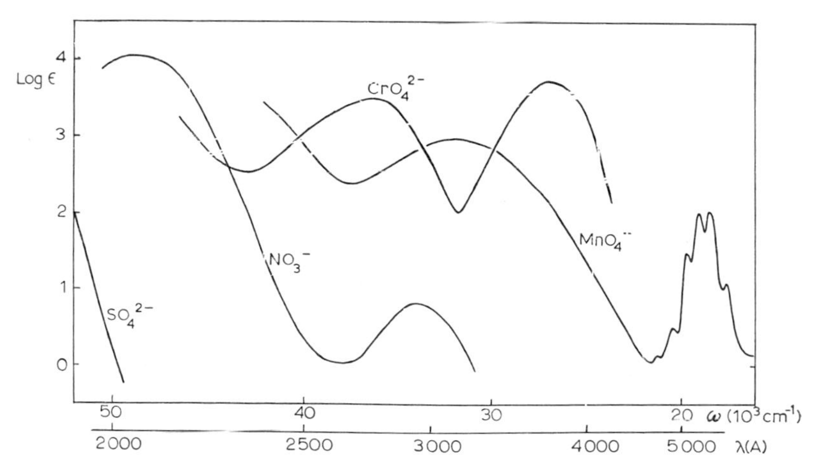

Fig. 129. Absorption spectra of SO_4^{2-}, CrO_4^{2-}, MnO_4^- and NO_3^-.

In the spectrum of the permanganate ion structure is discernible which can be attributed to vibration. The absorption curve can be analysed into eleven equally spaced bands conforming to $\omega = 17{,}313 + n.797$ cm^{-1} in the solid state. In ethyl acetate solution the constant frequency difference is 785 cm^{-1} and in aqueous solution 744 cm^{-1}.

References p. 287–289

Intermolecular Perturbations

In this final section of the chapter we note rather briefly some of the spectral changes which make manifest the effect of environment upon molecular energy levels.

Change of state

A fairly general phenomenon associated with the change of state from gas to liquid to solid is that of collision broadening. In a gas at low pressure the molecules are free to rotate and vibrate without much intermolecular disturbance. There are indeed collisions between gas molecules and kinetic theory indicates that the number of collisions suffered by one molecule of a gas at atmospheric pressure is of the order of 10^{+9} per second. On average, the time spent between collisions is 10^{-9} sec, and proportionately longer at lower pressures. The time between collisions is also lengthened by reducing the molecular velocity, *i.e.* at lower temperatures. Now the frequency of radiation associated with pure rotational change is generally in the range 10,000—100,000 Mc/s, or 10^{10}—10^{11} sec^{-1}. In the classical sense the time required for the transition is 10^{-10}—10^{-11} sec. This is not very much less than the time, 10^{-9} sec, between collisions which, since a collision may be assumed to alter the rotational state of the molecule, is just the mean lifetime Δt of the particular rotational state. In terms of the uncertainty principle (p. 75) the width of an energy level ΔE is given by $\Delta E \sim h/\Delta t$ (erg) or $1/\Delta t$ sec^{-1}. For a gas at atmospheric pressure, therefore, the width of the rotation levels, or the uncertainty inherent in their measurement, is of the order of 10^9 sec^{-1} compared with the value 10^{10}—10^{11} sec^{-1} of the transition. In practice microwave absorption spectra of gases are measured at pressures below 1 mm, say 10^{-3} atoms, when Δt may be 10^{-6} sec and hence $\Delta E = \Delta \nu \sim 10^6$ sec^{-1} = 1000 kc/s. In fact a typical line width in spectra observed at 10^{-2} mm pressure is 100 kc/s.

Considerable loss in resolution occurs when the pressure increases. Microwave spectroscopy is not usually attempted under those conditions since much of its value lies in the fine structue of rotational lines. Rotational structure appears in vibrational spectra measured usually at somewhat higher pressures, and then the width of individual lines may approach the separation between successive lines. So long as it remains say at 10% of that separation it may be less than the spectral slit width of the instrument (p. 47). When, however, the pressure broadening exceeds the separation between successive lines the effect is loss of band structure, and ultimately a change in band shape. This effect has been deliberately studied[77] (also Ref. 99, p. 241) with a view to gaining information about intermolecular forces. It is also used to obtain reliable estimates of total band intensity when the instrument resolution is insufficient to give good values of the intensities of individual rotational lines.

This discussion will serve to illustrate a general principle: that the sharpness of a spectral line depends upon the degree of definition of the energy levels between which the transition occurs, and hence upon the lifetime of the relevant states. In the liquid state rotational levels are generally eliminated, the collisions being such that it is no longer possible to define a rotational state. The effect on vibrational spectra is then to narrow the band on going from gas to liquid since the rotational ,,wings'' are shed. The vibrational transition itself is also subject to collision broadening and generally it is found that vibrational spectra are sharpened by reduction in temperature since this reduces the frequency and the energy of the collision.

Change of symmetry

One effect of change of state is to alter the symmetry of the molecular environment. Benzene provides a good example of this. In the vapour phase the high symmetry of the molecule, point group D_{6h}, is such that only a small number of the 30 normal modes is permitted to appear in the infra-red and Raman spectra. In fact, there should be only 20 distinct frequencies since 20 of the normal modes are doubly degenerate. Four frequencies should be active in the infra-red and inactive in the Raman: and seven different frequencies should be active in the Raman but not in the infra-red, and the remaining nine modes should be inactive in both. There should be no coincidences since the molecule has a centre of symmetry. This situation is indeed found in the spectrum of the vapour: see Fig. 130.

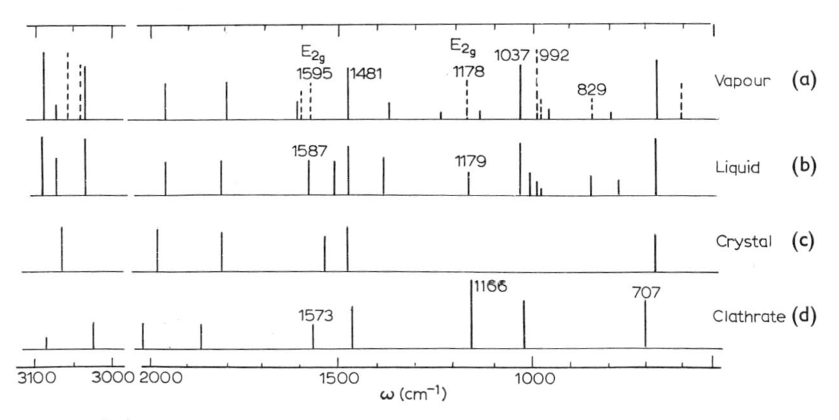

Fig. 130. Infra-red and Raman spectra of (a) benzene vapour, and the infra-red spectra of (b) benzene liquid, (c) crystal, and (d) in $Ni(CN)_2 \cdot NH_3 \cdot C_6H_6$.

References p. 287–289

On liquefaction the molecule finds itself under the influence of perturbing forces with no symmetry: furthermore the perturbations do not average out to an effective spherical symmetry (as they do in the n.m.r. spectra of liquids) during the time taken over a transition. The symmetry selection rules may therefore be expected to break down. It does not seem that the perturbation is sufficient to split the degeneracy of the E levels but it evidently does permit transitions to appear in the infra-red spectrum of the liquid which are only Raman-active in the gas (at wave numbers 829, 992, 1178, 1595 cm^{-1}).

In crystalline benzene a particular symmetry of crystal field is imposed upon the molecule and new selection rules can be deduced[78, 79] which affect the activity of the various modes. In the clathrate compound, $Ni(CN)_2 \cdot NH_3 \cdot C_6H_6$, the symmetry of the field affecting the molecule is very similar to that in crystalline benzene but, naturally, the relative strengths of the perturbations in the two cases will be very different[80]. The effect on the spectra is shown in Fig. 130.

Mention should also be made here of the technique which has recently been used to isolate ions from their normal environment. For example crystals of potassium bromide which have been pulled from a melt, containing a small amoung of cyanate as impurity, effectively isolate the NCO^- ion and enable its vibrational spectrum to be observed as very sharp absorption bands[81].

Solvent effects in electronic spectra

Solution often shifts the spectrum of a molecule to longer wavelengths, or lower frequencies, with respect to the vapour spectrum (*e.g.* Fig. 93, p. 222). This so-called *red shift* is greater in solvents of high dielectric constant. A qualitative explanation is that the charge displacement of the upper state requires less energy in a dielectric than in vacuum. Where a blue shift is observed it is usually associated with an $N \rightarrow Q$ transition involving a non-bonding orbital in the ground state. Here it is assumed that the solvation energy is greater in the ground state than in the upper state, and thus lowers the ground state with respect to the upper.

More specific solvent effects can occur such as those which determine the colour of iodine and bromine in various solvents. This particular system is merely one example of a large number of organic systems in which complex formation can occur[82]. It is generally agreed that the intense new band which occurs at 3000 Å in the spectrum of solutions of iodine in benzene or the very intense absorption bands given by a solution of, say, trinitrobenzene in aniline are due to charge-transfer transitions. There is some debate as to the nature of the ground state. It is probable that in many cases the complex is itself a charge-transfer complex, an ionic association A^+B^-, the spectral transition being essentially to an upper state AB[83]. Alternatively it has been suggested[84] that the two states are those which could be described by mixing of wave functions which individually would describe a "no-bond" condition A B, ψ_0, and an ionic state A^+B^-, ψ . Linear combination of two such orbitals $(a\psi_0 + b\psi_i)$ and $(a\psi_0 - b\psi_i)$ leads to energy levels E_+ (ground state) and E_- (upper state), between which the charge-transfer transition cocurs. A small value of b suggests a small contribution of ψ_i to the ground state and a large contribution to the upper: a large value of b implies that A^+B^- predominates in the ground state. Where solutions which give rise to such spectra show electrical conductivity, as do trinitrobenzene and diethylamine in ethanol[85], it seems probable that the ions exist to some extent free (or solvated) in solution and that ion-pairs A^+B^- give the observed spectra.

Ion pairs are recognized in many other systems and contribute a particular form of intermolecular perturbation. They have been discussed on p. 266. We exclude from the present discussion the more profound solvent effects which are essentially chemical reactions; particularly those which involve proton transfer to solvent (see p. 247). In these cases an entirely new molecule is produced with its own spectrum. However, it may not always be possible to distinguish this from more subtle effects on purely spectroscopic grounds.

Solvent effects in vibration spectra

Solvent shifts in infra-red and Raman spectra are now well-known phenomena. A theoretical intrepretation in terms of the dielectric constant of the solvent, later modified to use the refractive index, gives the relative frequency shift from the frequency ω_v of a band in the vapour as

$$\frac{\Delta\omega}{\omega_v} = C\,\frac{\varepsilon - 1}{2\varepsilon + 1} = C'\,\frac{n^2 - 1}{2n^2 + 1} \tag{4.18}$$

Equation (4.18) is known as the Kirkwood-Bauer-Magat (KBM) relationship and has been used[86] as a basis for detecting more specific interactions, notably hydrogen bonding. Thus the frequency shift for the N—H band of pyrrole is very much larger in the solvents pyridine and acetone than in the non-polar solvents which are presumed to define C′ in the KBM relationship (4.18). It is now recognised that the KBM equation is inadequate to account for solvent shifts even in non-polar solvents and in the absence of the more specific effects. Indeed it has been suggested (Ref. 99, p. 216) that there is no detectable spectroscopic difference, save in degree, between the effect of pyridine on pyrrole, which might be written as

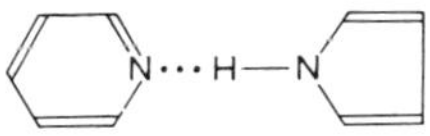

N···H—N

and the effect of benzene on pyrrole, Fig. 131 shows how the variation of the solvent effect on the N—H frequency of indole is a continuous one.

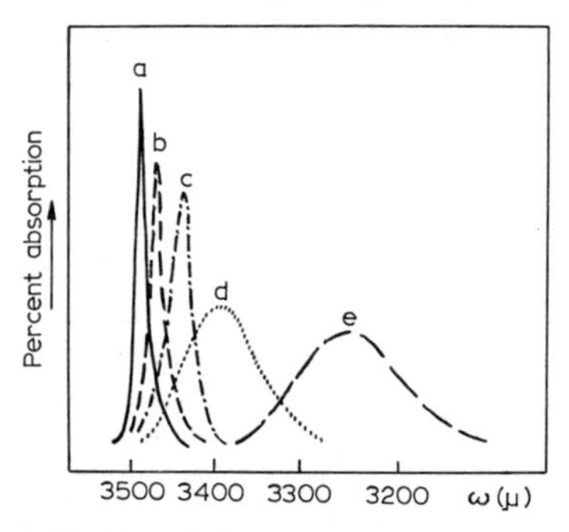

Fig. 131. N-H stretching band of indole in (a) hexane, (b) carbon disulphide, (c) benzene (d) methyl thiocyanate, (e) N,N-dimethyl acetamide.

It therefore appears that, in these solvent shifts, dielectric constants play little part and that the effects are local ones involving dipolar association (of which the hydrogen bond is a special case). Solvent shifts have been observed[87, 88] in bands other than those assigned to O—H stretching and bending (for example, C=O S=O, and N=O stretch) and they are, naturally, much smaller.

Hydrogen bonding

Despite the above considerations infra-red spectra have been widely used to study hydrogen bonding. Most of the work has been concerned with the shift in N—H and O—H stretching frequencies in hydrogen-bonded association of the sort indicated in Fig. 131. The magnitude of the shift in a system X—H... Y has been correlated with the acidity of various X—H and with the basicity of various Y since the proton transfer X^- . . . H—Y^+ may be regarded as the end product of the association. The magnitude of the frequency shift in the X—H stretching frequency in solid structures containing X—H . . . Y has also been correlated[89] with the X . . . Y distance: the very short O . . . O distance of 2.46 Å in crystalline maleic acid, for example, is correlated with a shift $\Delta\omega$ of nearly 2000 cm^{-1}.

Characteristic of a displaced X—H stretching band due to hydrogen bonding is the width of the band, as in Fig. 131. The spectra of crystalline organic compounds containing intermolecular hydrogen bonds generally show a very broad band extending from 3500 to 2500 cm^{-1}. Liquid alcohols and amines also show a broad band, not always extending as far as the crystal bands and sometimes composite with a suggestion of more than one kind of hydrogen bond. Dilution of these liquids with a non-polar solvent such as carbon tetrachloride leads to a diminution of the broad band and a growth of a sharp band at higher frequency, the ,,free" O—H or N—H frequency.

Bending, or X—H deformation, frequencies also appear to be shifted by hydrogen bonding but to higher frequency. If the hydrogen bond is attached to carbonyl oxygen, as is often the case, there is a shift in the carbonyl frequency to lower wavelengths. A prominent example is the dimerization of carboxylic acids thus:

$$R-C\begin{matrix} \nearrow O\cdots H-O \searrow \\ \searrow O-H\cdots O \nearrow \end{matrix}C-R$$

The characteristic C=O frequency is 1710 cm^{-1}, compared with the monomer frequency 1760 cm^{-1}. The O—H deformation in this structure is evidently[90] strongly mixed with the C—O stretch, giving bands at 1420, 1300 and 935 cm^{-1}.

A recent symposium[91] gives a good indication of the diversity of work done on hydrogen bonding.

Solvent effects in nuclear magnetic resonance

Most pronounced among the solvent shifts in nuclear magnetic resonance is the shift of proton resonance to lower applied fields when a hydrogen bond is formed. The resonance assigned to OH in the proton resonance spectrum of pure ethanol (Fig. 113, p. 253) is in the low field position ($\delta = {}^{-}0.5 \times 10^{-6}$ relative to water) appropriate to hydrogen bonding between ethanol molecules. Dilution of the ethanol with carbon tetrachloride moves the OH peak progressively through the CH_2 and CH_3 peaks to a limiting value $\delta = 4.1 \times 10^{-6}$: in the vapour the OH peak also appears on the high-field side ($\delta = 4.3 \times 10^{-6}$) of the CH_3 peak. Other examples are discussed by Schneider and by Pople (Ref. 91, pp. 55, 71; Ref. 105, Chap. 15). Solvent effects of this type can be observed in n.m.r. spectroscopy which might well be too weak to be detected otherwise; as, for example[92], the changes of the acetylenic hydrogen resonance in $CH_2Cl \cdot C \vdots CH$ in various solvents.

Chemical exchange

In n.m.r. spectroscopy chemical exchange is a factor which might be classed as a solvent effect where in other spectra it is only noted as a change of absorption with time. It frequently happens that significant n.m.r. spectra can only be obtained at low temperatures. A case in point is that of SF_4 (p. 259): on warming the sample the two peaks which indicated different pairs of fluorine atoms broadened and finally merged into one peak. Rapid exchange of the fluorine atoms between the two different positions is clearly

References p. 287–289

responsible for that effect; and a rough activation energy of 4 kcal/mole was calculated for the process.

Attempts to observe internal hydrogen bonding in aqueous solutions of highly alkylated succinic acids (I) were foiled by exchange. Whereas the

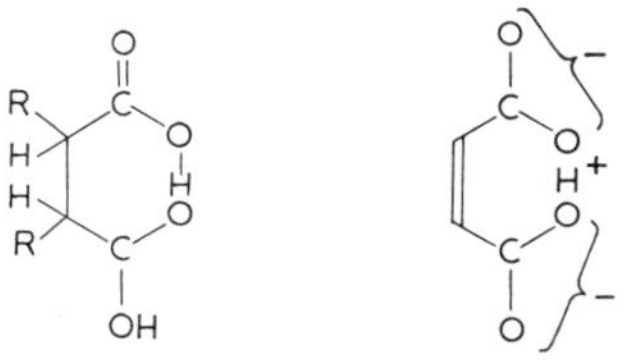

(I) (II)

materials dissolved in dimethylsuphoxide showed[93] a proton resonance at very low fields, in water no such line could be seen and could well be lost if exchange of protons with the solvent was rapid enough to reduce the residence time of any one proton to something comparable with the time of the n.m.r. transition. The same is true of the hydrogen maleate ion (II) and it seems that direct evidence as to the continued existence of the internal hydrogen bond in aqueous solution must rather come from infra-red spectra, where the transition time is very much shorter[94, 95].

The effect of chemical exchange on n.m.r. spectra can, of course, be put to good use in measuring rates of reaction, particularly protonation reactions. In an aqueous solution of methylammonium chloride[96], for example, at high enough hydrogen-ion concentration the CH_3-proton signal shows sharp quadruplet structure due to coupling with the NH_3^+ protons; the water peak is sharp and the NH_3^+-proton signal shows a triplet due to coupling with ^{14}N. As the hydrogen concentration is decreased, the quadruplet structure merges into a single sharp line showing that the protons are not retained in the NH_3^+-group long enough to permit the CH_3-protons to distinguish different spin orientations. From the concentrations known and the time required for the transition (given by the resonance frequency) it is possible to assess rate constants for the protolytic exchange. The NH_3^+-proton signal disappears as hydrogen-ion concentration decreases, as expected, and the water peak is also affected. It therefore appears that water is involved in the exchange reaction.

REFERENCES

[1] K. W. Hausser, R. Kuhn, A. Smakula, *Z. physik. Chem.*, B 29 (1935) 384.
[2] K. W. Hausser, R. Kuhn, G. Seitz, *Z. physik. Chem.*, B 29 (1935) 391.
[3] K. W. Hausser, R. Kuhn, *et al.*, *Z. physik. Chem.*, B 29 (1935) 363—454 (six papers).
[4] R. Kuhn and A. Winterstein, *Helv. Chim. Acta*, 11 (1928) 87, 116, 123, 427; 12 (1929) 493, 899.
[5] A. Maccoll, *Quart. Revs. (London)*, 1 1947) 16.
[6] G. N. Lewis, and M. Calvin, *Chem. Revs.*, 25 (1939) 273.
[7] E. Hückel, *Z. Elektrochem.*, 43 (1937) 752, 827.
[8] J. A. A. Ketelaar, *Chemical Constitution*, Elsevier, 1958, p. 265.
[9] C. A. Coulson, *Valence*, Oxford Univ. Press, 1953, p. 223.
[10] V. Prelog, *Helv. Chim. Acta*, 31 (1948) 555.
[11] G. A. Swan, *J. Chem. Soc.*, (1949) 1720; (1958) 2038.
[12] R. B. Woodward and B. Witkop, *J. Am. Chem. Soc.*, 71 (1949) 379.
[13] R. B. Woodward, W. J. Brehn, A. L. Nelson, *J. Am. Chem. Soc.*, 69 (1947) 2250.
[14] H. L. Holmes and Sir Robert Robinson, *J. Chem. Soc.*, (1939) 603.
[15] W. R. Brode, *Chemical Spectroscopy*, 2nd ed., Wiley, 1943.
[16] Symposium on Color and the Electronic Structure of Complex Molecules, *Chem. Revs.*, 41 (1947) 201—419.
[17] *International Critical Tables*, Vol. V, McGraw-Hill, 1929.
[18] *Landolt-Bornstein Tables*, 6th ed., Vol. I, part 3, Springer, 1951, pp. 78—358.
[19] *American Petroleum Institute, Project 44*, Carnegie Inst. Technol., 1953.
[20] *A. S. T. M. Absorption Spectra* (on punched cards), American Society for Testing Materials.
[21] K. V. Dimroth, *Angew. Chem.*, 52 (1939) 545.
[22] R. A. Friedel and M. Orchin, *Ultraviolet Spectra of Aromatic Hydrocarbons*, Wiley, 1951.
[23] R. A. Morton, *Absorption Spectra of Vitamins and Hormones*, 2nd ed., Hilger, 1942.
[24] H. M. Hershenson, *Ultraviolet and Visible Absorption Spectra, Index* 1934—54, Academic Press, 1956.
[25] J. Baddiley, J. G. Buchanan, P. A. Smith, D. H. Hayes, *J. Chem. Soc.*, (1958) 3743.
[26] J. J. Armstrong, J. G. Buchanan, F. E. Hardy, personal communication.
[27] N. B. Colthup, *J. Opt. Soc. Am.*, 40 (1950) 397.
[28] A. D. Cross, *Introduction to Practical Infra-red Spectroscopy*, Butterworth, 1960.
[29] E. R. Lippincott and M. C. Tobin, *J. Am. Chem. Soc.*, 73 (1951) 4990.
[30] E. E. Aynsley and W. A. Campbell, *J. Chem. Soc.*, (1958) 3290.
[31] L. A. Woodward and A. A. Nord, *J. Chem. Soc.*, (1956) 3721.
[32] L. A. Woodward, G. Garton, H. L. Roberts, *J. Chem. Soc.*, (1956) 3723.
[33] L. A. Woodward, *Phil. Mag.*, 18 (1934) 823.
[34] G. M. Bennet, J. C. D. Brand and G. Williams, *J. Chem. Soc.*, (1946) 869.
[35] P. O. Kinell, I. Lindquist, M. Zackrissen, *Acta Chem. Scand.*, 13 (1959) 190.
[36] C. H. Townes and A. L. Schawlow, *Microwave Spectroscopy*, McGraw-Hill, 1955.
[37] R. D. Spence, *J. Chem. Phys.*, 23 (1955) 1166.
[38] G. E. Pake, *J. Chem. Phys.*, 16 (1948) 327.
[39] R. E. Richards and J. A. S. Smith, *Trans. Faraday Soc.*, 47 (1951) 1261.
[40] C. M. Deeley and R. E. Richards, *J. Chem. Soc.*, (1954) 3697.
[41] P. T. Ford and R. E. Richards, *Discussions Faraday Soc.*, 19 (1955) 193.
[42] H. S. Gutowsky, *Discussions Faraday Soc.*, 19 (1955) 248.
[43] L. H. Meyer, A. Saika, H. A. Gutowsky, *J. Am. Chem. Soc.*, 75 (1953) 4567.
[44] W. E. Noland, J. H. Cooley, P. A. Mcveigh, *J. Am. Chem. Soc.*, 81 (1959) 1209.
[45] E. L. Muetterties and W. D. Phillips, *J. Am. Chem. Soc.*, 79 (1957) 322.
[46] F. A. Cotton, J. W. George and J. S. Waugh, *J. Chem. Phys.*, 28 (1958) 994.
[47] S. Alexander, *J. Chem. Phys.*, 28 (1958) 358.
[48] J. N. Schoolery, *Discussions Faraday Soc.*, 19 (1955) 215.

[49] W. C. PRICE, *Chem. Revs.*, 41 (1947) 257.
[50] A. PULLMAN and B. PULLMAN, *Discussions Faraday Soc.*, 9 (1950) 46.
[51] L. G. S. BROOKER, F. L. WHITE, R. H. SPRAGUE, S. G. DENT and G. VAN ZANDT, *Chem. Revs.*, 41 (1947) 325.
[52] L. P. HAMMETT, *Physical Organic Chemistry*, McGraw-Hill, 1940, Chap. 7.
[53] H. GILMAN (Ed.), *Organic Chemistry*, Vol. III, Wiley, 1953, p. 19.
[54] H. H. JAFFE, *Chem. Revs.*, 53 (1953) 191.
[55] M. ST. C. FLETT, *Trans. Faraday Soc.*, 44 (1948) 767.
[56] P. J. KRUEGER and H. W. THOMPSON, *Proc. Roy. Soc. (London)*, A 250 (1959) 22.
[57] L. E. ORGEL, *An Introduction to Transition Metal Chemistry*, Methuen, 1960.
[58] T. M. DUNN, in J. LEWIS and R. G. WILKINS (Eds.), *Modern Coordination Chemistry*, Interscience, 1960, Chap. 4.
[59] J. W. LINNETT, *Discussions Faraday Soc.*, 26 (1958) 7.
[60] F. E. ILSE and H. HARTMANN, *Z. physik Chem.*, 197 (1951) 239.
[61] O. G. Holmes and D. S. CLURE, *J. Chem. Phys.*, 26 (1957) 1686.
[62] H. HARTMANN and H. MÜLLER, *Discussions Faraday Soc.*, 26 (1958) 49.
[63] L. E. ORGEL, *Quart. Revs. (London)*, 8 (1954) 422.
[64] J. FRANCK, H. KUHN and G. ROLLEFSON, *Z. Physik*, 43 (1927) 155.
[65] R. HILSCH and R. W. POHL, *Z. Physik*, 57 (1929) 145; 59 (1930) 812; 64 (1930) 606.
[66] M. SMITH and M. C. R. SYMONS, *Discussions Faraday Soc.*, 24 (1957) 206.
[67] M. EIGEN, *Discussions Faraday Soc.*, 24 (1957) 25.
[68] H. FROMHERZ and K.-H. LIH, *Z. physik. Chem.*, A 153 (1931) 321.
[69] A. I. BIGGS, M. H. PANCKHURST and N. H. PARTON, *Trans. Faraday Soc.*, 51 (1955) 802, 806.
[70] M. LINHARD and M. WEIGL, *Z. anorg. Chem.*, 266 (1951) 49.
[71] F. S. DAINTON, *J. Chem. Soc.*, (1952) 1533.
[72] M. LINHARD, *Z. Elektrochem.*, 50 (1944) 224.
[73] J. H. B. GEORGE, J. A. ROLFE and L. A. WOODWARD, *Trans. Faraday Soc.*, 49 (1953) 375.
[74] P. L. GOGGIN and L. A. WOODWARD, *Trans. Faraday Soc.*, 56 (1960) 1591.
[75] H. W. CRANDALL, *J. Chem. Phys.*, 17 (1949) 602.
[76] J. P. MATHIEU and M. LOUNSBURY, *Discussions Faraday Soc.*, 9 (1950) 196.
[77] B. VODAR, *Proc. Roy. Soc. (London)*, A. 255 (1960) 44.
[78] R. D. MAIR and D. F. HORNIG, *J. Chem. Phys.*, 17 (1949) 1236.
[79] S. ZWERDLING and R. S. HALFORD, *J. Chem. Phys.*, 23 (1955) 2221.
[80] E. E. AYNSLEY, W. A. CAMPBELL AND R. E. DODD, *Proc. Chem. Soc.*, (1947) 210.
[81] J. A. A. KETELAAR and F. N. HOOGE, *European Molecular Spectroscopy Conference*, Freiburg i. Br., 1956.
[82] L. J. ANDREWS, *Chem. Revs.*, 54 (1954) 713.
[83] J. WEISS, *J. Chem. Soc.*, (1942) 245.
[84] R. S. MULLIKEN, *J. Am. Chem. Soc.*, 74 (1952) 811.
[85] R. E. MILLER and W. F. K. WYNNE-JONES, *J. Chem. Soc.*, (1959) 2375.
[86] M. L. JOSIEN and N. FUSON, *J. Chem. Phys.*, 22 (1954) 1169, 1264.
[87] L. J. BELLAMY and R. L. WILLIAMS, *Proc. Roy. Soc. (London)*, A 255 (1960) 22.
[88] W. C. PRICE, W. F. SHERMAN and G. R. WILKINSON, *Proc. Roy. Soc. (London)*, A 255 (1950) 5.
[89] K. NAKOMOTO, M. MARGOSHES and R. E. RUNDLE, *J. Am. Chem. Soc.*, 77 (1955) 6480.
[90] D. HADŽI and N. SHEPHARD, *Proc. Roy. Soc. (London)*, A 216 (1953) 247.
[91] D. HADŽI (Ed.), *Hydrogen Bonding — a report on the Ljubljana Symposium 1957* Pergamon Press, 1959.
[92] R. E. RICHARDS, *Proc. Roy. Soc. (London)*, A 255 (1960) 72.
[93] L. EBERSON and S. FORSÉN, *J. Phys. Chem.*, 64 (1960) 767.
[94] S. FORSÉN, *J. Chem. Phys.*, 31 (1959) 852.
[95] R. E. DODD, R. E. MILLER, W. F. K. WYNNE-JONES, *J. Chem. Soc.*, in the press.
[96] E. GRUNWALD, A. LOEWENSTEIN and S. MEIBOON, *J. Chem. Phys.*, 27 (1957) 630.

For most of the contents of this Chapter further reference should be made to:

[97] W. WEST (Ed.), *Chemical Applications of Spectroscopy*, Interscience, 1956, which contains sections on microwave and radiofrequency spectroscopy (W. Gordy); infra-red and Raman spectra with particular reference to group-frequency correlations (R. N. Jones and C. Sandorfy); electronic spectra (A. B. F. Duncan and F. A. Matsen) and fluoresence (W. West).

Reports of two conferences organised by the Hydrocarbon Research Group of the Institute of Petroleum in 1954 and 1958 reflect the progress and variety of interest in spectroscopy in the last decade; the are:

[98] G. SELL (Ed.), *Molecular Spectroscopy*, Institute of Petroleum, 1955.

[99] E. THORNTON and H. W. THOMPSON (Eds)., *Molecular Spectroscopy*, Pergamon Press, 1959.

These may be compared with the Faraday Society discussion of 1950 which gives a fair cross section of the use of spectroscopy in chemistry:

[100] Spectroscopy and Molecular Structure, *Discussion Faraday Soc.*, (1950), no. 9.

Restricted to the correlation of infra-red frequencies with molecular structure:

[101] L. J. BELLAMY, *Infra-red Spectra of Complex Molecules*, 2nd ed., Methuen, 1959, provides a critical and comprehensive account. The word *complex* in the title may mislead: it does not refer to complexes in the sense of either addition compounds or coordinated polyatomic ions.

Electronic spectroscopy of organic molecules and discussion of many examples of its use in organic chemistry is the main feature of

[102] A. E. GILLAM and E. S. STERN, *Electronic Absorption Spectroscopy*, 2nd ed., Arnold, 1958.

Nuclear magnetic resonance spectroscopy is the subject of many recent reviews and monographs of which the following are particularly useful:

[103] E. R. ANDREW, *Nuclear Magnetic Resonance*, Cambridge Univ. Press, 1956, is concerned mainly with the solid state and

[104] J. D. ROBERTS, *Nuclear Magnetic Resonance*, McGraw-Hill, 1959,

[105] J. A. POPLE, W. G. SCHNEIDER and H. J. BERNSTEIN, *High-resolution Nuclear Magnetic Resonance*, McGraw-Hill, 1959, deal with n.m.r. spectroscopy in liquids and gases, giving a full account of the theory and many examples of applications.

Further:

[106] Microwave and Radiofrequency Spectroscopy, *Discussion Faraday Soc.*, (1955), no. 19,

[107] *Chemical Society Symposium, Bristol 1958*, Chemical Society, 1958, contain a number of valuable reviews as well as more specific papers.

Chemical applications of spectroscopy are to be found in papers, in all the general chemical journals. Where the emphasis is spectroscopic, papers are likely to be found in

Spectrochimica Acta,

Journal of Molecular Spectroscopy,

Optics and Spectroscopy (*U.S.S.R.*), translation of *Optika i Spektroskopiya*.

Chapter 5

INTENSITY

A spectrum is a graph of intensity of radiative energy against wavelength. In the foregoing chapters we have assumed that such a graph is made up of a number of narrow lines at various wavelengths (or frequencies) marked by maxima in the graph and we have been mainly concerned to account for those frequencies in terms of transitions between energy levels. Now we must examine the factors which govern the intensities of the maxima, or the amount of energy which is absorbed or emitted in any transition.

Three factors determine the intensity. They are (a) the number, or concentration, of absorbing or emitting species, (b) the relative number of atoms or molecules found in the state from which the transition originates, and (c) the intrinsic transition probability. This chapter is devoted to consideration of each of these factors in turn, and some of their applications.

Concentration Dependence

The effect of concentration on spectral intensity has obvious analytical applications both in emission spectra and in absorption.

Emission spectrophotometry

A great deal of work has gone into the tabulation of atomic spectral lines to enable the identification of elements in flame, arc and spark spectra. For qualitative analysis a fair idea of the elements present in the sample may be gained from measurement of lines to an accuracy of 1 Å. In Brode's tables[46], in which principal lines of elements are listed by wavelength, there are generally less than half-a-dozen elements whose lines appear in any one interval of 1 Å. From the tabulation of principal lines by elements it is then possible to see what other lines of comparable intensity ought to appear in the spectrum for each of the suspected elements. Measurement of wavelength to 0.1 Å or 0.01 Å clearly improves the specificity and in practice the observation of two lines is sufficient for the unambiguous identification of an element.

The question of intensity is relevant to qualitative analysis in so far as a particular line may be due to a strong transition in an element present in low concentration, or to a weak transition in an element present in greater concentration. Very early in the history of spectroscopy atomic lines were classified according to their relative intensities: those few lines which remain observable as the concentration is decreased are the strongest lines generally resulting from S—P transitions. They are variously termed persistent lines, residual lines, or *raies ultimes*, and clearly they are the lines to look for in attempting to establish the presence of an element.

Quantitative analysis by means of emission spectra can range from simple flame photometry to the use of large and expensive grating polychromators (p. 19). In each case the emission of a particular line of a particular element under standard conditions is more-or-less proportional to the concentration of the element in the sample. It is important that the operating procedure should be rigorously standardized and calibrated. For a solution it is often convenient to spray the solution into an oxy-acetylene flame: for solids, electric arcs and sparks are usually more convenient. In either case the most careful standardization of procedure cannot eliminate source fluctuations during measurement and it is therefore necessary to use an internal standard.

For example, it may be required to determine the percentage in steel of carbon (using 1657 Å in the fluorite range) or manganese (using 2933 Å in the normal ultraviolet range). Standard calibration samples are made up (also containing such other elements as are likely to be present) and the arc spectrum is either photographed or recorded photoelectrically with a polychromator having pre-set exit slits. The intensity of the relevant line is then compared with the intensity of an appropriate iron line (1713 Å in the first case, 2880 Å in the second) and the ratio of the intensities is plotted against the percentage concentration of the test element. In the photoelectric instruments there can be provision for an automatic computation of the ratio and printing out on a strip of paper. The virtue of the internal standard is that it suffers precisely the same fluctuations as other elements in the sample. Which element is chosen as internal standard naturally depends upon the nature of the mixture.

The sensitivity of spectral lines has been defined in various ways. What matters for accuracy of determination is the concentration sensitivity of the line, or the rate of change of intensity with concentration. Generally an element present in concentration greater than 0.01 % can be determined with an accuracy of $\pm$ 2 %; that is, the standard deviation in a concentration result of say 500 p.p.m. is $\pm$ 10 p.p.m. Often considerably better accuracy is achieved. The absolute sensitivity or limit of detection may be given as the concentration below which no line can be observed or, better, as the concentration at which the standard deviation becomes greater than $\pm$ 25 %. Limits vary considerably: beryllium and copper can be estimated to that accuracy down to about 5 p.p.m.; other elements perhaps only to 100—300 p.p.m. Many elements have been detected down to 0.5 p.p.m. but with considerably greater error than 25 %. In flame emission photometry absolute sensitivities are quoted in terms of molarity and vary from 10^{-3} to 10^{-6} M.

Atomic absorption spectroscopy

A recent development[1] in elementary analysis is also a form of flame photometry: it differs in that the atoms in a flame are detected by absorption rather than by emission. A hollow cathode lamp (p. 14) generates a very sharp line quite specific to the element from which it is constructed. This light is passed through the flame and is absorbed only if the same element is present in the flame. In this application the flame merely acts as a vapourizer and it is not required, as it is in emission flame photometry, to excite the atoms to higher energy levels. There is in principle no need for any other dispersion element since the source is strictly monochromatic: nevertheless a spectrophotometer may be used as a convenient detector. The result is recorded as a percentage absorption or optical density and an internal standard is not required. Sensitivities are generally better, sometimes as much as 10 times better, than emission flame photometry.

Absorption spectrophotometry

For quantitative analysis of substances which absorb radiation at any wavelength it is necessary, and usually sufficient, to make measurements of the ratio of transmitted energy, I, to incident energy, I_0. The instrument may be a simple photometer which uses filters to select a broad band of wavelength or it may be one incorporating a high-resolution monochromator. I/I_0 may be measured simply by noting two meter readings, the first with the sample cell in the beam, the second after it has been removed. Whatever the procedure it can, with careful standardization, be used to prepare a calibration graph of I/I_0 against concentration. This may then be used for finding the concentration of a test solution by measurement of I/I_0. Absorptiometric analysis along these lines is widely applied to the estimation of metals by their coloured complexes in solution as well as to quantitative analysis for many organic molecules in both ultraviolet and infra-red regions. Provided the substance concerned is the only one which absorbs under the conditions of the experiment the method is very straightforward: sometimes a calibration may be devised to take into account small amounts of other absorbing materials. Naturally greater selectivity is obtained with instruments employing prisms or gratings than with those which use filters. It will often be found that the calibration graph approaches linearity if $\log_{10}(I_0/I)$

is plotted against concentration but the method, as so far described, is quite empirical and the calibration cannot be transferred from one instrument to another.

Beer's law

There is, however, a theoretical basis for transference of results of intensity measurements under certain circumstances. We have seen (p. 47) that, for monochromatic light passing through a homogeneous medium of thickness l in which the absorbing species is present in molar concentration c the following relationship should hold:

$$D \equiv \log_{10} (I_0/I) = \varepsilon lc \tag{5.1}$$

D, the optical density or absorbance, is defined by the first equation. The linear dependence of optical density upon concentration and path length is generally known as Beer's law.

Equation (5.1) can be simply derived from the consideration of the chance that a molecule will absorb a quantum if it encounters one. We can define an absorption cross-section, σ, of the molecule such that any quantum falling in that area is absorbed. We cannot define precisely the actual molecular cross-section, but it may be expected to be of the order of 10^{-16} cm². Hence a value of σ of 10^{-20} cm² would suggest a chance that 1 in 10^4 encounters with a quantum results in absorption. Thus σ expresses the intrinsic probability of the occurrence of transition, and will be different for each transition. Now, if the molar concentration is c (mole l^{-1}) the number of molecules per square centimetre of area perpendicular to the incident light in a small element of path length dl, is $10^{-3}\,Nc\mathrm{d}l$, where N is Avogadro's number. The absorbing area is therefore $\sigma 10^{-3} Nc\mathrm{d}l$ cm² out of a total area of 1 cm². $\sigma 10^{-3} Nc\mathrm{d}l$ is thus the fraction of light absorbed and we may write

$$\mathrm{d}I/I = -\,\sigma 10^{-3} Nc\mathrm{d}l \tag{5.2}$$

Integration between I_0 and I on the left and 0 and l on the right gives

$$\ln (I_0/I) = \sigma 10^{-3} Ncl \tag{5.3}$$

This is the same form as (5.1) and identical with it if

$$\varepsilon = 0.434 \sigma 10^{-3} N \tag{5.4}$$

Thus ε is a measure of molar absorbing cross-section. (If c were measured in moles cm^{-3}, ε would then have units cm² moles^{-1} appropriate to this interpretation.) It also expresses the intrinsic probability of a transition; a question to which we return later in the chapter.

The instrumental factors which are involved in determination of ε, or in making proper use of Beer's law have been discussed in Chapter 1. Evidently, if the instrument slit range is small enough compared with the band width, equation (5.1) may be expected to hold at any wavelength of the band and particularly at the wavelength, λ_{max}, at which the optical density is a maximum, D_{max}. Failing this, we must look for derived quantities from which we can construct an equation which is linear in concentration.

A particular shape may be assumed for the absorption band, whose apparent height, D'_{max}, and apparent half-height width, $\Delta\omega'_{\frac{1}{2}}$, may be converted by the method outlined on p. 50 to true D_{max}, ε_{max} and $\Delta\omega_{\frac{1}{2}}$. Then, according to (1.35):

$$D_{max} \cdot \Delta\omega_{\frac{1}{2}} = \text{constant} \cdot cl \tag{5.5}$$

Thus it should be possible to plot the corrected $D_{max} \cdot \Delta\omega_{\frac{1}{2}}$ against cl and obtain a straight line. Other means of correction for slit width have also been given[2,3]. An interesting fact which emerges from these considerations is that Beer's law may still hold for the apparent D'_{max}, under conditions which make D'_{max} only 90% of the true D_{max}. The conditions for analytical use of Beer's law, provided they are standardized, are thus less stringent than those required for measurement of absolute intensity.

An alternative measure of intensity is the integrated area of a band. In terms of the above correction for the effect of a finite slit the true integrated intensity, A, of a band is given by:

$$A = 2.303 \int_0^\infty \varepsilon \mathrm{d}\omega = 2.303 \frac{\pi}{2} \varepsilon_{max} \cdot \Delta\omega_{\frac{1}{2}} \tag{5.6}$$

The measured area of a band, if the apparent value $\varepsilon' = D'/cl$ is plotted against ω, is the apparent integrated intensity B, defined by

$$B = 2.303 \int \varepsilon' \, \mathrm{d}\omega \tag{5.7}$$

It is found in practice that B is much less sensitive to the effect of slit width than is ε'_{max}. Hence an alternative analytical procedure for finding c is to measure total band area under a curve of D against ω (without dividing by cl):

$$\text{total area} = \int D \, \mathrm{d}\omega = 0.434 \, B \cdot cl \tag{5.8}$$

The linear relationship (5.8) between the measured band area and concentration should hold under less exacting conditions than the linear relationship between D'_{max} and c. A calibration graph of B against c can be prepared which stands the best chance of being linear.

Analysis of mixtures

For analytical purposes any of the above methods must be tested on the particular instrument to be used. It may be that overlapping bands make it difficult to define the area of one band and enforce the use of peak optical densities. Such overlapping will in any case require corrections to be made taking account of the additivity of optical density (p. 49). Lothian (Ref. 47, Chap. 5) gives details and equations for handling D_{max} values of multi-component mixtures.

In the analysis of gaseous mixtures the effect of pressure broadening (p. 280) on the apparent intensity must be borne in mind. It is evident, for example, in the analysis of mixtures of CF_3H and C_2F_6, two difficultly separated products of the reaction of CF_3 radicals with hydrocarbons. The peak intensity of the absorption band at 1150 cm^{-1} due to CF_3H not only depends upon the partial pressure of CF_3H but also upon the proportion of C_2F_6 present in the mixture. One way to meet this difficulty is to add a large excess of some non-absorbing gas, say dry air, to a standard pressure.

Chemical Equilibria

It is now generally recognized that, under proper experimental conditions and taking account of stray light effects and reflection losses, Beer's law is valid in homogeneous media. Apparent deviations from Beer's law, or from corresponding linear dependence of Raman scattering intensity upon concentrations, can then be ascribed to chemical effects. The intensities can be used to assess equilibria provided that there are wavelengths at which the absorptions due to the participating species are markedly different. Spectrophotometry has the advantage as a method of studying equilibria that it can obtain a measure of the concentration of a participating species which cannot be isolated from solution. There is the consequent disadvantage that such species cannot be directly weighed to enable a calibration curve to be made and an extinction coefficient to be obtained. It is therefore necessary to resort to extrapolation to obtain a value of ε and it is not always clear what is the best extrapolation.

Basically the problem is concerned with the difference between a thermodynamic equilibrium constant in terms of activities and the corresponding expression in terms of concentrations. In concentrated solutions of electrolytes long range electrostatic interactions may blur the definition of species to which an absorption may have been assigned, but for the most part it can be taken that such an absorption identifies a particular species and measures its concentration. Intensities of vibrational spectra most certainly measure concentration. If, however, a linear extrapolation is to be used to obtain an unknown extinction coefficient the concentrations have to be corrected by the appropriate activity coefficients, which rely upon particular theoretical assumptions.

Some examples may be given to illustrate the difficulties and to show what information can be obtained. One general point should first be strongly emphasized. Whereas for analytical purposes with non-reacting materials the absorption spectra are almost independent of temperature, in studies of chemical equilibria it is really essential that the cell compartments should be maintained at constant temperature since equilibrium constants may vary considerably with temperature.

Determination of acid-base ionization constants

The use of indicators is a method of determining the concentration of hydrogen ions which is essentially absorptiometric in nature. An acid HA dissociates in water thus:

$$HA + H_2O \rightleftharpoons H_3O^+ + A^-$$

To this system we add a conjugate acid–base pair, $HX—X^-$, in such small concentration that the system is not appreciably altered. Under these conditions the equilibrium

$$HX + H_2O \rightleftharpoons H_3O^+ + X^-$$

measures the hydrogen ion concentration as

$$[H_3O^+] = K^1 \frac{[HX]}{[X^-]}$$

where K_1 is the acidity constant of the conjugate pair, activity coefficients being set at unity for the moment. If HX and X^- are of different colours then a change in $[H_3O^+]$, which alters the ratio $[HX]/[X^-]$ leads to a change in colour. The conjugate pair is said to act as an *acid–base indicator*.

The colour change may be followed[4] absorptiometrically as in Fig. 132, which shows the absorption spectrum of bromophenol blue in solutions of pH from 3.0 to 5.4. The variation of peak optical density is also plotted against pH. The spectrum suggests that the peak at 5900 Å is due solely to the con-

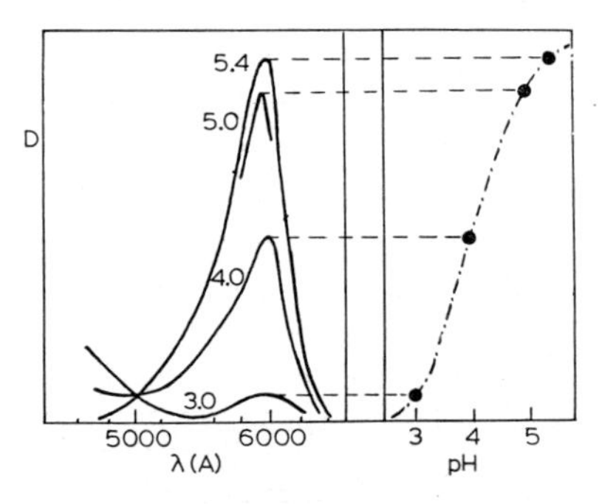

Fig. 132. Absorption spectrum of bromophenol blue at various pH. Variation of peak optical density with pH.

jugate base, X^-, of bromophenol blue (since the peak is reduced by increasing the acidity) and the HX absorption evidently occurs somewhere below 4500 Å. There is no problem of overlap in this case: therefore we may write $[X^-] = D/\varepsilon l$ where D and ε are the optical density and extinction coefficient at the maximum. If the total concentration of bromophenol blue is c then $[HX] = c - [X^-]$ and

$$[H_3O^+] = K_1 \left(\frac{c}{[X^-]} - 1\right) = K_1 \left(\frac{c \cdot \varepsilon l}{D} - 1\right) \tag{5.9}$$

or

$$\mathrm{pH} = \mathrm{p}K_1 - \log_{10} \left(\frac{c \cdot \varepsilon l}{D} - 1\right) \tag{5.10}$$

This gives the shape of the curve of D against pH shown in Fig. 132. Clearly such a curve, which can be obtained empirically without knowledge of K_1 or ε by measurement in various buffer solutions, can then be used to determine pH and the degree of dissociation of a weak acid HA. Naturally the numerical values only hold good for a particular solvent and cannot be simply transferred to another.

Conversely, this experiment provides the information about the nature of the indicator conjugate pair itself. Equation (5.10) can be fitted to the curve of D against pH to obtain values of ε and of K_1. In that equation $D \neq c \cdot \varepsilon l$, since c refers to the total concentration, $[HX] + [X^-]$. But if the pH is made sufficiently large so that all the indicator is present as X^-, then $c = [X^-]$ and $D = c \cdot \varepsilon l$, and hence ε is found. Then at the value of D corresponding to $\frac{1}{2}c \cdot \varepsilon l$, $\mathrm{pH} = \mathrm{p}K_1$.

There is no restriction to the visible region when spectrophotometry is employed and much simpler acid–base pairs can be examined, or used as indicators, when the range is extended into the ultra-violet. A good example of this is the absorption spectrum of *p*-nitrophenol from 3000—4400 Å in various buffer solutions. Fig. 133 shows the reciprocal movement of two absorption bands with change in pH. The band at 3170 Å clearly belongs to the base in the conjugate pair defined by

$$HO \cdot C_6H_4NO_2 \rightleftharpoons H^+ + O \cdot C_6H_4NO_2^-$$

In aqueous solution:

$$H_2O + HO \cdot C_6H_4 \cdot NO_2 \rightleftharpoons H_3O^+ + O \cdot C_6H_4 \cdot NO_2^-$$

and the acid dissociation constant of *p*-nitrophenol is given by

$$\mathrm{p}K_a = \mathrm{pH} - \log \frac{\alpha}{1 - \alpha} - \log \frac{\gamma_{H^+} \gamma_{A^-}}{\gamma_{HA}} \tag{5.11}$$

In this equation K_a is now the thermodynamic ionization constant, the γ are activity coefficients, and pH is defined in terms of hydrogen-ion concentration, *viz.* $\mathrm{pH} = -\log_{10} [H_3O^+]$. H_3O^+, of course, stands for the hydrogen

References p. 328–329

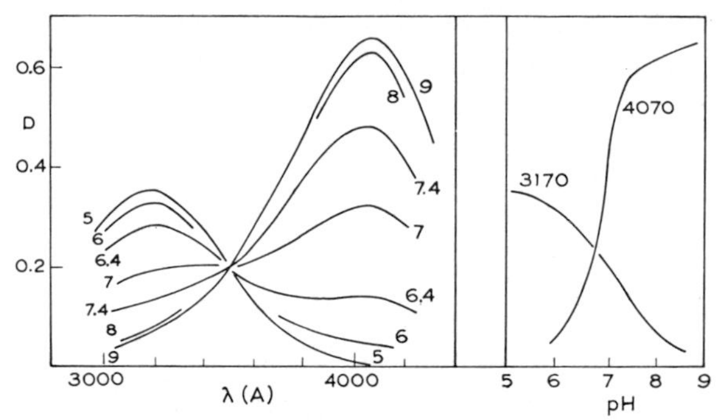

Fig. 133. Ultraviolet absorption of *p*-nitrophenol, $3.6 \times 10^{-5} M$, in buffered aqueous solutions. Variation of D_{3170} and D_{4070} with pH.

ion, whatever its degree of hydration. HA and A stand for $HO \cdot C_6H_4 \cdot NO_2$ and $O \cdot C_6H_4 \cdot NO_2^-$.

The term $\alpha(1 - \alpha)$ is the one with which the spectroscopic data are primarily concerned. α is the degree of dissociation of HA. If Beer's law is valid, and if there is no overlap of bands,

$$D_{4070} = \varepsilon_A \alpha c l \qquad (5.12)$$
$$D_{3170} = \varepsilon_{HA}(1 - \alpha) c l$$

where c is the total concentration of *p*-nitrophenol. Hence

$$\frac{\alpha}{1-\alpha} = \frac{D_{4070}}{D_{3170}} \frac{\varepsilon_{HA}}{\varepsilon_A} \qquad (5.13)$$

The extinction coefficients are obtained by measuring the limiting values of D_{4070} at high pH ($\alpha \to 1$) and of D_{3170} at low pH ($\alpha \to 0$). The values obtained are $\varepsilon_{HA} = 9720$ (3170 Å) and $\varepsilon_A = 18{,}330$ mole^{-1} l cm^{-1} (at 4070 Å). At intermediate pH, α can be obtained from the measured values of D_{4070} or D_{3170} and hence pK_a from equation (5.11). From D_{4070} $pK_a = 6.99$ (an average of nine values lying between 6.95 and 7.05).

From D_{3170} the evaluation of α is complicated by a small overlap correction. It is evident from the curves at high pH in Fig. 133 that the 4070 Å band makes a small contribution to the absorption at 3170 Å: in fact ε'_A (at 3170 Å) $= 1390$ mole^{-1} l cm^{-1}. Therefore (5.12) must be modified to:

$$D_{3170} = [\varepsilon_{HA}(1 - \alpha) + \varepsilon'_A \alpha] c l \qquad (5.14)$$

With this correction α, and therefore pK_a, may be obtained from measurements of D_{3170} at intermediate pH and this also gives $pK_a = 6.99$.

The values of pK_a so obtained clearly depend upon the pH of the buffer solutions used and the equation used to relate activity coefficients to ionic strength. To discuss the evaluation of the term $\log(\gamma_{H^+} \gamma_{A^-}/\gamma_{HA})$ would go beyond the scope of this book. However, we may note that theoretical or semi-empirical equations may be used to evaluate the γ terms and hence pK_a: or, conversely, the pK_a obtained from e.m.f. or conductance measurements may be compared with spectrophotometric data to test the range of application of activity coefficient formulae.

The *p*-nitrophenol absorption in Fig. 133 also illustrates a common phenomenon in pH dependence of absorption curves. There is a point at 3500 Å at which the optical density is evidently independent of pH and therefore independent of α. This point, at which all the curves intersect, is called the *isobestic point*. For the two component system it can be seen that the optical density D^i at the particular wavelength is the sum of contributions from ε_{HA} and $\varepsilon^i{}_A$, thus:

$$D^i = [\varepsilon^i{}_A \alpha + \varepsilon^i{}_{HA}(1 - \alpha)]cl \tag{5.15}$$

The term in brackets can only be independent of α if

$$\varepsilon^i{}_A = \varepsilon^i{}_{HA} = \varepsilon^i \tag{5.16}$$

and hence

$$D^i = 2\varepsilon^i \cdot cl \tag{5.17}$$

This gives another variation on the determination of α. ε^i is determined from D^i at known concentration c, and likewise ε_A from D_{4070} at high pH. Then:

$$\alpha = \frac{D_{4070} \cdot 2\varepsilon^i}{D^i \cdot \varepsilon_A} \tag{5.18}$$

Alternatively, where *p*-nitrophenol is to be used as a pH indicator, the ratio D_{4070}/D^i is independent of concentration and hence the calibration plot of the ratio against pH does not require accurate knowledge of c.

The absence of an isobestic point indicates a more complex system. It may be that there are more than two species involved in the equilibrium. Alternatively there may be a medium effect in which the change of acidity, which causes the change in α, also causes changes in ε_A and ε_{HA} and thus makes the value ε^i, at which ε_A and ε_{HA} are equal, dependent upon pH. This second alternative is illustrated in the absorption spectrum of aceto-

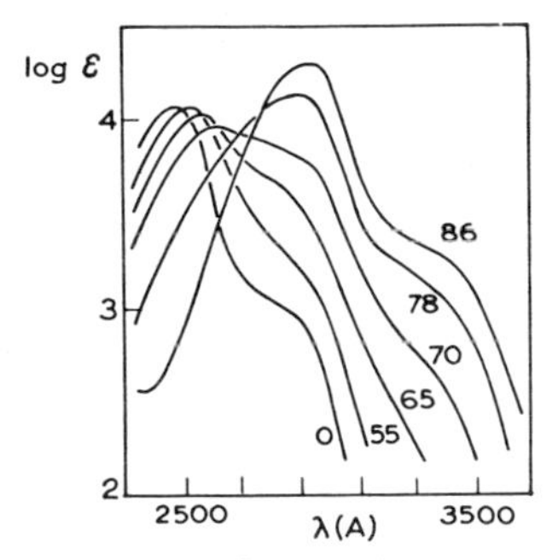

Fig. 134. Absorption spectrum of acetophenone in aqueous sulphuric acid as solvent. The numbers on the curves indicate the solvent composition (% H_2SO_4).

phenone[7] in aqueous sulphuric acid as solvent, varying from 0—96 % H_2SO_4 (Fig. 134). The acetophenone is evidently behaving as a very weak base according to the equation

$$C_6H_5 \cdot CO \cdot CH_3 + H_2SO_4 \rightleftharpoons C_6H_5 \cdot \overset{+}{C}OH \cdot CH_3 + HSO_4^-$$

and the maximum at about 2950 Å corresponds to the cation. However, the change of medium results in band shifts and the isobestic point is lost. One method of dealing with this situation in calculating an ionization constant is to shift all the curves laterally to a common point of intersection; that is, to create an arbitrary isobestic point.

In cases where it is not possible to alter the conditions sufficiently to find the maximum value of optical density an extrapolation procedure such as the following must be adopted. A weak organic acid is so weak in ethanol solution that it is not possible, in the equilibrium

$$HA + OEt^- \rightleftharpoons A^- + HOEt$$

to reach the situation in which virtually all the substance is present as A^-. If we suppose that only A^- absorbs at a given wavelength and its extinction coefficient is ε_A we may again write $D = \varepsilon_A \cdot \alpha cl$, but now we cannot reach the limiting value of D when $\alpha \rightarrow 1$. If, however, the equilibrium constant, in ethanol solution, is:

$$K = \frac{[A^-]}{[HA][OEt^-]} \cdot \frac{\gamma_A}{\gamma_{HA}\,\gamma_{OEt}} \tag{5.19}$$

and we set $\gamma_{HA} = 1$ and $\gamma_A = \gamma_{OEt}$, it is readily shown that, at a concentration c of the conjugate pair HA and A^-:

$$\frac{cl}{D} = \frac{1}{\varepsilon_A} + \frac{1}{\varepsilon_A K} \cdot \frac{1}{[OEt^-]} \tag{5.20}$$

Thus there should be a linear relationship between cl/D and $1/[OEt^-]$, with intercept $1/\varepsilon_A$ and slope $1/(\varepsilon_A - K)$. In fact excellent straight lines have been published[8] for such weak acids as 2,4-dinitroaniline in ethanol for which $K = 4.773$. Consequently $pK_a = 18.46$, referring to the equilibrium in ethanol

$$HA + EtOH \rightleftharpoons A^- + EtOH_2^+.$$

Quantitative Raman studies

There are as yet very few applications of quantitative Raman spectroscopy to the study of chemical equilibria. They are mainly confined to nitric and sulphuric acids. Earlier work with photographic recording of line intensities was rendered very difficult by the effect of band broadening on the intensity of blackening of a photographic plate. Nevertheless dissociation constants for nitric acid of 24 and 21 mole l^{-1} (at 25°C) were obtained by that method, which have since been verified by photoelectric Raman methods. The development of photoelectric recording should make a big contribution to studies of chemical equilibria in solution[12]. In favourable cases species can be detected and their concentrations measured down to about 0.1 *M*. From the point of view of comparison with other equilibrium data or for application of limiting-law extrapolation 0.1 *M* is a relatively high concentration. On the other hand the Raman intensity unequivocally measures concentration and therefore provides a good basis for the study of activity coefficients.

One good example[13] is the sulphuric acid–water system. Using the Raman lines at 910 cm^{-1} for H_2SO_4, 980 cm^{-1} for SO_4^{2-} and 1040 cm^{-1} for HSO_4^- it was first verified with ammonium hydrogen sulphate that the Raman intensity was proportional to molarity and then that in sulphuric acid the observed molarity of HSO_4^- was the same as that obtained by subtracting observed values for H_2SO_4 and SO_4^{2-} from the total molarity of the acid. Then it was shown that the actual concentrations of the three species in sulphuric acid varied with total molarity as in Fig. 135. This is direct chemical information of a kind which can hardly be obtained in any other way. It tells us that at no time does the concentration of SO_4^{2-} ions rise above 2*M*. At a total molarity of 14*M* virtually all the sulphur is present as HSO_4^- ions,

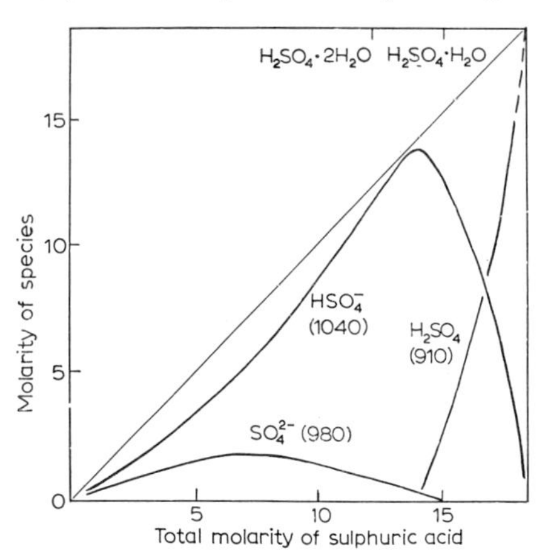

Fig. 135. Concentrations from Raman intensities of SO_4^{2-}, HSO_4^- and H_2SO_4 in sulphuric acid.

even though there are only about $1\frac{1}{2}$ water molecules present for each H^+ both to form H_3O^+ and to act as solvent. Above a total molarity of about 15.3*M* there is insufficient water for the completion of the process

$$H_2O + H_2SO_4 \rightleftharpoons H_3O^+ + HSO_4^-$$

and thus the cross-over of the curves for HSO_4^- and H_2SO_4 represents the change in the equilibrium as concentration increases.

References p. 328–329

Since the Raman intensities do in fact measure concentration directly they provide a means of investigating activity coefficients. For example in the equilibrium

$$HSO_4^- + H_2O \rightleftharpoons H_3O^+ + SO_4^{2-}$$

the thermodynamic equilibrium constant

$$K_2 = \frac{[H_3O^+][SO_4^{2-}]}{[HSO_4^-]} \cdot \frac{\gamma_{H^+}\gamma_{SO_4^{2-}}}{\gamma_{HSO_4^-}} \tag{5.21}$$

is known. The ratio of the concentrations is found directly from Raman measurements. Hence the ratio of activity coefficients may be determined and compared with the theory of ionic interaction.

Ionic association

The acid–base function discussed above is a particular example of ionic association. In other ionic associations between cation and anion there are additional difficulaties which can be seen from the following example.

Aqueous cupric perchlorate gives an absorption at about 2500 Å, attributed to the free (hydrated) cupric ion. The band is considerably enhanced and shifted to longer wavelengths by the presence of sulphate ions. An assignment to specific ionic interaction, as in the formation of an ion pair, is always more convincing if a new band appears. In this case no new band can be seen and it might be possible in principle to explain the observation as a medium effect on a single species whose extinction coefficient is affected by the environment. However, the change in extinction coefficient is from about 60 to over 200 mole^{-1} l cm^{-1} and it is more probable that there is an ion-pair band at about the same wavelength which obscures the free ion absorption. It is assumed that there is an equilibrium

$$CuSO_4 \rightleftharpoons Cu^{2+} + SO_4^{2-}$$

and a dissociation constant

$$K = \frac{[Cu^{2+}][SO_4^{2-}]}{[CuSO_4]} \cdot \gamma_\pm^2 \tag{5.22}$$

where $\gamma_\pm$ is the mean activity coefficient of the cation or anion. We may assume that $CuSO_4$ and Cu^{2+} are the only two species absorbing at the chosen wavelength with extinction coefficients ε_1 and ε_0 respectively.

If the ion pair is formed by mixing solutions of cupric perchlorate (assumed completely dissociated) to molarity a and sodium sulphate to molarity b and if it is further assumed that the additional non-absorbing ion pairs $NaSO_4^-$ and HSO_4^- are formed (in the presence of a small addition of perchloric acid to prevent hydrolysis) then:

$$K = \frac{(a-x)(b-x-y_1-y_2)}{x} \gamma_\pm^2 \tag{5.23}$$

where $x = [CuSO_4]$, $y_1 = [NaSO_4^-]$, $y_2 = [HSO_4^-]$. The optical density is:

$$\begin{aligned} D &= \varepsilon_1 xl + \varepsilon_0(a - x)l \\ &= \varepsilon_0 al + (\varepsilon_1 - \varepsilon_0)xl \end{aligned} \tag{5.24}$$

With some rearrangement it follows that

$$\frac{K}{(\varepsilon_1 - \varepsilon_0)\gamma_\pm^2} = \frac{abl}{D - \varepsilon_0 al} - \frac{b}{\varepsilon_1 - \varepsilon_0} + \frac{x + y_1 + y_2}{\varepsilon_1 - \varepsilon_0}\left(1 - \frac{a}{x}\right) \tag{5.25}$$

$\varepsilon_0 al$ is the measured optical density of a solution of cupric perchlorate, of molarity a, in which copper is entirely present as free ion. Alternatively, $D - \varepsilon_0 al$ is obtained directly by comparing the mixed solution with the perchlorate solution. If a and b are varied in such a way as to keep the ionic strength constant, the plot of $abl/(D - \varepsilon_0 al)$ against b has a limiting slope (at low b) of $1/(\varepsilon_1 - \varepsilon_0)$ and an intercept of $K/(\varepsilon_1 - \varepsilon_0)\gamma_\pm^2$. Uncertainty arises from the need to add sodium perchlorate in order to adjust the ionic strength and from the fact that y_1 and y_2 contribute to the ionic strength. $\gamma_\pm$ may be estimated from the Debye-Hückel equation, or a modification of it. Knowing ε_1, x may be calculated by (5.24) from measurements on solutions at various ionic strengths. The unknowns in (5.25) or (5.23) are now K, $\gamma_\pm$, y_1 and y_2. y_1 and y_2 are estimated by an approximation procedure, using known values of association constants. Various values of a parameter in the equation for the activity coefficient give various values of $\gamma_\pm$ for each ionic strength and hence a set of K values at different ionic strength for each chosen value of the parameter. The correct K is taken to be the one which remains constant with change in ionic strength. The value of K obtained by this method (at 25°C) is 4.7×10^{-3} mole l^{-1}.

An alternative approach to the same equilibrium is to use copper sulphate alone. In this case the dissociation constant can be written in terms of the degree of dissociation, α, thus:

$$K = \frac{\alpha^2 c}{1 - \alpha}\gamma_\pm^2 \tag{5.26}$$

where c is the total molar concentration of the copper sulphate. The apparent optical density at the chosen wavelength is

$$\begin{aligned} \varepsilon_{app} &= \varepsilon_0\alpha + \varepsilon_1(1 - \alpha) \\ &= \varepsilon_1 - (\varepsilon_1 - \varepsilon_0)\alpha \end{aligned} \tag{5.27}$$

As before, ε_1 is not known and therefore α cannot be calculated from the observed ε_{app}. The procedure adopted was to calculate $\gamma_\pm$ from three possible formulae and then to adjust K until the corresponding values of α (by 5.26) gave the best linear plot of ε_{app} against α. The interesting fact that emerges is that for each of the chosen formulae for $\gamma_\pm$ a set of K, ε_1 and ε_0 values could be found, each set equally consistent with the observations. They were (at 25°C):

References p. 328–329

d	4.3	10	14	Å
10^3K	8.0	4.0	3.5	mole l^{-1}
ε_1	353.9	222.2	205.4	mole^{-1} l cm^{-1}
ε_0	62.6	62.3	61.9	

The uncertainty in the dissociation constant is obvious. The first line gives d, the distance of closest approach of the free ions used in the formula for $\gamma_{\pm}$. It is a somewhat moveable boundary separating free ions from ion-pairs and, so far as the spectrophotometric data go, there is some degree of arbitrariness in defining the boundary. Since the choice of d in such models of ionic association is of some significance, it is interesting to compare conductimetric and cryoscopic values of K, each of which also depends upon the assumed value of d. The best agreement with the spectrophotometric K values is when $d = 10$ Å and $K = 3.8 \times 10^{-3}$ (cryoscopic, 0°C) and 4.0×10^{-3} mole l^{-1} (conductimetric, 25°C).

It is not always possible to measure the quantity $\varepsilon_0 al$ or $\varepsilon_0 \alpha$ corresponding to the unassociated cation since, especially in non-aqueous or mixed solvents of low dielectric constant, it is likely that all ions are associated in some form. Such is the case with the thiosulphates of sodium, potassium and magnesium in 50 % ethanol[16].

A more favourable example of quantitative measurement of ionic association is one in which the absorption due to the complex appears at a new maximum, well removed from the band due to the free ion. This is the case with the lead halide complexes, PbX^+, and, in the case of the bromide and iodide, further new bands appear due to the complex PbX_2. Values of the dissociation constant for

$$PbX^+ \rightleftharpoons Pb^{2+} + X^-$$

at 18°C in water and in 40 % aqueous methanol are:

	water	aqueous methanol
$PbCl^+$	2.57×10^{-2}	1.84×10^{-3}
$PbBr^+$	1.42×10^{-2}	0.96×10^{-3}
PbI^+	1.2×10^{-2}	1.2×10^{-3}

We have already referred to some spectroscopic studies on ionic association in connection with the qualitative evidence for such association. Some of the papers referred to include quantitative data. An important early paper which is relevant to this section and the next deals with the ion pairs $FeCl^{2+}$, $FeBr^{2+}$, and $Fe(OH)^{2+}$ as well as the higher associations involving a number of ligands.

Stoicheiometry of complex ions

A complex ion is formed by the attachment of one or more ligands A to a central cation M and we require to know which of the possible complexes MA_n is stable. Solutions of M and A of equal molarity, m, are mixed in a series of volume ratios $(1 - x) : x$ and the optical density at a particular wavelength is plotted against x. In the absence of any interaction

$$D_0 = \varepsilon_M(1 - x)m + \varepsilon_A xm \tag{5.28}$$

where ε_M and ε_A are the extinction coefficients of M and A respectively. If, on the other hand, there exists an equilibrium

$$M + nA \rightleftharpoons MA_n$$

and the equilibrium concentrations are c_M, c_A and c it follows that:

$$c_M = (1 - x)m - c$$

$$c_A = xm - nc \tag{5.29}$$

$$Kc = c_M c_A^n \tag{5.30}$$

The first two equations (5.29) are stoicheiometric requirements and (5.30) defines the stability constant of the complex.

If the extinction coefficient of the complex at the chosen wavelength is ε then

$$D = l(\varepsilon_M c_M + \varepsilon_A c_A + \varepsilon c) \tag{5.31}$$

and hence, introducing (5.28) (5.29) and (5.30),

$$\Delta = D - D_0 = l(\varepsilon - \varepsilon_M - n\varepsilon_A) \tag{5.32}$$

The measured quantity Δ is plotted against x and we wish to know what information can be gained from maxima or minima in the curve. It follows from (5.32) that, unless by chance $\varepsilon = \varepsilon_M + n\varepsilon_A$, $d\Delta/dx = 0$ when $dc/dx = 0$. Differentiation of (5.29) and (5.30) with respect to x and setting $dc/dx = 0$ gives $- dc_M/dx = dc_A/dx = m$ and

$$x_0 = \frac{n}{1 + n} \quad \text{or} \quad n = \frac{x_0}{1 - x_0} \tag{5.33}$$

where x_0 is the value of x at which $dc/dx = 0$.

D_0 may be calculated from the known values of ε_M and ε_A or Δ may be directly measured by comparing the absorption of the mixture in a 1 cm cell with that of the two solutions separately mixed with water in the same ratio in two 1 cm compartments of a double cell. Δ is plotted against x for a number of wavelengths in order to avoid the situation, which may occur at one wavelength, that $\varepsilon = \varepsilon_M + n\varepsilon_A$. A single maximum or minimum occurring in all the curves at a common value of x_0 indicates the formation of a single complex, in which $n = x_0/(1 - x_0)$. It can be shown by further differentiation that a minimum occurs when $\varepsilon < \varepsilon_M$ and a maximum when $\varepsilon > \varepsilon_M$.

An example[19] of this straightforward behaviour is given by potassium chromate and hydrochloric acid (Fig. 136). In this case $x_0 = 0.5$ and $n - 1$, a result which is in agreement with the equilibrium:

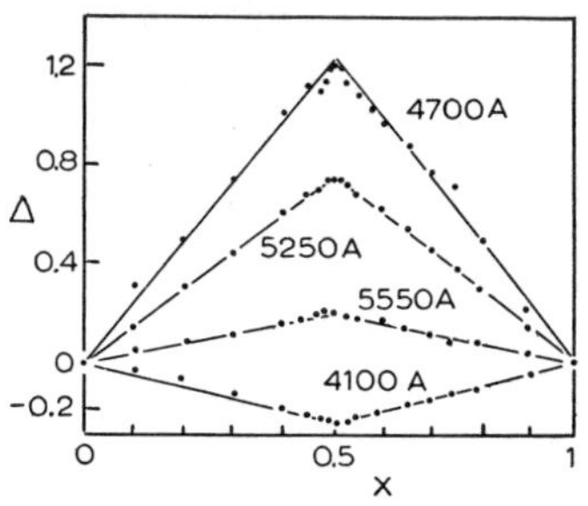

Fig. 136. Method of continuous variations. Variation of Δ with x when x ml of 0.1 M aqueous K_2CrO_4 is added to $(1-x)$ml of 0.1M aqueous HCl.

$$2CrO_4^{2-} + 2H_3O^+ \rightleftarrows 2HCrO_4^- + 2H_2O \rightleftharpoons Cr_2O_7^{2-} + 3H_2O$$

A similar example[20] of 1 : 1 complex (or ion pair) formation is shown by the occurence of maxima at $x_0 = 0.5$ on mixing uranyl perchlorate with sulphuric acid and with nitric acid:

$$UO_2^{2+} + HSO_4^- + H_2O \rightleftharpoons UO_2SO_4 + H_3O^+$$
$$UO_2^{2+} + NO_3^- \rightleftharpoons UO_2NO_3^+$$

In the first example the position of the maximum is equally compatible with the formation of $UO_2HSO_4^+$ but this is excluded by the dependence upon pH of the value of Δ which, from (5.33), is a measure of the concentration of the complex.

This so-called method of continuous variations was first described[21] in 1928 and has since been extended[18] to include cases where $n > 1$ and where more than one species is present. Curves of Δ against x for different wavelengths may show maxima at different values of x. Thus nickel sulphate and ethylenediamine give $x_0 = 0.483$ at 6220 Å, 0.67 at 5780 Å and 0.76 at 5300 and 5450 Å. These are sufficiently close to 1/2, 2/3 and 3/4 to indicate the complexes $[Ni\ en]^{2+}$, $[Ni\ en_2]^{2+}$ and $[Ni\ en_3]^{2+}$. Similar results are obtained for Ni^{2+} and *o*-phenanthroline. For cupric ions and ammonia there is evidence for $[Cu(NH_3)_2]^{2+}$ as well as the more familiar $[Cu(NH_3)_4]^{2+}$.

Stability constants

The more formidable problem presented by complex formation is that of stability constants. Basically the problem is the same as before, how to determine extinction coefficients of species which are present only in solution along with a number of other similar species also with unknown extinction coefficients. If a complex is sufficiently stable and its formation is virtually complete at high ligand concentration it is possible to find its extinction coefficient or to determine it by a short extrapolation. However, it is not uncommon for as many as six ligands to attach themselves successively to the central atom or successively to replace another. Methods of handling complicated systems of this type have been described[22–24]. If a metal ion M forms a series of complexes with a ligand A

$$\mathrm{M} + \mathrm{A} \rightleftharpoons \mathrm{MA}$$
$$\mathrm{MA} + \mathrm{A} \rightleftharpoons \mathrm{MA_2}$$
$$\cdots\cdots\cdots$$

we may define a set of stability constants for each complex

$$\beta_1 = \frac{[\mathrm{MA}]}{[\mathrm{M}][\mathrm{A}]}; \quad \beta_2 = \frac{[\mathrm{MA_2}]}{[\mathrm{M}][\mathrm{A}]^2}; \quad \cdots\cdots \tag{5.34}$$

If we measure absorption at a wavelength at which the metal ion and the complexes absorb ($\varepsilon_0, \varepsilon_1, \varepsilon_2 \ldots$ for M, MA, MA_2) but not the free ligand A, then the optical density is given by

$$\begin{aligned} D/l &= \varepsilon_0[\mathrm{M}] + \varepsilon_1[\mathrm{MA}] + \varepsilon_2[\mathrm{MA_2}] + \ldots + \varepsilon_n[\mathrm{MA_n}] \\ &= [\mathrm{M}]\{\varepsilon_0 + \varepsilon_1\beta_1[\mathrm{A}] + \varepsilon_2\beta_2[\mathrm{A}]^2 + \ldots + \varepsilon_n\beta_n[\mathrm{A}]^n\} \end{aligned} \tag{5.35}$$

The total concentration of metal c_M is given by

$$\begin{aligned} c_M &= [\mathrm{M}] + [\mathrm{MA}] + [\mathrm{MA_2}] + \ldots + [\mathrm{MA_n}] \\ &= [\mathrm{M}]\{1 + \beta_1[\mathrm{A}] + \beta_2[\mathrm{A}]^2 + \ldots + \beta_n[\mathrm{A}^n]\} \end{aligned} \tag{5.36}$$

If the mean, or measured, optical density is defined as

$$\bar{\varepsilon} - D/c_M \cdot l \tag{5.37}$$

it follows that:

$$\bar{\varepsilon} = \frac{\varepsilon_0 + \varepsilon_1\beta_1[\mathrm{A}] + \varepsilon_2\beta_2[\mathrm{A}]^2 + \ldots + \varepsilon_n\beta_n[\mathrm{A}]^n}{1 + \beta_1[\mathrm{A}] + \beta_2[\mathrm{A}]^2 + \ldots + \beta_n[\mathrm{A}]^n} \tag{5.38}$$

Defining extinction coefficient differences

$$\Delta\bar{\varepsilon} = \bar{\varepsilon} - \varepsilon_0, \qquad \Delta\varepsilon_1 = \varepsilon_1 - \varepsilon_0, \qquad \Delta\varepsilon_2 = \varepsilon_2 - \varepsilon_0, \ldots \tag{5.39}$$

then:

$$\Delta\bar{\varepsilon} = \frac{\Delta\varepsilon_1\beta_1[\mathrm{A}] + \Delta\varepsilon_2\beta_2[\mathrm{A}]^2 + \ldots + \Delta\varepsilon_n\beta_n[\mathrm{A}]^n}{1 + \beta_1[\mathrm{A}] + \beta_2[\mathrm{A}]^2 + \ldots + \beta_n[\mathrm{A}]^n} \tag{5.40}$$

Evidently the quantity $\Delta\bar{\varepsilon}/[\mathrm{A}]$ which can be measured for a number of solutions with different ligand concentrations tends to $\Delta\varepsilon_1\beta_1$ as $[\mathrm{A}] \to 0$. The process can be extended as follows:

$$\begin{aligned} &\text{put } f_1 = \Delta\bar{\varepsilon}/[\mathrm{A}], && \lim f_1 = a_1 = \Delta\varepsilon_1\beta_1 \\ &\text{put } f_2 = (f_1 - a_1)/[\mathrm{A}], && \lim f_2 = a_2 = \Delta\varepsilon_2\beta_2 - a_1\beta_1 \\ &\text{put } f_3 = (f_2 - a_2)/[\mathrm{A}], && \lim f_3 = a_3 = \Delta\varepsilon_3\beta_3 - a_1\beta_2 - a_2\beta_1 \\ &\text{put } f_4 = (f_3 - a_3)/[\mathrm{A}], && \lim f_4 = a_4 = \Delta\varepsilon_4\beta_4 - a_1\beta_3 - a_2\beta_2 - a_3\beta_1 \end{aligned} \tag{5.41}$$

and so on. The extrapolation in each case is to $[\mathrm{A}] = 0$.

Now let $y = 1/[\mathrm{A}]$. Substituting in (5.40) and multiplying numerator and denominator by y^n gives

$$\Delta\varepsilon = \frac{\Delta\varepsilon_n\beta_n + \Delta\varepsilon_{n-1}\beta_{n-1}y + \ldots + \Delta\varepsilon_1\beta_1 y^{n-1}}{\beta_n + \beta_{n-1}y + \ldots + \beta_1 y^{n-1} + y^n} \tag{5.42}$$

References p. 328–329

and extrapolation procedures can be devised with (5.42) as they were for (5.40). Thus:

$$\lim \Delta\bar{\varepsilon} = b_1 = \Delta\varepsilon_n$$

Now put

$$g_2 = (\Delta\bar{\varepsilon} - b_1)/y \quad \lim g_2 = b_2 = (\Delta\varepsilon_{n-1} - \Delta\varepsilon_n)\beta_{n-1}/\beta_n \tag{5.43}$$

and so on. The extrapolation here is to $y = 0$.

The operation of these extrapolations can now be illustrated with the simple system in which two complexes are formed, *viz.* MA and MA_2. It might be applied, for example, to $HgCl^+$ and $HgCl_2$: $PbCl^+$ and $PbCl_2$. D is measured for a number of solutions in which c_M is constant and [A] is varied. $\bar{\varepsilon}$ is evaluated, ε_M is known from absorption in the absence of A, and hence $\Delta\bar{\varepsilon}$. $\Delta\bar{\varepsilon}$ is plotted against y and f_1 against [A]; extrapolations to zero in each case give b_1 and a_1. g_2 and f_2 may now be evaluated and plotted against y and [A] respectively, extrapolation to zero giving b_2 and a_2. It then follows from (5.41) and (5.43) that:

$$\beta_1 = \frac{a_1 b_1 - a_2 b_2}{a_1 b_2 + b_1^2}; \quad \beta_2 = \frac{a_2 b_1 + a_1^2}{a_1 b_2 + b_1^2} \tag{5.44}$$

and

$$\Delta\varepsilon_1 = a_1/\beta_1; \qquad \Delta\varepsilon_2 = b_1 \tag{5.45}$$

Spectrophotometric determinations of stability constants are based upon methods of extrapolation such as these. It depends upon the complexity of the system how much of the formal treatment is needed. An example[25] of its use is the evaluation of stability constants of chromium acetate complexes. Tables of stability constants of complexes are available[22–24, 26]: they quote values obtained by all methods, including spectrophotometric.

Kinetic Applications

The concentration dependence of absorption intensities has obvious analytical application in the study of reaction rates. It may be that the reaction is slow enough or can be quenched so that samples can be withdrawn and analysed at leisure. Absorptiometry may play its part in such analysis but no new features are involved. Alternatively the reaction may be too fast for this procedure. In this case spectrophotometry permits the change in concentration of reactants to be followed *in situ*. There are many ways in which this can be done and many chemical reactions to which the technique has been applied: only a few examples need be mentioned here.

Fast reactions in solution are often followed by the tube flow method originally described by Hartridge and Roughton[27] to follow the reaction between oxygen and haemoglobin. Two solutions of reactants are driven by syringes down separate tubes into a special mixing chamber and the mixed fluid then flows down an observation tube. At various distances down this tube, representing different times after mixing, the concentration of a reactant or a product may be measured by various methods, including spectrophotometry. Alternatively[28] the flow may be stopped suddenly and the subsequent changes observed continuously at one position. The same object of varying the reaction time while observing at one position is achieved by accelerated flow. Systems which have been treated in this way include[29] the catalytic decomposition of hydrogen peroxide with catalase and peroxidase; the decomposition of oxyhaemoglobin (optical density measured at 4300 Å); the proton transfer reaction between trinitrotoluene and ethoxide ion in ethanol; the rate of reaction of carbon monoxide with myoglobin.

An alternative *capacity flow* method[30] measures the steady state concentration of a reactant or product in a large reaction vessel into which reactants are flowing at a controlled rate. The rate of formation of the complex $FeS_2O_5^+$ has been measured in[31] this way and the extinction coefficient obtained for what is ordinarily a transient violet intermediate.

An early and classical application[32] of spectrophotometry in the determination of steady state concentrations is that illustrated in Fig. 137. The vessel contains iodine molecules whose concentration is obtained by meas-

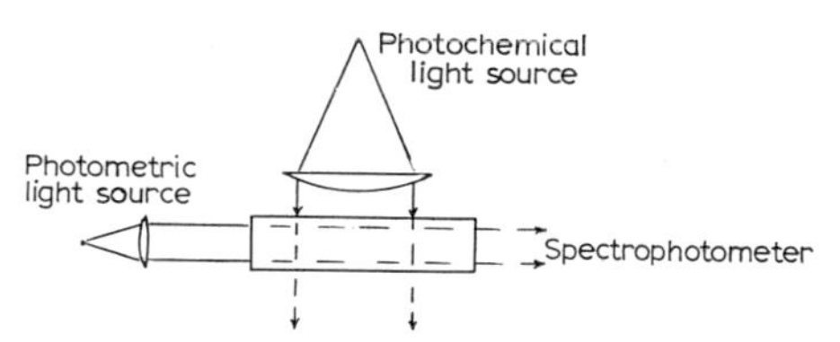

Fig. 137. Measurement of change of steady state concentration of undissociated iodine molecules on irradiation.

urement of optical density. When the photochemical irradiation is turned on iodine atoms are formed. The balance between rate of formation of iodine atoms and rate of recombination determines a new steady state concentration of iodine molecules, observed as a new optical density. From this experiment information was obtained about the effect of various inert gases on the rate of recombination of iodine atoms.

Other studies[33–34] which make ingenious use of photoelectric methods are those of the various reactions of the oxides of nitrogen which involve changes in concentration of NO_2. Fig. 138 shows a typical result of an experiment in which the concentration of NO_2 is decreasing. The light used to detect the NO_2, selected either by filters or by a quartz monochromator, is chopped by a rotating shutter which acts as a time marker. The decrease in the absorbing species NO_2, with time results in an increase in the photoelectric signal ob-

References p. 328–329

tained when the shutter is opened. The trace obtained on a cathode ray oscilloscope screen is as shown in Fig. 138 (photoelectric signal increases downward) and from it can be deduced the rate of change of NO_2 concentration.

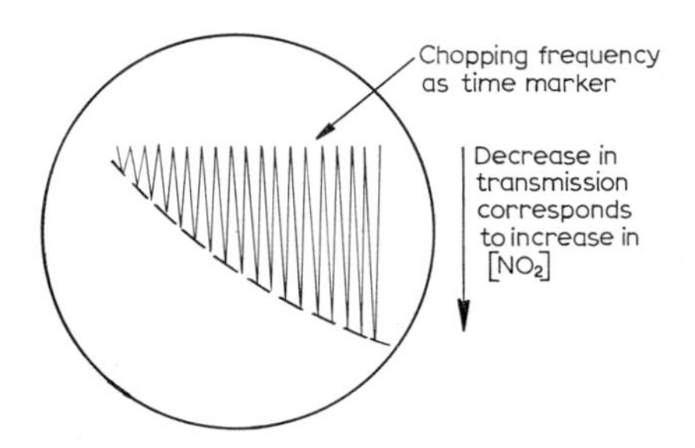

Fig. 138. CRO presentation of change of absorption with time due to increase in concentration of an absorping reaction product; *e.g.*, NO_2.

An important factor in all methods of following rapid changes in concentration by means of photoelectric or similar detectors is the rate of response of the detector. Thus the infra-red region has not been much used partly because thermal detectors cannot be made to respond to changes occurring in less than 10^{-2} sec: indeed most infra-red spectrometers use interrupted radiation (in order to amplify the signal) at 12—16 sec^{-1}. Photoconductive infra-red detectors can be used with an interruption frequency of 800—1000 sec^{-1} and thus reactions involving small changes in 10^{-2} sec should be measurable. Photoelectric cells have a much faster response and can be used in the ultra-violet and visible region. Photography is capable of recording exposures of the order 10—100 μsec provided the source intensity is sufficient, as is the case when a flash discharge is used. It need hardly be said that the time constant of the detector is of no importance in measurement of steady state concentrations.

Flash photolysis

An important kinetic technique which relies upon spectrophotometry is flash photolysis. It has not only yielded important new information about reaction mechanism: it has also enabled many free radicals to be identified and characterized spectroscopically[35]. Among these, for example, the methyl radical has been shown to be planar and the formyl radical, CHO, to be linear in the ground state but bent in excited states.

The advantage of flash photolysis can be seen against the background or normal photolysis. If acetone vapour at 50 mm pressure and 125°C is irradiated with the full light from a 100 watt mercury arc the steady state methyl radical concentration is about 2×10^{-12} mole cm^{-3} or a partial pressure of 5×10^{-5} mm. In the flash photolysis the procedure is to induce photochemical decomposition in a very short time. The discharge of energy of the order of 10,000 J through a gas discharge tube may last for 5×10^{-5} sec and radiate within that period 10^{21} quanta of useful ultra-violet light. The effect on an absorbing compound in an adjacent reaction vessel is almost complete dissociation into radicals which immediately begin to react or recombine. Thus momentarily the result of irradiating 50 mm of acetone is probably about 20 mm of methyl radicals and 20 mm of acetyl radicals. A second flash discharge of much lower power and flash duration can be made to provide sufficient light for photographing the absorption spectra of the

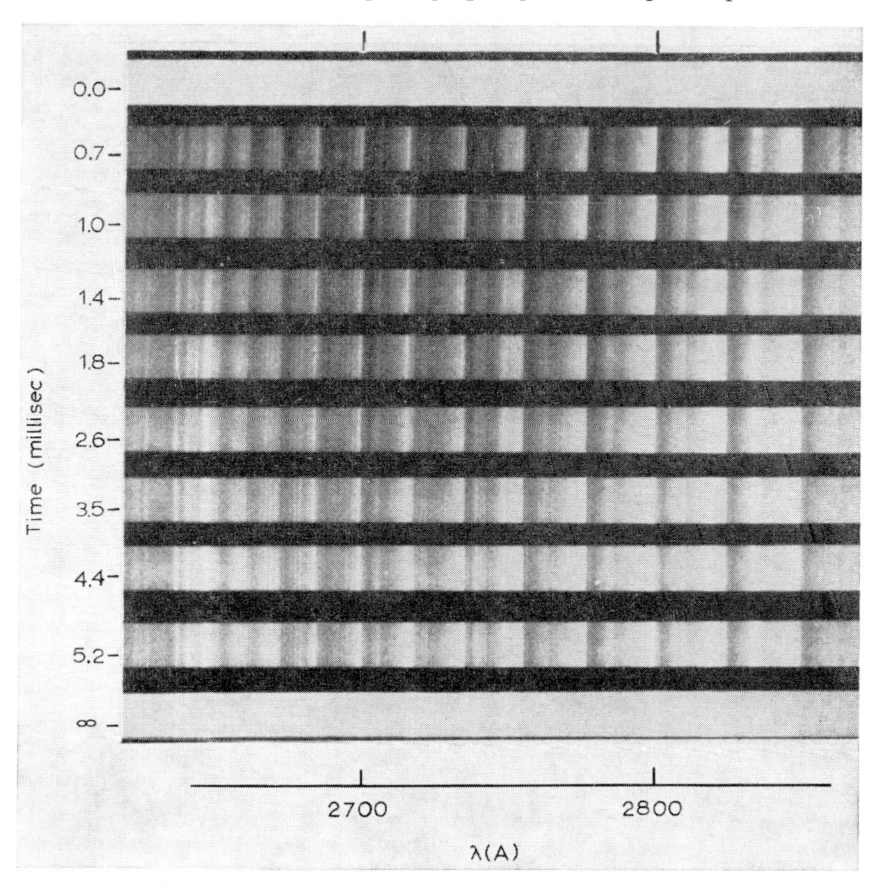

Fig. 139. Absorption spectrum of the ClO free radical at various times after flash irradiation of a chlorine–oxygen mixture.

References p. 328–329

radicals at a predetermined time after the photolysis discharge. In this way a series of absorption spectra is obtained of the reaction products at various times after the initiation. The spectra lead to information about structures and electronic states of the free radicals concerned: the rates of change of their intensities give detailed kinetic information.

One of the first examples of this technique was the detection of the ClO radical when mixtures of chlorine and oxygen were subjected to flash irradiation. There is no overall reaction but evidently a series of radical reactions such as:

$$Cl_2 + h\nu \rightarrow 2Cl$$
$$2Cl + O_2 \rightarrow 2ClO$$
$$2ClO \rightarrow Cl_2O_2 \rightarrow Cl_2 + O_2$$

It can be seen from Fig. 139 that the species responsible for absorption is at maximum concentration about 1 msec after the flash. From the simplicity of the spectrum the species is evidently a diatomic molecule and, in the circumstances, ClO is the only possibility. Fig. 140 shows the derived curve of the variation of ClO concentration with time. [ClO] must in this case be expressed as a product of a concentration term and the unknown extinction coefficient: that is to say, the optical density is taken as a measure of the concentration. Rate constants thus contain the unknown factor ε. Nevertheless the shape of the curve in Fig. 140 indicates a bimolecular disappearance of ClO and thus provides evidence as to mechanism. Fig. 140 shows analysis of the observed variations of [ClO] into a formation curve and a decay curve: it also shows on a different scale the variation with time of the intensity of the photochemical flash.

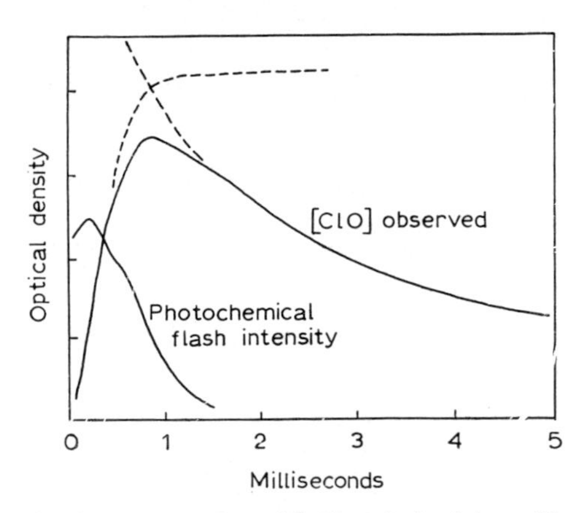

Fig. 140. Variation of ClO concentration with time derived from Fig. 139. The intensity of photochemical flash is on a different scale.

Where the absorption spectra of the radicals are required for their own sake photographic recording with a secondary spectro-flash is the only possibility. For purely kinetic purposes, however, it is an advantage to be able to use a photoelectric detector to follow the change in absorption. For this purpose a continuous low-intensity monochromatic source is used, set at an appropriate wavelength and a curve such as that in Fig. 140 is obtained on a cathode ray oscilloscope in one experiment. By this means the iodine recombination reaction has been studied[38] with results which confirm and extend those obtained by the steady state method described above.

Population of Levels

We have noted already (p. 81) the fact that there is a distribution of molecules among levels in a manner which depends upon the energy differences between the levels. Thus, if there are two levels separated by ΔE, the number of molecules n_i in the upper state i and the number n_j in the lower state j will be in the ratio:

$$\frac{n_i}{n_j} = \frac{g_i}{g_j}\, e^{-\Delta E/kT} \tag{5.46}$$

where g_i and g_j are the statistical weights or degeneracies of the two states. Equation (5.46) is the appropriate form of the Boltzmann distribution law. Table 21 gives numerical values of $e^{-\Delta E/kT}$ for various ΔE and various temperatures.

TABLE 21

NUMERICAL VALUES OF $e^{-\Delta E/kT}$ FOR VARIOUS ΔE AND VARIOUS TEMPERATURES

	N.m.r.	Rotational		Vibrational		Electronic
ΔE	0.001 (30 Mc/s)	1 (30,000 Mc/s)	10	100	1000	10,000 cm^{-1}
	1.24×10^{-7}	1.24×10^{-4}	1.24×10^{-3}	1.24×10^{-2}	0.124	1.24 *eV*
	2.86×10^{-6}	2.86×10^{-3}	2.86×10^{-2}	0.286	2.86	28.6 kcal/mole
1°K	0.99855	0.237	5.6×10^{-7}	2.9×10^{-63}	4×10^{-626}	10^{-6254}
10°K	0.99986	0.866	0.237	5.6×10^{-7}	2.9×10^{-63}	4×10^{-626}
100°K	0.999986	0.986	0.866	0.237	5.6×10^{-7}	2.9×10^{-63}
300°K	0.9999952	0.9952	0.953	0.619	8.2×10^{-3}	2.0×10^{-21}
1000°K	0.9999986	0.99855	0.986	0.866	0.237	5.6×10^{-7}

Note: $k = 1.38 \times 10^{-16}$ erg molecule^{-1} degree^{-1}
$= 0.696$ cm^{-1} molecule^{-1} degree^{-1}
$R = 1.98 \times 10^{-3}$ kcal mole^{-1} degree^{-1}

The relative population of levels given by (5.46) is for a system in thermal equilibrium. Electrical or other non-equilibrium excitation artificially and temporarily populates the upper levels: transitions reverting to the lower levels give rise to emission spectra. The Boltzmann distribution does not, therefore, have any direct influence on emission intensities. On the other hand, transitions in absorption spectra originate in levels whose populations are determined by (5.46). Hence absorption intensities are partly determined by the Boltzmann factor.

Nuclear spin Zeeman levels are the most closely spaced levels with which we are concerned in this book. $\Delta E/kT$ is of the order of 10^{-3} even at 1°K and is about 5×10^{-6} at room temperature. The relative population of, say, the two spin states of a proton in a magnetic field of 5000 gauss (30 Mc/s) at 100°K is 0.999986. In other words for every 2 million molecules there are 14 more in the lower state than in the upper. It is a very small proportion but nevertheless it amounts to an excess in the lower state of 3×10^{17} molecules in a gram molecule of material.

With so small a proportion of molecules in excess in the lower state there is a real possibility that the absorption in a sufficiently strong radiation field, may lead to *saturation*. In optical absorptiometry it is tacitly assumed that the excess population of the lower of two states involved in a transition is not significantly disturbed by absorption. In magnetic resonance the rate of absorption of quanta sufficient for the effect to be observed may be comparable with the rate at which the upper state can rid itself of its artificial excess of molecules.

Relaxation and resonance absorption

The process of relaxation to equilibrium after the distribution between two levels has been disturbed by absorption may be treated simply as follows. Suppose the upper level contains n_2 molecules and the lower contains n_1, and let the probability of a transition upwards be p_{12} and of transition downwards be p_{21}. At equilibrium

$$n_{2e}p_{21} = n_{1e}p_{12} \tag{5.47}$$

and

$$\frac{p_{21}}{p_{12}} = \frac{n_{1e}}{n_{2e}} = e^{\Delta E/kT} = 1 + \frac{\Delta E}{kT} \tag{5.48}$$

since $\Delta E/kT$ is very small. Away from equilibrium the rate of change of n_1 and n_2 is governed by the two probabilities, thus

$$-\frac{dn_2}{dt} = \frac{dn_1}{dt} = n_2 p_{21} - n_1 p_{12}$$

or

$$\begin{aligned}\frac{d}{dt}(n_1 - n_2) &= -(2n_1 p_{12} - 2n_2 p_{21}) \\ &= -(2n_1 p_{12} - 2n_{1e} p_{12} - 2n_2 p_{21} + 2n_{2e} p_{21})\end{aligned}$$

Or, putting p as a mean value for p_{12} and p_{21}, $\Delta n = n_1 - n_2$ and

$$\Delta n_e = n_{1e} - n_{2e}; \quad \frac{d\Delta n}{dt} = -2p(\Delta n - \Delta n_e) \tag{5.49}$$

Thus, the rate of return to equilibrium is proportional to the departure from equilibrium.

The solution to (5.49) is

$$\Delta n - \Delta n_e = (\Delta n - \Delta n_e)_0 \, e^{-t/\tau_1} \tag{5.50}$$

where

$$\tau_1 = \frac{1}{2p} \tag{5.51}$$

According to (5.50) the difference between the excess population and its equilibrium value is a maximum $(\Delta n - \Delta n_e)$ immediately after a perturbation, at $t = 0$, and is reduced by a factor e after a time $t = \tau_1$.

τ_1 is of particular importance in nuclear magnetic resonance. In that context it is called the *spin-lattice relaxation time* since it is a measure of the rate at which the excited spin system can relax to its thermal equilibrium, with other degrees of freedom, such as lattice vibration. τ_1 varies considerably depending upon the ease and mode of coupling between spin and other motion: in liquids it usually lies between 10^{-2} and 10^2 sec, increasing, for example, as viscosity decreases.

The process of resonance absorption may now be examined in more detail. In the presence of radiation of the appropriate frequency and density ρ there is an equal chance $P\rho$ of a molecule in the upper state being induced to emit or of a molecule in the lower state to absorb. The net absorption of energy is therefore $n_2 P\rho - n_1 P\rho = \Delta n P\rho$. Equation (5.49) now becomes

$$\frac{d\Delta n}{dt} = -\frac{(\Delta n - \Delta n_e)}{\tau_1} - 2\Delta n P\rho \tag{5.52}$$

and the steady state is given by $d\Delta n/dt = 0$ or

$$\Delta n = \frac{\Delta n_e}{1 + 2\tau_1 P\rho} \tag{5.53}$$

and the net absorption is

$$\Delta n P\rho = \frac{\Delta n_e P\rho}{1 + 2\tau_1 P\rho} \tag{5.54}$$

The net absorption is merely $\Delta n_e P\rho$ provided $2\tau_1 P\rho$ is small. If $2\tau_1 P\rho$ becomes large either because of a long spin-relaxation time or a strong radiation field (high intensity) then $\Delta n P\rho$ tends to $\Delta n_e/2\tau_1$: this is the limit of saturation in which the net absorption is determined by the rate of relaxation and is independent of $P\rho$. In that condition Lambert's law relating the attenuation to the thickness of the sample does not apply.

References p. 328–329

Rotational states

The energy differences between rotational states are larger than the Zeeman splittings involved in nuclear spin resonance, but they are still small enough to permit comparable populations of neighbouring levels at normal temperatures (Table 21). There are, however, many rotational levels and it is therefore appropriate to give the Boltzmann distribution over many states. If the energy of the Jth state is E_J and its statistical weight g_J then the number of molecules N_J of that state, relative to the total, is

$$\frac{N_J}{N} = \frac{g_J \, e^{-E_J/kT}}{\sum_J g_J \, e^{-E_J/kT}} \tag{5.55}$$

The denominator in this expression is the partition function: if the expressions $g_J = 2J + 1$ and $E_J = J(J + 1)hB$ (p. 187) are substituted it becomes the rotational partition function. When the summation is approximated by an integral it becomes kT/hB. Thus the distribution over rotational states is:

$$\frac{N_J}{N} = \frac{(2J + 1)hB}{kT} \, e^{-hJ(J+1)B/kT} \tag{5.56}$$

The Boltzmann exponential factor in (5.56) decreases with increasing J but the degeneracy $2J + 1$ increases. There is, in consequence, a maximum which can readily be shown to occur at

$$J_{\max} + \tfrac{1}{2} = \sqrt{\frac{kT}{2Bh}} \tag{5.57}$$

For HCl ($B = 10.32 \text{ cm}^{-1}$) at 300°K N_J/N has values, calculated from (5.56), indicated in Fig. 141. $hB/kT = 0.049$, so that $kT/2Bh = 3.2$ and by (5.57)

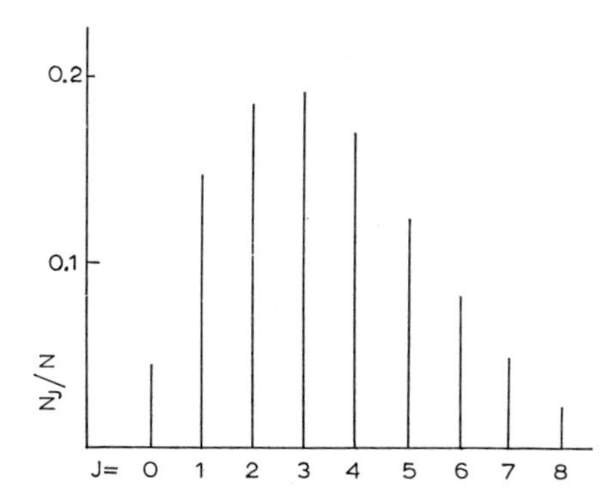

Fig. 141. Values of N_J/N for the first nine values of J for HCl at 300°K.

$J_{\max} = 2.7$. This is in agreement with the two largest values of N_J/N at $J = 2$ and 3 in Fig. 141.

In pure rotational spectra in the microwave region we are concerned with transitions between adjacent J levels whose relative populations are comparable. To determine the effect of the relative population upon intensity, we can evidently apply principles similar to those involved in deriving equation (5.54) but with slight modification. The statistical weights of the rotational levels are responsible for the fact that for small J the relative population *increases* with increasing energy. Thus the $J = 1$ level for HCl at 300°K has $N_J/N = 0.147$ while $J = 0$ has $N_J/N = 0.049$. Clearly if the probability of an upward transition were equal to that of a downward transition there could be no net absorption of energy. In fact, it can be shown that for the transition $J \to J + 1$ the probability (at unit radiation intensity) is $P/g_J = P/(2J + 1)$ and for $J + 1 \to J$ it is $P/g_{J+1} = P/(2J + 3)$: this favours absorption over emission by a factor of 3 for $0 \to 1$ transitions, and a factor of 5/3 for $1 \to 2$ transitions, and so on. Taking this factor into account there is always a net absorption from a lower level to a higher: *i.e.* PN_J/g_J is always larger than PN_{J+1}/g_{J+1}.

Saturation can occur in microwave absorption as it can in nuclear magnetic resonance. It depends upon the rate of decay of excess population in upper J states and the mechanism for decay in this case is by collisions between molecules. The excess rotational energy is thereby converted to kinetic energy. Since pure rotational spectra are generally observed at as low pressures as possible, with consequent reduction in the number of collisions, saturation is frequently observed, especially at high microwave intensities. From a practical point of view, the occurrence of saturation can affect quantitative analysis by microwave absorption. It can also set a lower limit to the width of observed microwave lines since there is a limit to the lowering of pressure before saturation occurs.

In the rotational structure of vibrational transitions the effect of population is different again. In this case the upper rotational levels are well removed from the lower and unless the vibrational frequency is particularly low the population of upper states may then be ignored. The probability of each transition being assumed to be the same the intensities of the lines in a rovibrational band should follow the same pattern as the distribution of molecules between levels. Thus the relative intensities in the resolved hydrogen chloride band should follow Fig. 141; this is seen to be the case on comparing Fig. 87 (p. 204).

In this context the effect of certain nuclear spin states in molecules containing symmetrically placed nuclei with $I = 0$ is to put $g = 0$ for odd J levels, *i.e.* those levels have no population. This explains the missing lines in the rovibrational spectrum of CO_2 (p. 205). Alternating intensities in other symmetrical molecules arise in a similar way, the nuclear spin states de-

References p. 328–329

termining different statistical weights for alternate rotational states.

It should be clear from Table 21 that changes in temperature have a marked effect upon spectra such as these where $e^{-\Delta E/kT}$ is not close to either 0 or 1. Naturally reduction of temperature favours the lower rotational levels. Where the $0 \rightarrow 1$ transition is sought a low temperature is used, not just in order to increase the population of $J = 0$ relative to $J = 1$ but also to increase the population of both relative to the higher levels.

Vibrational and electronic absorption spectra

The magnitude of ΔE in vibrational and electronic transitions is such that the population of the upper level may generally be ignored by comparison with the lower. Thus to a first approximation any population effect arises from distribution among lower state levels as, for example, in the intensity variation of rotational lines in rovibrational bands.

There are circumstances in which thermally excited vibrational levels are encountered and it is therefore necessary to take account of the population of the upper state in exact calculations from intensities of vibrational bands, especially those of lower frequency. Pure rotational spectra are sometimes observed for upper vibrational states. Difference bands and "hot" bands originate from excited vibrational states. $\omega_1 - \omega_2$ in the infra-red absorption of SO_2 (Fig. 66 p. 164) is an example of a difference band. "Hot" bands are often observed as distortions in unresolved infra-red bands or as extra structure in resolved rovibrational bands. They arise from transitions $v = 1 \rightarrow 2$ and in the harmonic approximation should occur at precisely the same frequency as $v = 0 \rightarrow 1$. Anharmonicity leads to a slight displacement. The transition is allowed by the selection rules equally with the fundamental and thus hot bands are only weak because of the low population of the $v = 1$ levels. For this reason it is not possible to observe sequences (p. 79) in absorption. In emission there is no such limitation.

Raman spectral transitions may originate in the states $v = 0$ or $v = 1$ (Fig. 38 p. 79). Those appearing on the long wavelength or low frequency side of the exciting line are the so-called Stokes lines and originate in the $v = 0$ levels. The anti-Stokes lines on the high frequency side of the exciting line are very much less intense than these Stokes lines since they originate in the less populated $v = 1$ levels.

Franck-Condon Principle

The relative intensities of molecular spectra, both in absorption and emission, may be interpreted in terms of a mechanical principle suggested by Franck, and subsequently put in quantum mechanical terms by Condon.

The elementary statement of the principle is that *movement of the nuclei is negligible during the time taken by an electric transition.* The idea is obviously closely related to the Born-Oppenheimer approximation (p. 76) in which the various motions of a molecule are considered to be separable.

Absorption

The application of the principle follows in a simple manner. Fig. 142 shows typical upper and lower vibronic levels of a diatomic molecule in which $r'_e = r''_e$, $r'_e > r''_e$ and $r'_e \gg r''_e$. Absorption spectra originate in $v'' = 0$, since we may assume the thermal population of other v'' levels to be low. An electronic transition in which the nuclei do not move is represented by a vertical line on the diagram. Then, from the Franck-Condon principle, the most probable transition from any particular initial state will be that represented by a vertical line raised from the most probable configuration in the initial state. Which vibrational level is the final state in the most probable transition depends upon the relative positions of the two potential energy curves: it will be that level for which the configuration probability density is a maximum at the same value of r as the most probable configuration of the initial state. Except in the level $v = 0$, the most probable con-

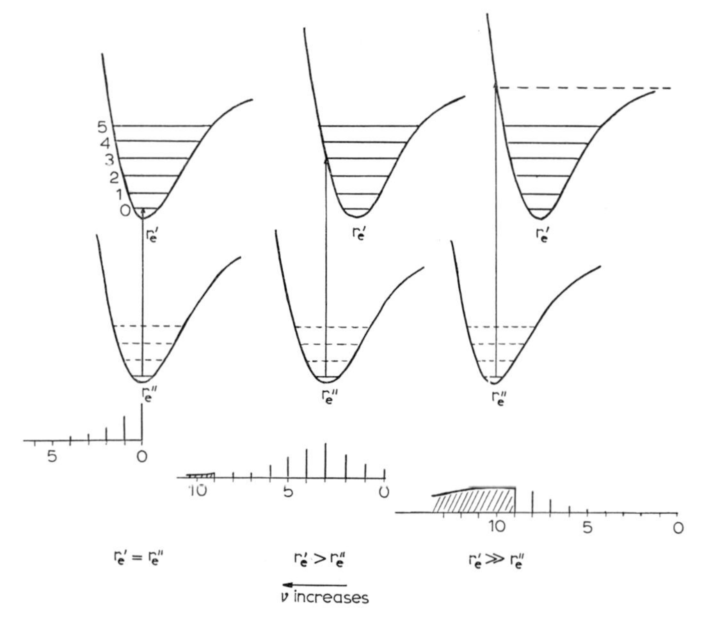

Fig. 142. Rovibronic transitions and absorption spectra illustrating the Franck-Condon principle. Only the most probable transition is shown in each case.

References p. 328–329

figuration approximates to the classical turning point; that is, to the value of r at which the horizontal energy level intersects the potential energy curve. Thus the level which corresponds to the point of intersection of the vertical line and the potential curve will be the final level in the most probable transition.

Vibronic transitions are not restricted to $\Delta v = \pm 1$ and a single progression is to be expected in the absorption spectrum. If $r'_e = r''_e$ the most prominent member of the progression is the first for which $v' = 0$, and the intensity falls off to higher frequency. For $r'_e > r''_e$ the most prominent band will be one for which $v' > 0$, say $v' = 3$, and the intensity is lower in bands at lower and higher frequencies. The situation depicted for $r'_e \gg r''_e$ is one in which the vertical line meets the upper potential energy curve at a point above the dissociation limit. Thus the maximum intensity occurs in the continuum. The first member of the progression is probably too weak to be observed and intensity increases towards the limit (for simplicity the convergence is not shown in Fig. 142).

Emission

In emission spectra of diatomic molecules the transitions may be from any vibrational level in the upper state to vibrational levels in the lower state. The variation of intensity between progressions based upon different v' will depend upon the mode of excitation. The upper vibrational levels may be fairly evenly excited, or there may be preferential excitation of one v' level. We will suppose that there are roughly the same number of molecules excited to each of the v' levels and illustrate the effect of the Franck-Condon principle on the relative intensity within each of the first four progressions (progression of v'' on $v' = 0$, 1, 2, and 3 in turn). Fig. 143 shows a typical disposition of potential energy curves. The progression originating in $v' = 0$ has its most intense band at $v'' = 1$ since the vertical line meets the lower curve at that level. For each of the remaining progressions, originating in levels for which $v' = 0$, we find the most probable transition by dropping perpendiculars from the ends of the horizontal lines representing the vibrational states. The justification for this is that the classical oscillator spends more time at the turning points than at intermediate positions. Maximum intensities therefore occur at two points in each progression, as shown.

If the pairs of numbers v'' at which maxima occur are plotted against v' as in Fig. 144 the points lie approximately on a parabola which is symmetrical about the diagonal $v' = v''$. This so-called *Condon parabola* is wider the greater the difference between r'_e and r''_e: when $r'_e = r''_e$ it degenerates into a straight line.

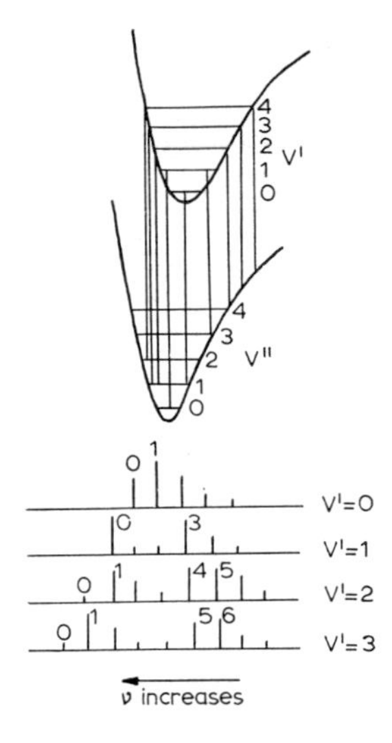

Fig. 143. Four progressions in an emission spectrum illustrating the Franck-Condon principle.

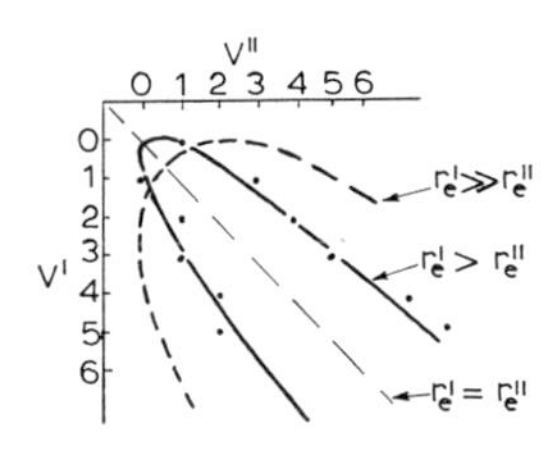

Fig. 144. Condon parabolas. The full curve corresponds to the situation in Fig. 143.

It may be noted that the occurrence of one maximum only in the $v' = 0$ progression is consistent with the quantum mechanical interpretation of vibrational states, wherein the lowest state is least like the classical vibration. For $v' = 0$ there is only one most probable configuration — when r is close to r'_e — and hence only one most probable transition.

Dissociation processes: photochemistry

The Franck-Condon principle assists qualitative discussion of dissociation processes, and is the basis of photochemistry. The situation shown in Fig. 142 in which $r'_e \gg r''_e$ is a simple initiation of photochemical dissociation. If the molecule is irradiated with light of frequency corresponding to the transition shown it will be immediately raised to the upper electronic state. The potential energy curve in that state is such that the nuclei experience strong repulsion at a value of $r' = r''_e$. Since the repulsion energy is greater than the dissociation limit the motion of the nuclei occurring after the electronic change will lead to dissociation of the excited molecule, the two fragments moving apart with kinetic energy equal to the excess of excitation energy over the dissociation energy. Essentially the same process occurs when the upper state has a wholly repulsive potential energy curve.

References p. 328–329

A large number of photochemical reactions are initiated by primary processes which are known to be direct dissociations. Notably it is in this way that the halogens dissociate and that the photosensitive reaction between chlorine and hydrogen is initiated. The two lowest lying excited states of bromine are illustrated in Fig. 52 (p. 130). The electronic term for the $^3\Pi_u$ state is 13814 cm^{-1} (relative to the ground state $^1\Sigma_g$): the corresponding wavelength is 7250 Å. Since the state has only a shallow minimum, wavelengths a little shorter than 7250 Å are sufficient to induce dissociation into normal unexcited atoms. Hydrogen, on the other hand, does not dissociate under irradiation even with quanta more than sufficient in energy to dissociate it. The mercury line at 2537 Å is sufficiently energetic (112.5 kcal/gram molecular quantum: $D(H_2) = 103.2$ kcal/mole) but hydrogen does not absorb at wavelengths above 1040 Å. The repulsive state which would dissociate into normal hydrogen atoms (Fig. 52) is a $^3\Sigma_u$ state and the prohibition against singlet-triplet transitions operates strictly in hydrogen where it does not in bromine. However, when hydrogen containing mercury vapour is irradiated with ultraviolet light at light at 2537 Å from a low pressure mercury arc the mercury vapour absorbs 112.5 kcal/mole and transfers the energy to the hydrogen molecule by collision. This process of dissociation is an example of *photosensitized decomposition.*

Excitation to an upper state may be insufficient to initiate decomposition (as in Fig. 142, case $r'_e > r''_e$). Then the molecule will normally lose energy by fluorescence. During the lifetime of the excited state (about 10^{-8} sec) it may suffer collision with another molecule with the following possible consequences: (i) degradation to thermal energy, (ii) the energy of collision may be just sufficient to make up that required for the molecule to dissociate, (iii) reaction with the collision partner, (iv) energy transfer to sensitize the decomposition of the collision partner.

A related phenomenon is *predissociation.* The name is given to the development of diffuse regions in the middle of discrete band systems. In absorption the diffuse region appears before the genuine continuum. One way in which this can occur is shown in Fig. 145. A transition A yields a discrete band but at higher frequencies a transition such as B to the state α gives the molecule such energy that on vibration it attains a configuration close to the configuration of the state β at the same energy. A radiationless jump to the repulsive state β leads to dissociation. So far as the absorption spectrum is concerned the effect of the coincidence of energy in states α and β at the same nuclear configuration is to broaden the vibrational levels in the state α, by mixing with the continuum of the state β. Transitions in the region of B are broadened and may appear quite diffuse. On the other hand the broadening may be slight, with the effect only showing up as an anomalous drop in

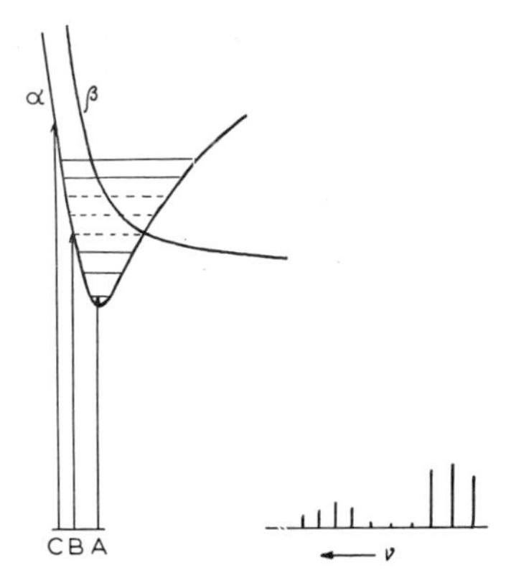

Fig. 145. An example of predissociation.

intensity. In any case with the situation illustrated in Fig. 145 higher frequency transitions such as C again show discrete vibrational bands. The Franck-Condon principle is equally applicable to radiationless transitions and there is clearly less chance of such a transition from the higher vibrational levels excited by the quantum C.

There are other situations in which predissociation can occur and in polyatomic molecules the opportunities for radiationless transitions are much greater. Some such process must be involved when an aldehyde or a ketone absorbs by a transition essentially assigned to the carbonyl group yet dissociates by breaking a C—C link.

Absolute Intensity

The remaining factor governing the intensity of an observed spectrum is the probability that a transition will occur. There are two possibilities envisaged in dipole interaction between a quantum system and a radiation field. They are: (a) spontaneous emission, the probability of the occurrence in unit time of a spontaneous transition from an upper state 2 to a lower state 1 with emission of radiation of wave number ω being

$$S_{21} = \frac{64\pi^4\omega^3}{3h} |R^{12}|^2$$

and (b) induced emission and absorption, the probability per molecule of a transition in unit time from state 1 to 2, or vice versa, in the presence of radiation of wavenumber ω and density $\rho(w)$ being

$$P_{12}\rho(\omega), \text{ where } P_{12} = P_{21} - \frac{8\pi^3}{3h^2c} |R^{12}|^2 \qquad (5.58)$$

References p. 328–329

S_{21} and P_{12} are the so-called *Einstein probabilities* for spontaneous emission and for induced emission and absorption (usually denoted A and B but these symbols have been used for integrated absorption intensities). $|R^{12}|$ is the *transition moment*, or the matrix element of the variable electric dipole moment, in respect of states 1 and 2.

Selection rules

We have given examples of the form which $|R^{12}|$ takes for dipole transitions between vibrational (p. 134) and rotational states (p. 188) and shown that selection rules follow in two ways. In the first place the symmetry of the two states 1 and 2 may ensure that $|R^{12}|$ is zero: in this case the transition is forbidden by a symmetry selection rule. Secondly, $|R^{12}|$ may be zero in vibrational states because the vibration is assumed to be either mechanically harmonic or electrically harmonic. The resulting harmonic oscillator selection rule is less rigorous than a symmetry selection rule.

For electronic states involving many non-interacting electrons it can be shown that the overall dipole transition moment for any conceivable transition is always zero except in those cases in which only one electron changes its state. One-electron transitions are subject to selection rules which are essentially based upon the quantum mechanical equivalent of conservation of angular momentum: $\Delta J = 0, \pm 1$ but $J = 0 \nleftrightarrow 0$; $\Delta M = 0, \pm 1$ but $\Delta M \neq 0$ for $\Delta J = 0$; and the frequently invoked rule that $\Delta S = 0$. The latter is the prohibition against change in multiplicity, against singlet-triplet transitions, which holds so long as resultant spin S can be defined. The proof of these rules depends upon the demonstration that the appropriate transition moment vanishes.

The rule that $\Delta J = 0, \pm 1$ is rigorous for dipole transitions in atoms and molecules. So also is the Laporte selection rule which forbids transitions from even to even states or from odd to odd states. Even and odd states are those which remain unchanged or change sign on reflection through a centre of symmetry (inversion). The transition moment for a dipole transition between two electronic states 1 and 2 contains integrals such as

$$\int \psi_1{}^* x \psi_2 \mathrm{d}x$$

where ψ_1 and ψ_2 are atomic or molecular orbitals. If ψ_1 and ψ_2 are centrosymmetric and are either both even or both odd their product is even and the integrand (in view of the factor x) is odd with respect to the inversion, when $-x, -y, -z$ is substituted for $+x, +y, +z$. Since a change of coordinates cannot change the value of the integral that value must be zero when the integrand is odd. Thus the transition moment is zero for transitions even → even, odd → odd, but not for even → odd. In particular, transitions between terms of the same electron configuration (p. 116) are not allowed. Octahedral $d-d$ transitions are also forbidden but the rule is evidently transgressed in complex ions (p. 266).

Absolute intensity in absorption

In this chapter and in Chapter I we have used two apparently contradictory measures of intrinsic absorption intensity. One is the extinction coefficient, ε, which has dimensions (concentration × length)$^{-1}$ and the other is the integrated band intensity A (equation 5.6) having the dimensions of extinction coefficient × wavenumber. There is no objection to this provided the two quantities are recognized as alternative, but not interchangeable, measures of intensity. The choice between the two when they are to be used relatively is governed by convenience, by instrumental factors, and ultimately by empirical tests of the sufficient validity of equations such as (5.1), (5.5), or (5.6). However, in relating experimental quantites to transition probabilities it is evident that there is no arbitrary choice and that the integrated band intensity is the appropriate quantity. This may be obvious but it is as well to examine the point in a little more detail since at least one monograph derives a relationship in terms of ε and then, without any change in dimensions, replaces ε by $\int \varepsilon \mathrm{d}\omega$.

Hitherto it has been assumed that it is in principle possible to measure an intensity of transmitted energy $I(\omega)$ at a wavenumber, ω. Not only does instrument resolution generally prevent this but so also does the finite width of the spectral line. Various factors contribute to line broadening (p. 280) by reducing the lifetime, and therefore increasing the uncertainty in the energy, of the upper state. In the last resort the natural line width arises from spontaneous and induced radiative degradation of the excited state and in this case there are theoretical reasons[39] for assuming that the intensity function has the Lorentz shape:

$$I_{\text{abs}}(\omega) = \frac{1}{\omega_0^2} \cdot \frac{a}{(\omega - \omega_0)^2 + b^2} \tag{5.59}$$

where maximum absorption occurs at $\omega = \omega_0$ with intensity $a/\omega_0 b^2$ and the width of the line at half maximum intensity is $2b$.

It is clear, therefore, that at any wavenumber ω, the measured quantity is $I(\omega)\delta\omega$, where $\delta\omega$ is a wavenumber interval which may be assumed small enough in what follows for the effect of absorbing path length to be constant over the range $\delta\omega$. We now follow an argument somewhat similar to that used on p. 293. It is assumed that the light incident on a sample has intensity

I_0, constant over the whole line or band. In the wavenumber range $\delta\omega$ the intensity at distance x through the absorbing sample is $I(\omega, x)\delta\omega$ and at distance $x + \mathrm{d}x$ it is $I(\omega, x + \mathrm{d}x)\delta\omega$. The change in intensity is

$$- \mathrm{d}I(\omega, x)\delta\omega = [I(\omega, x + \mathrm{d}x) - I(\omega, x)]\delta\omega \tag{5.60}$$

Now if the radiation density is $\rho(\omega, x)$ and the difference in population (molecules cm^{-3}) of the two states of interest is Δn_{12} the net rate of transitions in unit volume over the range $\delta\omega$ is $\rho(\omega, x)P_{12}\Delta n_{12}\delta\omega/\Delta\omega$ where $\delta\omega/\Delta\omega$ divides the $P_{12}\Delta n$ for the whole band of width $\Delta\omega$ into equal contributions. Each transition reduces the energy by $hc\omega_{12}$ (where ω_{12} is the central absorption wavenumber and thus an average for the whole band, assuming $\omega_{12} \gg \Delta\omega$) and thus in unit cross section over a path length $\mathrm{d}x$

$$- \mathrm{d}I(\omega, x)\delta\omega = hc\omega_{12}\rho(\omega, x)P_{12}\Delta n_{12}(\delta\omega/\Delta\omega)\mathrm{d}x \tag{5.61}$$

Since the flux intensity, I, and the radiation density are related by $I = c\rho$ (5.61) becomes

$$- \mathrm{d}I(\omega, x)\delta\omega/I(\omega, x) = h\omega_{12}P_{12}\Delta n_{12}(\delta\omega/\Delta\omega)\mathrm{d}x \tag{5.62}$$

Integrating between $x = 0$ and 1 and $I = I_0$ and $I(\omega, l)$ we have

$$\ln [I_0/I(\omega, l)]\delta\omega = 2.303\varepsilon(\omega)Cl \cdot \delta\omega = h\omega_{12}P_{12}\Delta n_{12}l\delta\omega/\Delta\omega \tag{5.63}$$

where $\varepsilon(\omega)$ is the molar decadic extinction coefficient and C is the molar concentration. Making $\delta\omega$ infinitesimal, integrating over the whole band, and using (5.58):

$$A = 2.303 \int \varepsilon(\omega)\mathrm{d}\omega = \frac{\Delta n_{12}}{C} h\omega_{12}P_{12} = h\omega_{12}P_{12}10^{-3}N = 10^{-3}N \frac{8\pi^3}{3hc} \omega_{12}|R^{12}|^2 \tag{5.64}$$

if N is Avogadro's number and it can be assumed that essentially all molecules are in the ground state. Thus the integrated absorption intensity is related to the transition moment, or matrix element, for the two states 1 and 2.

For a gas it is customary to use $k(\omega)$, defined for a gas at a pressure of 1 atm at 0°C ($C = 1/22.4$) as

$$k(\omega) = \frac{1}{l} \ln\left(\frac{I_0}{I}\right) = \frac{2.303}{22.4} \varepsilon(\omega) = 0.1028\varepsilon(\omega) \tag{5.65}$$

Then

$$\int k(\omega)\,\mathrm{d}\omega = h\omega_{12} P_{12} \Delta n_{12} = \frac{8\pi^3\Delta n_{12}}{3hc} \omega_{12} |R^{12}|^2 \tag{5.66}$$

When the measured intensity is of a complete rovibrational band of a gaseous molecule the right hand side of (5.66) has to be summed over all the rotational transitions taking due account of Δn for each transition. The result is found to be

$$\int k(\omega)\,\mathrm{d}\omega = \frac{n\pi}{3c} \left[\left(\frac{\partial\mu_x}{\partial q}\right)^2 + \left(\frac{\partial\mu_y}{\partial q}\right)^2 + \left(\frac{\partial\mu_z}{\partial q}\right)^2\right] \tag{5.67}$$

where n is the total number of molecules per unit volume at 0°C and 1 atm pressure and the differentials are the rates of change of dipole moment in the three directions x, y and z with respect to the normal coordinate of the particular vibration considered. In a sufficiently symmetrical molecule only one of the derivatives differs from zero and is therefore calculable by (5.67).

If the dipole moment of a molecule can be expressed as the vector sum of bond moments, then the $\partial\mu/\partial q$ can be expressed in terms of bond moments and the rate of change of those moments with bond extension or change in bond angle. For further discussion of these and related matters see Coulson[40].

An alternative mode of expressing experimental intensities is in terms of *oscillator strength or f-value.* This quantity comes from dispersion theory and it can be shown that

$$f = \frac{8\pi^2\omega mc}{3he^2}\,|R^{12}|^2 = \frac{mc^2}{\pi e^2 \Delta n_{12}} \int k(\omega)\mathrm{d}\omega$$
$$= 4.20 \times 10^{-8} \int k(\omega)\,\mathrm{d}\omega = 4.32 \times 10^{-9} \int \gamma\varepsilon(\omega)\mathrm{d}\omega \qquad (5.68)$$

Δn_{12} is, for electronic transitions, usually the total number of molecules per cm^3. e and m are the charge and mass of the electron and γ is a factor (of the order of $\frac{1}{2}$) introduced to allow the use of data from liquids and solutions, where the experimental quantity is usually $\varepsilon(\omega)$.

It can further be shown that, over all transitions, $\Sigma f = Z$, the total number of electrons. For transitions of a single electron $\Sigma f = 1$ and therefore the value of f for the strongest transition may approach 1. In that case $\int \varepsilon(\omega)\mathrm{d}\omega$ may approach 5×10^8. The corresponding value of ε_{max} depends upon the band width (equation 5.6): if $\Delta\omega_{\frac{1}{2}} \sim 1000$ cm^{-1}, $\varepsilon_{max} \sim 3 \times 10^5$.

The characteristic feature of charge-transfer spectra is their great intensity. If, as a simple approximation, a transition involves a charge redistribution amounting to one electron moving half an internuclear distance the transition moment is effectively $\frac{1}{2}er$. Then $f = 2\pi^2 mc\omega r^2/3h$ and is greater the higher ω and the larger r. For a typical distance of 1 Å and a typical frequency of 50,000 cm^{-1} (2000 Å) $f \sim 0.075$. The N—V transition ($^3\Sigma_g^- - {}^3\Sigma_u^-$) of the oxygen molecule gives rise to the first absorption by oxygen at 1800—1900 Å and may be regarded as a charge transfer from the ground state to the antibonding state represented as a hybrid of O^+O^- and O^-O^+. An approximate calculated f value is 0.44 and the observed is 0.193. The corresponding ε_{max} is about 5×10^3.

For further discussion see Mulliken[41,42].

Empirical correlations

It has been common practice in ultraviolet spectrophotometry to relate molecular structure not only to wavelengths of absorption maxima but also

References p. 328–329

to their maximum molar extinction coefficients. The correlation between structure and intensity has been well recognized. It is only comparatively recently that correlations between molecular structure and intensity have been sought in the infra-red. The intensity of carbonyl bands has been used[43] to characterize steroids, and many other examples are known in which the intensity of a band associated with a particular group is affected by the environment of the group in the molecule. Changes of band intensity have also been correlated with change of solvent[44]. However, it cannot yet be said that the influence either of solvent or of structure on absorption intensities is well-understood: it presents one of the outstanding problems[45] in chemical spectroscopy.

REFERENCES

1 A. WALSH, *Spectrochim. Acta*, 7 (1955) 108.
2 S. BRODERSEN, *J. Opt. Soc. Am.*, 44 (1954) 22.
3 J. B. WILLIS, *Australian J. Sci. Research*, 4 A (1951) 173.
4 W. R. BRODE, *J. Am. Chem. Soc.*, 46 (1924) 581.
5 A. I. BIGGS, *Trans. Faraday Soc.*, 50 (1954) 800.
6 R. A. ROBINSON in W. J. HAMER (Ed.), *The Structure of Electrolyte Solutions*, Wiley, 1959, p. 253.
7 L. A. FLEXSER, L. P. HAMMETT and A. DINGWALL, *J. Am. Chem. Soc.*, 57 (1935) 2103.
8 R. S. STEARNS and G. W. WHELAND, *J. Am. Chem. Soc.*, 69 (1947) 2025.
9 J. CHEDIN, *Ann. chim. (Paris)*, [11] 8 (1937) 243.
10 O. REDLICH, *Z, physik. Chem.*, A 182 (1938) 42.
11 O. REDLICH and J. BIGELEISEN, *J. Am. Chem. Soc.*, 65 (1943) 1883.
12 T. F. YOUNG, L. F. MARANVILLE and H. M. SMITH, in W. J. HAMER (Ed.), *The Structure of Electrolyte Solutions*, Wiley, 1959, p. 35.
13 T. F. YOUNG, *Record Chem. Progr. (Kresge-Hooker Sci. Lib.)*, 12 (1951) 81.
14 W. D. BALE, E. W. DAVIES and C. B. MONK, *Trans. Faraday Soc.*, 52 (1956) 816.
15 W. G. DAVIES, R. J. OTTER and J. E. PRUE, *Discussions Faraday Soc.*, 24 (1957) 103.
16 G. O. THOMAS and C. B. MONK, *Trans. Faraday Soc.*, 52 (1956) 685.
17 A. I. BIGGS, M. H. PANKHURST and H. N. PARTON, *Trans. Faraday Soc.*, 51 (1955) 802, 806.
18 E. RABINOWITCH and W. H. STOCKMAYER, *J. Am. Chem. Soc.*, 64 (1942) 335.
19 W. C. VORSBURGH and G. R. COOPER, *J. Am. Chem. Soc.*, 63 (1941) 437.
20 R. H. BETTS and R. K. MICHELS, *J. Chem. Soc.*, (1949) S 286.
21 P. JOB, *Ann. chim. (Paris)*, [10] 9 (1928) 113.
22 K. B. YATSIMIRSKII and V. P. VASIL'EV, *Instability Constants of Complex Compounds*, Pergamon Press, 1960.
23 A. K. BABKO, *Fiziko-khimicheskiĭ analiz kompleksnykh soedineniĭ v rastvorakh*, (*Physico-chemical Analysis of Complex Compounds in Solutions*), Kiev: Izdatel. Akad. Nauk Ukr. S.S.R, 1955.
24 K. B. YATSIMIRSKII, *Zhur. Neorg. Khim.*, 1 (1956) 2306.
25 K. B. YATSIMIRSKII and T. I. FEDOROVA, *Zhur. Neorg. Kjhim.*, 1 (1956) 2301;
26 *Stability Constants, Pt. 1, Organic Ligands*, Chem. Soc. (Spec. Publ.), 1957; *Pt. 2, Inorganic Ligands*, Chem. Soc. (Spec. Publ.), 1958.
27 H. HARTRIDGE and F. J. W. ROUGHTON, *Proc. Roy. Soc. (London)*, B 94 (1923) 336.
28 F. J. W. ROUGHTON and B. CHANCE, in S. F. FRIESS and A. WEISSBERGER (Eds.), *Investigation of Rates and Mechanisms of Chemical Reactions*, Interscience, 1953.

[29] *Discussions Faraday Soc.*, 17 (1954) 120—139.
[30] B. Stead, F. M. Page and K. G. Denbigh, *Discussions Faraday Soc.*, 2 (1947) 263.
[31] F. M. Page, *Trans. Faraday Soc.*, 49 (1953) 635.
[32] E. Rabinowitch and W. C. Wood, *J. Chem. Phys.*, 4 (1936) 497.
[33] H. L. Johnston and D. M. Yost, *J. Chem. Phys.*, 17 (1949) 386.
[34] H. L. Johnston, *Discussions Faraday Soc.*, 17 (1954) 14.
[35] D. A. Ramsay, *Advances in Spectroscopy*, Vol. I, Interscience, 1959, p. 1.
[36] G. Porter, *Proc. Roy. Soc. (London)*, A 200 (1950) 284; *Discussions Faraday Soc.*, 9 (1950) 60.
[37] G. Porter and F. J. Wright, *Discussions Faraday Soc.*, 14 (1953) 23.
[38] K. E. Russell and J. Simons, *Proc. Roy. Soc. (London)*, A 217 (1953) 271.
[39] W. Heitler, *Quantum Theory of Radiation*, 2nd ed., Oxford Univ. Press, 1944, p. 118.
[40] C. A. Coulson, *Spectrochim. Acta*, 14 (1959) 161.
[41] R. S. Mulliken, *J. Chem. Phys.*, 7 (1939) 1420.
[42] R. S. Mulliken and C. A. Rieke, *Repts. Progr. Phys.*, 8 (1941) 231.
[43] R. N. Jones, E. Augdahl, A. Nickson, G. Roberts and D. J. Whittingham, *Ann. N. Y. Acad. Sci.*, 69 (1957) 38.
[44] H. W. Thompson, *Spectrochim. Acta*, 14 (1959) 145.
[45] Discussion reported in *Proc. Roy. Soc. (London)*, A 255 (1960) 1—80.
[46] W. R. Brode, *Chemical Spectroscopy*, 2nd ed., Wiley, 1943,
contains 250 pages of tables and charts, mainly of atomic spectral lines. There is also an extensive bibliography relating to spectrochemical analysis and to other tables of spectral lines.
[47] G. F. Lothian, *Absorption Spectrophotometry*, 2nd ed., Hilger and Watts, 1958,
deals with instrumentation, and practical limitations on accuracy, as well as giving a number of examples of the application of absorptiometry.
The theory of intensity is covered in many of the monographs dealing with individual regions: these are to be found in the bibliographies of previous chapters.

INDEX